TRANSPORT
OF DELIGHT

Series on Technology and the Environment

JONATHAN RICHMOND

TRANSPORT
OF DELIGHT

THE MYTHICAL CONCEPTION OF
RAIL TRANSIT IN LOS ANGELES

THE UNIVERSITY OF AKRON PRESS

AKRON, OHIO

Manufactured in the United States of America

First edition 2005

09 08 07 06 05 5 4 3 2 1

LIBRARY OF CONGRESS CATALOGING-IN-PUBLICATION DATA

Richmond, Jonathan.
    Transport of delight : the mythical conception of rail transit in Los Angeles / Jonathan Richmond.
       p. cm. — (Series on technology and the environment)
    Includes bibliographical references and index.
    ISBN 1–884836–94–1
    1. Street-railroads—California—Los Angeles Metropolitan Area.   2. Local transit—California—Los Angeles Metropolitan Area.   I. Title. II. Technology and the environment (Akron, Ohio)
TF920.R53 2004
338.4'6'0979494—dc22
2004007247

The paper used in this publication meets the minimum requirements of American National Standard for Information Sciences—Permanence of Paper for Printed Library Materials, ANSI Z39.48—1984. ∞

To the memory of my mother

Contents

List of Illustrations

Maps

Lis

PLEASE EXCUSE ME — I HAVE TO BE SOMEONE MUST Fly INTERESTING TALK PLEASE RING ME I G Py T. 0117884160

List of Tables

Train of Thought

I must admit it: I like trains. I've doted on trains as far back as I can remember. My first career aspiration was to be a guard on London's Underground: the power of train doors hissing open and shut mesmerized me, and I dreamed of pushing the buttons that remotely operated them.

It didn't take long to graduate to higher things: next I wanted to be a train driver. On weekends I would take countless trips to nowhere-in-particular just for the joy of riding trains. I read the British Rail timetable for pleasure, memorizing alarmingly large parts of it.

The day of truth came when I needed to be in New York for an early meeting. Short of time, I forsook AMTRAK for Boston's Logan Airport and the air shuttle. Enjoying trains for themselves is all very well, but when I needed to actually go somewhere, I found I behaved with disdain for the amorous attributes of the technology.

I nonetheless became intrigued about how conceptions of technology—including the romantic ones I had myself experienced—might influence practical decisions in unsuspected ways. Although this book concludes that rail is a bad option for Los Angeles, based on how people actually choose how to travel, it is not primarily about the argument between building railways or busways. Whether buses or trains are "better" does not change the main focus of this work, which is about how we think.

While this book explores how we conceive of technology, it is written in the hope that we learn not so much about technology as about ourselves: If we can appreciate how we understand things, it will give us new options to see them differently. If what follows leaves you thinking about things differently, then I will have succeeded.

Acknowledgments

This book, which started life as a dissertation at Massachusetts Institute of Technology, involved often lengthy interviews with large numbers of people in 1985 and 1986; they are all to be especially thanked. They are listed in Appendix A, with the exception of those who requested anonymity.

Some people provided extra-special help. Burke Roche, Deputy to the late Supervisor Kenneth Hahn, was not only generous with his time, but provided myriad leads to other fruitful interviews and sources of information. Were he still with us, I would also wish to show my appreciation to Supervisor Hahn for breaking his usual rule of declining academic interviews. I remember the twinkle in his eyes as he told me that seeing me was the only way to relieve his secretary of my daily requests for a meeting.

Supervisor Michael Antonovich and his then-deputy, the late Tom Silver helped secure difficult-to-obtain data in addition to granting an interview. The supervisor's invitation to the opening of the Blue Line made for a day I shall never forget, and I also much appreciated the opportunity to later work for him as transportation policy adviser. Michael Lewis, deputy to former Supervisor Peter Schabarum, also gave considerable help. The late Supervisor Baxter Ward participated in one of the lengthiest interviews for this project, and I learned a great deal from it.

At the former Los Angeles County Transportation Commission (LACTC), Ed McSpedon greeted a potential critic with a remarkable sense of humor, and even wrote a detailed personal letter providing information I had requested just days before the Blue Line opened, a display of professionalism and kindness beyond the call of duty.

The Southern California Rapid Transit District (RTD) Library (now the Los Angeles County Metropolitan Transportation Authority (MTA) Library) was the site of much of my effort to piece together documentation on the historical and political development of rail transit in Los Angeles County. The late Dorothy Peyton Gray gave me much personal attention, enabling me to track down and learn from the obscure but illuminating elements in the extraordinary collection she developed. Tom Rubin, then at RTD, provided heroic support in obtaining elusive data and supplying much insight into the financial aspects of transit operations. I also made use of the library of the Automobile Club of Southern California in Los Angeles, and received much valuable assistance from Jim Ortner.

One member of staff of Southern California Association of Governments (SCAG) showed me particular indulgence, although naming of this person is not possible since all SCAG contacts were on a basis of not for attribution. George Frank of the San Diego Association of Governments helped me learn about transit development in San Diego, while everyone at the Metropolitan Transit Development Board (MTDB), San Diego was also helpful. Many others, in Seattle, Portland, San Francisco, Santa Clara, Orange County, Vancouver, and Washington, D.C., gave generously of their time, too.

I was received warmly by the community members I met in Watts, and enjoyed hospitality from Freita Shaw-Johnson. She told me a great deal about her community—as did the wonderful "Sweet" Alice Harris, Grace Payne, and several others—creating a picture of life in Watts that quite stripped away any stereotypes. Diane Watson, who represented the Watts community as a state senator in Sacramento, showed a remarkable devotion to her constituents, which I admired and appreciated as much as the time she generously gave me.

Most recently several staff at the MTA helped me update this work, and my thanks go to all of them, and to former Chief Operating Officer Allan Lipsky, in particular, for making the arrangements. There are too many others who provided interviews and other assistance to mention; thank you to one and all.

The late Professor Donald Schön was my thesis supervisor. Our many lengthy discussions were invariably enlightening, and sharpened my approaches to inquiry; he at times changed my thinking where few others would have succeeded, and his criticism led to substantial improvements as one draft gave way to another. Thanks, as well, to Professors Ralph Gakenheimer and Joseph Weizenbaum, who were also on my doctoral committee at MIT.

Professor Martin Wachs, then at the University of California, Los Angeles, served as an external committee member. Not only did he show an ability to uncomplainingly dredge through my work with a toothcomb; he deluged me with a continual stream of articles, clippings, suggestions, enthusiasm, and encouragement. Unhesitating to answer the most unreasonable request for assistance, I cannot thank him enough.

Daniel Brand, Vice-President of Charles River Associates in Boston supplied a valuable critique of my work on forecasting in chapters 4 and 5. SCAG also furnished comments on my forecasting work.

The intellectual influence of the late Professor C. West Churchman permeates every element of this work. I first encountered him as a privileged member of "West's seminar" at the University of California, Berkeley, an experience which left none of its participants unchanged. Professor Churchman taught that there is no event without ethical connotations, and that consideration of ethics must come prior to any action. That's a lesson never to be forgotten. A number of long meetings with West during the course of my research were significant in shaping its progress. Professor Melvin Webber, also of UC, Berkeley, provided many helpful suggestions, too, while Professor William Garrison, yet another Berkeley influence, constantly reminded me that if I could never achieve the myth of objectivity, that I could, at least, be subjective in fair and professional ways.

Professor Alan Altshuler supervised my master's thesis, teaching me the difference between an "assertion" and a "proof." His influence was significant in sharpening my approach to research, and it lingers on. I am grateful to Alan Altshuler as well as to Diana ben Aaron, Alan Evans, Raphäel Fischler, Sir Peter Hall, Doug Hart, Martin Krieger, Barry Surman, and Martin Wachs for more recent comments and editorial suggestions during the period between dissertation and book.

Professor Yossi Sheffi, then Head of the Transportation Systems Division of the Department of Civil Engineering at MIT, showed he believed in my project from the start, and provided unceasing support and encouragement. The Center for Transportation Studies at MIT provided fellowship support as the end of the tunnel finally came into sight. Its then Director, Professor Joseph Sussman, also approved the purchase of software I needed for my work, in addition to helping with conference expenses. I am especially grateful to both Yossi and Joe for putting together an emergency financial package which enabled me to graduate. The administrative support of Professor Frank Perkins, formerly Dean of the Graduate School at MIT, was also much appreciated.

I began turning thesis into book while a visiting professor at the University of Sydney; thanks to Professor David Hensher, Director of the Institute of Transport Studies, for hosting and encouraging me. Most recently, Alan Altshuler invited me to be a fellow at the A. Alfred Taubman Center for State and Local Government (which he directed) at Harvard University's John F. Kennedy School of Government, where I finished work on the book. He also generously gave me significant funding for a final research trip to Los Angeles, and for the costs of preparing graphic elements, indexing, and securing permissions. He was a constant source of advice and support over the last few years; my many thanks to Alan. I acknowledge, also, the insight and good humor of David Luberoff, executive director of the Rappaport Institute for Greater Boston at Harvard and the administrative assistance of Sandra Garron and Deborah Voutselas at the Taubman Center.

Jeff Stine, Curator of Engineering and Environmental History at the Smithsonian Institution acquired this book for the University of Akron Press. His many revealing comments on the draft manuscript were critical to developing it for final publication. Thanks also to Dorothy Hoffman for her expert indexing.

As concluding a doctoral dissertation is a time to look back, I also acknowledge the influence of my tutors at the London School of Economics, where I was an undergraduate: the succession of countless argumentative tutorials with John Martin, Ian Hamilton, and the late Kenneth Sealy, taught me how to question assumptions and, indeed, what criticism is all about. The beginnings of my interest in serious social inquiry are rooted at St. Paul's School, however, and particularly with my economics teacher, John Allport. His personal attention, encouragement of my interests, valuable guidance, and continual kindness will never be forgotten.

I was moved that fifty people turned up for my MIT doctoral defense. The final question of the defense—not supposed to come from an undergraduate, but that wouldn't discourage Will Glass-Husain whose nerve is happily redeemed by his brilliance—led me to change an element of my conclusion before submitting the thesis, proving that defenses can be more than ritual events.

David Kazdan, working on his dissertation at Case Western as I wrote mine at MIT, exchanged email with me several times a week. We chronicled our experiences and provided the mutual psychotherapy needed to finish. China Altman, Bob & Joanne Bonde, Sharon & David Bruce, Rory O'Connor, Fritz Cronheim—now sadly deceased—Peter Goldsbrough, Amy Gorin, Reuven Lerner, David Maltz, Rabi Mishalani, Eric Peyrard, Nana & Richard Prevett, John Purbrick, Roberta Ridley, Eric Starkman, Barry Surman, Michael Thouless & Yili Wu, and Adrian Wykes also supplied constant encouragement. *The Tech*, MIT's student newspaper, proved to be the best of places to hang out as well as learn the merits of journalistic writing styles.

My parents, Theo & Lee Richmond, sister Sarah, and brother Simon encouraged the madness of this never-ending enterprise over far too many years—my love to all of you—while my grandmother, Becca Souccar, never stopped reminding me that "tu dois finir." I am glad she was still alive when I graduated even if, unfortunately, she did not live to see the book.

This work is dedicated to the memory of my mother, Diane, for whom I say:

יִתְגַּדַּל וְיִתְקַדַּשׁ שְׁמֵהּ רַבָּא

CHAPTER 1

Introduction

Man is born free and is everywhere in trains.

—Tiresias, after Rousseau

A Problem of Agoraphobia

When Copernicus argued in 1543 that the earth rotates daily on its own axis and moves annually around a stationary sun, he was attacked by a Lutheran follower, Melanchthon, since "the eyes are witnesses that the heavens revolve in the space of twenty-four hours."[1]

Only through awareness of the shortcomings besetting the way we receive and deal with information do we stand a chance of finding a more ready path to understanding. But not only are we unaware; we do not seek to be more aware. We suffer, says Kenneth Boulding, from agoraphobia, "the fear of open spaces, especially open spaces in the mind." We identify with and are reassured by recognizable forms; we try to blot out the void and disorder of the unknown over which we have no control. Though one can only be wise, warned Harold Laski, "if he admits that his knowledge of the subject is mainly a measure of his ignorance of its boundaries," we delude ourselves into believing that we have successfully closed in on the essence of the subject under study in an effort to escape from the reality and consequences of our ignorance. Thus, says Russell Ackoff, "We usually try to reduce complex situations to what appear to be one or more simple solvable problems. This is sometimes referred to as 'cutting the problem down to size.' In so doing we often reduce our chances of finding a creative solution to the original problem."[2]

We behave most of the time as if we lived in bubbles.[3] The bubble provides a womblike sense of comfort and security, an uncertainty-controlled environment in which we can work, using the repertory of procedures—mediated by the universe of understandings—on the list of problems contained within the bubble's walls. We may peer out to glance at the outside world; it is seen, however, only from the perspective afforded by our particular bubble. Because the bubble has invisible walls, we are not even aware of being in a bubble.

The bubble enables us to tacitly live by a set of assumptions, rather than inviting us to critically appraise them; but the assumptions, of which we are so frequently unaware, guide the results we reach and influence the choices we make. Unquestioned norms and procedures lead planners and politicians to typically only examine "alternatives" laid down on a determinate path, rather than trace that path to its source to seek out more novel directions. As Alan Altshuler, writing on *The Urban Transportation System*, remarked,

> Analytic activities have tended overwhelmingly to focus on the appraisal, advocacy, and/or incremental adaptation of . . . technologies and services—which we term *preselected solutions*—rather than on laying bare the character of the problems generating demands for public action or searching with a fresh eye for effective remedial strategies. Paramount among the preselected solutions have been highway and transit improvements, and policy discussion has typically proceeded as if these were the only options available for addressing sources of dissatisfaction with the urban transportation system.[4]

When we fail to lay *bare the character* of problems and focus instead on uncritically *preselected* solutions, we are apt to see possible policy choices only in terms of these solutions; were we to range more broadly in our inquiry, we might come up with options which could serve us better. Why, then, do we constrain ourselves?

This book is a study about the failure of thought and its causes.[5] It starts with a bizarre decision: to construct a comprehensive rail passenger system in an environment where it appears incapable of providing real benefits. It continues with an account of the political and institutional actions that led to that decision. It then develops and applies a theory which generates conceptions of understanding to explain actions which otherwise seem strange. In so doing, it shows how the bubbles that constrain our thoughtfulness and inventiveness get constructed around myths which themselves depend on the understandings formed through our history, cultural background, and the experiences that make up our daily lives.

As Donald Schön pointed out: "Underlying every public debate and every formal conflict over policy there is a barely visible process through which issues come to awareness and ideas about them become powerful. The hidden process by which ideas come into good currency gives us the illusory sense of knowing what we must worry about and do." Schön complains that we tend to "disregard the less visible process and to accept the ideas underlying public conflict over policy as mysteriously given."[6] This work is about revealing that "hidden" process and understanding how its operation precludes the use of creativity in the process of policy formation. Although it includes a detailed case study in transportation planning, its goals are general, rather than specific. It aims to expose the deficiencies in our thinking out and acting on problems in the hope that such knowledge can help point the way to better planning.

The following case study reviews policy development and planning for light rail in Los Angeles, one of a number of American cities to have embraced rail transit in the 1970s and 1980s. Of special interest, project funding for the initial Blue Line, the focus of this work, was obtained from local sales tax revenue, rather than federal grants, discounting the notion that the project was initiated purely to bring federal dollars to Los Angeles. Chapter 3 evaluates the Los Angeles rail program. Processes of planning are then examined on two levels. Two chapters seek to expose and understand the methods employed by the professional working or consulting for government in the context in which they are used (chapters 4 and 5). The main emphasis in the rest of the study is on surfacing the (generally tacit) mechanisms by which politicians inform themselves and make decisions in the formulation of policy. A grasp of such processes is fundamental to an understanding of not only how transportation systems, but all planned systems, are adopted and developed.

Light Rail in Los Angeles

Several previous attempts to fund rail projects in Los Angeles in the 1960s and 1970s had failed. But, in 1980, Los Angeles County voters approved Proposition A, a measure to lower the base bus fare from eighty-five cents to fifty cents for three years, provide transportation funds for the discretionary use of local administrators, and—most importantly—to build a countywide system of rail rapid transit lines. Standing in front of a preserved "Red Car," which once served on the 1,164 miles of track of the Los Angeles interurban rail system, Los Angeles County Supervisor Kenneth

Hahn spoke at a 1981 press conference. "Now we need light rail transportation," he said. "It's a priority for the nation, let alone for Los Angeles. Just travel the freeways; see the jammed bumper-to-bumper freeways." Only 54 percent of voters had been in favor of the proposition and its half-cent tax increase, and delays resulted pending a favorable resolution of the legal question of whether a two-thirds majority vote was required for the proposition to become law. But on July 1, 1982, amidst much fanfare, bus fares came down and sales tax went up.

In July 1985, bus fares returned to their former levels. The Los Angeles County Transportation Commission (LACTC), sponsor of Proposition A, now found itself preoccupied with implementing its "pledge" to provide rail transit for county residents. The commission had by then selected a light rail approach (meaning that trains can operate in street traffic as well as on their own rights-of-way, unlike heavy rail subways, which run exclusively in dedicated rights-of-way), and had decided to proceed first with a line traveling south from downtown Los Angeles through the low-income minority communities of Watts and Compton to a terminal in Long Beach, the second-largest city in the county. This had been the route of the last of the "Red Car" interurban streetcars to pass out of existence, the culmination of a decline in the Red Car system underway by the 1920s. The line was opened on July 14, 1990. It has been operated since April 1, 1993 by the Los Angeles County Metropolitan Transportation Authority (MTA), which replaced the former LACTC and Southern California Rapid Transit District (RTD).

Rail transit plays an important role in many East Coast American cities because of the well-defined intense demands for travel between concentrated foci of employment and their surrounding suburbs. A relatively small number of high-capacity rail lines can effectively serve the needs of many commuters. In contrast, the low density and widespread distribution of both population and economic activity in Southern California generates a dispersed and complex pattern of transportation demands between a myriad of origins and destinations. This calls for service more similar to a telephone network (which connects anywhere to everywhere) than to rigid radial public transportation; this does not augur well for rail solutions.

Critics, particularly in the academic community, have found rail transportation inappropriate for Los Angeles and other dispersed western cities.

The degree of consensus reached among economists and planners who are opposed to rail in cities such as Los Angeles is remarkable. They believe that more could be gained from improvements to the existing bus system—which can provide direct service between a larger number of origins and destinations than can rail—and from better management of existing roads. This view is supported by the conclusions reached in chapter 3.

It is puzzling, then, that while basic questions as to the need for the revived streetcar system have been neglected, the project has been strongly supported on many fronts by both politicians and the public at large. There has been very little dissent. Why?

Politicians make decisions on public expenditures, and a study of this sort requires an analysis of the political aspects of decision-making. This book goes beyond conventional political analysis, however, unearthing the myths created by the subconscious world of the images, symbols, and metaphors that rule daily experience, and investigating the power of myths to steer the decision-making process.

Given the critical role urban railways played in the development of Los Angeles, this work first presents a historical study of the rise and fall of the Pacific Electric "Red Car" system. Understanding why the system arrived and went out of existence is helpful to an evaluation of the case for a light rail system of similar specifications today. The contextual information of a historical study also gives clues as to how people who remember the old system understand the case for light rail today.

Chapter 3 evaluates the case for light rail with a presentation of concepts which would have been available at the time of the decision to proceed with the Long Beach line. It also provides an analysis of the performance of this and other rail projects since their opening. The following chapter critically analyzes the forecasting methodology used to predict passenger demand for the Long Beach system, while chapter 5 explores the ethical dimensions of ways such work influences and supports political decision-making.

Chapter 6 then gives an account of the political and institutional circumstances that led to the decision to put Proposition A of 1980 on the ballot, and to select the Long Beach line as the first light rail route to be brought into operation. It also brings readers up to date on the politics of decision-making in Los Angeles following the commencement of rail operations.

If we have described by this stage how the elements of political power came together to enact a rail program, we have yet to understand what made rail seem appealing. Too often political analyses examine the interaction of interests and the outcomes of their struggles, but neglect to look at why interests have come to formulate their desires in particular ways. To trace the understandings which made rail attractive to decision makers, chapter 7 proposes a theory of myth, based on the human need for simplicity and abhorrence of complexity, and on the availability of a set of symbols, images, and metaphors which come together coherently to create a myth that acts with the power of truth.

The next three chapters test the power of the theory of myth to provide explanation by examining if its tenets explain concept formation and decision-making for rail transit in Los Angeles. Chapter 11 focuses on the attraction of light rail to the depressed communities of Compton and Watts, examining the power of rail's symbolic claims to provide access to opportunities as well as to places. The remaining two chapters synthesize mythical and political analyses and present conclusions, while an epilogue presents the scene on the Blue Line's opening day.

This study depends heavily on an extensive set of interviews conducted especially for this project, as well as on materials obtained from transcripts of political meetings and media sources. A total of 209 interviews were conducted (see appendix A for list of participants), 103 of them in Los Angeles County, the remainder in Orange County (California), Portland (Oregon), Sacramento, San Diego, San Francisco, San Jose, Seattle, Vancouver, B.C., and Washington, D.C.

Most interviews were conducted over the summers of 1985 and 1986, following the unanimous approval given by the Los Angeles County Transportation Commission for the final design and construction of the downtown Los Angeles–Long Beach line on March 27, 1985. Interview subjects included present and former elected officials and staff in a variety of organizations, as well as community representatives and five university professors. A particular emphasis was put on seeking out members and staff of the LACTC, which promoted Proposition A, but interviews were conducted at all levels of government including the Los Angeles regional organization, Southern California Association of Governments (SCAG) (which performed technical analysis for the Long Beach light rail Environmental Impact Statement), and city, state, and federal representatives and staffs. The interviews outside Los Angeles were conducted primarily to see if the mental processes at work in Los Angeles were also in evidence elsewhere.

A loosely structured questionnaire was used to guide discussion through key issues (see appendix B), but respondents were allowed freedom to explore their particular interests and perceptions. Interviews were taped, except where participants objected. Where confidentiality was agreed upon, the names of particular interviewees are not supplied. All interview material obtained at SCAG is used without attribution. In most other cases, names are given.

All interviews were transcribed from tape or notes (generating over three thousand pages), and forty core Los Angeles County interviews (and one other interview) were comprehensively keyworded, using the textbase program FYI 3000 and the textbase/hypertext software IZE, according to the subject in each chunk of text, responses given, and organization represented. The approach permitted precise data retrieval from a wide variety of angles, always bearing in mind the need to place individual excerpts in the context of the complete interview from which they were retrieved. Evaluating the interviews required a form of literary criticism, in which validity depends on the functioning of a "logic" according to the rules of the theory under test. Discussion of issues of interpretation and validity is included in chapter 7. All quoted material which appears without attribution in this text comes from the interviews conducted for this project.

This work was initially presented as an MIT doctoral dissertation in 1991.[7] The passage of more than a decade since then has provided evidence of the performance of the adopted rail systems, and also brought political changes. Some further research was conducted in Los Angeles and with recent documentary materials to bring this book up-to-date.

A Note on Rationality

Rationality, which according to the dictionary is the pursuit of reason, has given itself a bad name. Critics have contended that rationality has its limits but, in doing so, have smeared the very name of reason.[8] The problem is at least partly related to the fact that a particular type of reason—and one that is not, furthermore, always "reasonable"—has taken "rationality" and the concept of reason itself for its name. John Friedmann and Barclay Hudson state that rationalism "is predominantly concerned with how decisions can be made more rationally.... A decision (usually about the proper allocation of resources) will be called rational when it arrives at a single "best" answer to a *stated problem*" (my emphasis). Under this approach, "professional activity consists in instrumental problem solving made rigorous by the application of scientific theory and technique."[9]

Given a *stated problem*, the task of systematic inquiry in public affairs is to evaluate the benefits and costs of alternative solutions, and compute which is best. The question of what problem should be asked or how it gets framed— something not subject to quantification or to analytical reason—gets cast aside. Technical or analytical reason, which is often equated with "rationality" itself, treats problems mechanically by evaluating their parts in specific ways according to given criteria. Analytical rationality functions within a "bubble" of given definitions and cannot determine the validity of selecting which particular alternatives to evaluate. It cannot decide whether the criteria for evaluation are appropriate, or whether its own internal logic is helpful to the task.

The ability to follow given criteria to answer the stated problem is what makes the process "rational." As Churchman says,

> [A] whole rationalist economic theory was generated by the concept of a rational (economic) man; the theory is valid because of the way in which it defined rationality and not because it describes the realities of the human being. As we have seen, the same comment applies to academic model building. One defines an objective function to be thus and so. Then what "should be done" follows from the definition and has nothing to do with reality.[10]

The task of asking what problem should be addressed is one of the deepest functions of reason. It could be termed "reflective rationality," to give back the name of reason to one of reason's most essential tasks, and one which is sadly too often ignored.

Trying to Understand Narrowness

To West Churchman, following his teacher Edgar Singer, the most crucial ethical question is: Should a particular question be investigated at all? Asking one question rather than another is an ethical act, for answers to different questions imply the implementation of alternative sets of outcomes with divergent impacts on the lives of different people. Connected with the design of appropriate questions for inquiry is the "sweep-in" approach. Given an initial question, related issues need to be "swept in" to broaden inquiry to reach for root causes. When a planner finds a system like a prison or hospital in difficulty, for example, Churchman says, "The planner should search not for ways to make the prison or the hospital run more smoothly, but for the reasons why we have things like badly run prisons and hospitals."[11]

Churchman complains about the narrow definition of problems "which provide feasible boundaries to the ethical issues, which need no further defense, i.e., which stop the conversation." He calls for "an 'unbounded' systems approach which must include a study of humanity, not within a problem area, but universally." Churchman is firmly a rationalist; he believes in the power of reason. But his approach does not consist of applying a narrow set of criteria to a given "problem." Rather, it involves opening up the boundaries of inquiry, guided by ethical principles. It regards all systems as part of larger systems, with all parts given relevance only in relation to all other parts of all other systems. "Those of us who practice social science learn the hard way that there are no simple questions and that the process of addressing a specific question will eventually require answers to more and more questions."[12]

Questions of transportation are therefore also questions of urban form and the environment, of the history and identity of a community, of its distribution of resources, of employment, education, and race. What is the use, for example, of studying how to provide transportation "access" to a particular employment site if barriers to attaining employment are related to inadequate education or racial prejudice? Our role as planners should not be to take a "transportation problem" as given, but to reflect on its definition and open up inquiry to reach for the root causes of social problems which may lie far beyond the scope of an initial simple problem formulation. This is the art of "bubble bursting" and "reflective rationality."

We will not be seeing Churchman's idealized approach in operation in either technical analysis or political decision-making in Los Angeles: instead we shall observe the asking of only the narrowest of questions, the answers to which cannot effectively help cure the problems under study. This has disturbing implications for policy making and planning, and we therefore need to identify where and how things are going wrong. That is what this book is all about.

Figure 2-1. Henry E. Huntington at his San Marino home. (Featured in the book *La Reina* by Laurance Landreth Hill, published by Security Trust & Savings Bank in 1929, p. 105. (From Security Pacific National Bank Collection, Los Angeles Public Library, by permission)

CHAPTER 2

The Rise and Fall of the Pacific Electric: A Case Study in Technological Evolution and Displacement

It required no conspiracy to destroy the electric railways; it would, however, have required a conspiracy to save them.

—David Brodsly

"NOT ALL POETS EXPRESS THEMSELVES in words," wrote the *Electric Railway Journal* in its obituary notice to mark Henry Huntington's death on May 23, 1927. "The romantic and imaginative soul," the *Journal* declared, "found an outlet for his feelings in organizing and directing railroads," as well as through collecting art and philanthropy.

> He was famous in his younger days as a director of railroads and in his older days as a pre-eminent collector of rare manuscripts and a connoisseur of works of art. His genius for acquisition was no less noteworthy, whether the object of his desire was a railroad or a Rubens, and the same characteristics which insured success in his early business ventures and won for him the reputation of pioneer and genius stamped him as a distinguished collector of books and works of art and entrenched him firmly in more than local affections.[1]

Historical analysis is too rarely a part of social science or planning studies, but to appreciate how light rail might fit into in today's Los Angeles, we need to understand the rise and fall of its predecessor, interurban railroads. The turn-of-the-nineteenth-century interurban railroad sowed the dispersed patterns of Los Angeles. But, while the interurban gave birth to a low-density standard of living to which immigrants aspired, subsequent technological development provided a vehicle—the automobile—that an-

Figure 2.2. The Pacific Electric brought new residents to Long Beach. Pictured here is the first Pacific Electric car on the opening day of the Long Beach line, July 4, 1902. (From Security Pacific National Bank Collection, Los Angeles Public Library, by permission)

swered far better the Southern Californian dream of freedom of movement, and the interurban's reign was over.

A knowledge of this process tells us why a mature city with urban functions and travel patterns formed around the availability of the highway and the car cannot return to dependence on a displaced technology. It also illuminates the lingering associations that erroneously suggest that such a turning backward could be feasible or desirable. History shows that just when rail was being displaced by automotive technologies, a climate of misplaced optimism masked the inevitable decline of rail as the major component in the passenger transportation system. Recently, we have seen the same unrealistic hopes afoot again. If the rail system that created Los Angeles resonated with poetry, its dream lives on as part of modern-day popular folklore, even though the dream of the old cannot justify construction of the new. Despite the myth that the Red Cars of Los Angeles perished as the result of a conspiracy of automotive interests, the rail technology was in fact displaced because the new—automotive—technology served consumer needs better.

Beginnings

The first settlers of the pueblo *la Reyna de los Angeles* ("The Queen of the Angels"), founded on orders of California's Spanish governor, Felipe de Neve, were eleven families who had made the difficult trek a thousand miles from Sonora and Sinaloa, Mexico. They came to start a farming community, and it was formally established on September 4, 1781.[2] By the end of the decade, the population of the Pueblo was 139.[3] At the first federal census in 1850, it was 1,610. As was the rule in pre-industrial cities, the distance between home and work was limited by walking speed, and this constrained growth.[4]

Fundamental change arrived with the first transcontinental rail connection in 1876, augmented by a more direct connection with the east in 1881.[5] The first railroads generally followed the routes of Indian, Mexican colonial, and American stage trails, which in turn reflected the region's natural geography.[6] They provided the axis for the initial spurt in growth. While agricultural settlements such as San Bernardino, Riverside, Pasadena, El Monte, Pomona, and Long Beach came into existence in anticipation of the transcontinental railroad, development escalated in the 1880s, with more than one hundred towns established in Los Angeles County. "The railroads were not only the motivating factor in the boom, but the location of their lines influenced the alignment and provided the focus of the new subdivisions."[7] The population of the original Los Angeles community increased from six thousand to over fifty thousand between 1870 and 1890, while the Los Angeles County population as a whole went from fifteen thousand to around one hundred thousand. The newly created Orange County drew an additional thirty thousand settlers.[8]

The newcomers taking root in the developing Los Angeles were quite different from the relatively poor and unskilled immigrants arriving on the eastern seaboard. The new Angelenos were generally American-born, with many having originated in the midwest and bringing with them some means. Coming from a rural background, they shared an ideal of a low-density single family lifestyle long before automobility was to spread that concept far and wide.[9] Los Angeles' "geography of the ideal" presented its immigrants with the antithesis of the urban morphology of the traditional cities of the east.[10] Robert Fogelson says that those arriving in Los Angeles saw the eastern metropolis—congested, impoverished, filthy, immoral, transient, uncertain, and heterogeneous—as the receptacle for all European evils and as a contradiction to their ideal, which was represented by the spacious, affluent, clean, decent, permanent, predictable, and homogeneous characteristics of the residential suburb. Thus, they sought city outskirts which had the potential to reflect rural values rather than urban vices, and this choice was the basis for the extraordinary dispersal of Los Angeles.[11]

Los Angeles and the Interurban

In 1869, the first local rail line—between the Wilmington wharves and downtown Los Angeles—opened, followed in 1872 with a route linking downtown and Santa Monica. The first streetcar line began operations in 1874. Within thirteen years, there were forty-three separate franchises for lines operating in the City of Los Angeles. The first electric (as against horse, mule, or cable-driven) one arrived in 1887.[12] But if the railroad cutting frontiers from the east provided the major artery bringing migrant life to Los Angeles and—together with the early streetcar lines—laid the seeds for fulfillment of the suburban dreams of L.A.'s migrants, it was the in-

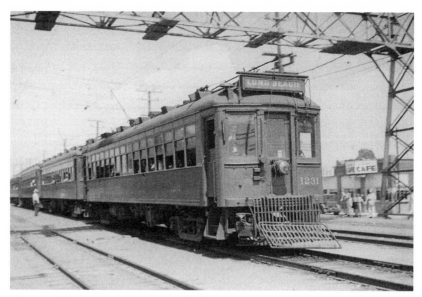

Figure 2-3. The Long Beach line was the last of the Pacific Electrics to go. (From Security Pacific National Bank Collection, Los Angeles Public Library, by permission)

terurban rail systems which spread new blood over a vast territory beyond that originally occupied by the pueblo.

As Radolph Karr points out, the interurban railroad was a hybrid: it operated on private rights-of-way like a steam railroad, but also on streets like a street railway. Stopping in many more places than a steam train, it provided far greater accessibility. The flexibility of operating in either single- or multiple-car formats also provided the option of much more frequent departures than was economical with a steam engine and set of coaches.[13] The higher quality service stimulated demand, which in turn justified further extensions in service. The greatest geographical growth occurred when the electric railways were most vigorously expanding, making Los Angeles the largest city in terms of area in the United States.[14] It was this potential for urban expansion, rather than a will to provide transportation services, that caused railways to grow. Railway development was a tool of real estate development rather than a response to transportation demand. Los Angeles landholders built railways between 1890 and 1910 specifically to profit from the low-density residential development favored by migrants to the Southland.

While Los Angeles had never developed a significant urban core, business and commercial functions nonetheless remained focused there until the dawn of the automobile. The railways allowed distant suburbs to be created and made attractive by the promise of ready connections to downtown, although their new residents were to frequently call for crosstown links which the railroads said were not financially viable. The railways were often mechanically unreliable and the subject of constant public complaints. Financially unsound as transportation companies, the railways rarely made money from operating rail service but instead catalyzed huge real estate profits.[15] As one prominent railwayman explained, "It would never do for an electric line to wait until the demand for it came. It must anticipate the growth of communities and be there when the home builders arrive—or they are very likely not to arrive at all, but to go to some other section already provided with arteries of traffic."[16]

Moses Sherman and Eli Clark began the first interurban operation— from Pasadena to downtown Los Angeles—on May 4, 1895, using portions of local lines between the two cities. On opening day, ten trolleys raced from one end of the line to the other.[17] But it was Henry Huntington who was to build the largest interurban network in the nation, operating over 1,164 miles of track by 1923 and connecting points over a hundred miles apart.[18] "Surely Californians, and especially sons of Los Angeles, are both grateful and proud that Henry E. Huntington had a hand in the making of the West," declared Huntington's obituary:

> For at the psychological hour in 1898 when the City of Los Angeles was cautiously courting prosperity along came the needed financier, who with foresight and courage projected a network of electric railways in the city which converted hamlets and villages into commercial and trading centers and made barren ground desired sites for homeseekers. Southern California, Henry E. Huntington and prosperity became a triumvirate the exploits of which will fill many pages of history.[19]

Huntington, together with a group of associates, incorporated the Pacific Electric Railway Company on November 10, 1901. The first line to be built—which was to also be the last interurban route to close and the first one to be resurrected under Los Angeles County's Proposition A—connected downtown Los Angeles with Long Beach. The line opened on July 4, 1902, with the Red Cars of the Pacific Electric bringing thirty thousand visitors.[20]

Huntington sold an estimated $3 million worth of land in connection with his rail enterprise.[21] His tactic was to run real estate and water utility companies in parallel with railway development. Huntington Land & Improvement would purchase land adjoining the planned tracks, which would then be subdivided and sold as suburban residential lots. "Indeed, Huntington integrated his undertaking so effectively that while Pacific Electric lost millions and Valley Water thousands, Huntington Land's earnings justified the entire investment."[22]

"The population of Long Beach, first community touched by the magic of the Big Red Cars, grew from a village of 2,200 residents to a city of nearly 18,000 in less than a decade. Most of its new residents arrived on the P. E.," reports Spencer Crump.[23] Watts, midway down the Long Beach line, was also transformed by the trolley, becoming a bedroom community with only minimal commercial activity of its own, but establishing itself as one of the most important centers of the rail network.[24] Huntington's promotions extended existing settlement peripheries, and also created thirteen new towns, twelve of them on the Pacific Electric.[25] This process, "built not a city, but a series of connecting villages."[26] Huntington's Red Cars traveled over sixty miles from downtown Los Angeles, passing through Covina, Claremont, Upland, Etiwanda, and Rialto en route to San Bernardino, Redlands, and Riverside.[27]

Laurence Veysey says that photographs of places like Hollywood in 1910 show crops and pastureland, with the rail line the only urban feature of the environment. When the Pacific Electric reached the hitherto remote San Fernando Valley in 1911, it brought about the substantial settlement of a previously uninhabited expanse of sagebrush. The rail stations themselves formed the original nucleus of some of the new satellite towns which owed their existence to their steel lifeline to Los Angeles.[28] The population of Los Angeles, which had stood at just over 100,000 in 1900, more than trebled by 1910.[29] The metropolitan area as a whole grew from 180,920 to 507,300 residents over the same period, a decade during which "the electric trolley held a virtual monopoly of interurban transport."[30] But trouble was already afoot for the Red Cars. As Karr documents, the only years after 1913 during which passenger operations made a profit were 1923 and some years during World War II.[31]

The Conspiracy Theory of Bradford Snell

Popular belief would have it that the "Big Red Cars" of Los Angeles, along with streetcar systems in other cities, were eliminated by a conspiracy of automotive, oil, and rubber interests to replace them first with buses, and ultimately with automobiles. The "conspiracy theory" draws largely on a paper produced by a staff attorney, Bradford C. Snell, for use during congressional hearings before the Subcommittee on Antitrust and Monopoly of the United States Senate Committee on the Judiciary.[32] Disenchantment with the road and the car was at its height when Snell released his report, and environmentalists who protested the damage done by the highway were calling for freeway projects to be withdrawn and replaced by rail transit. In 1973, the Federal Aid Highway Act allowed just this, if desired at the local level, and provided for 80 percent matching grants. The highway revolts in the air were part of a larger anti-establishment social movement, and Snell's presentation of "a study of the social consequences of monopoly" was timed right to receive a receptive audience in 1974.[33]

It is important to consider Snell's claims, for they lead to the conclusion that the loss of rail lines was a function of monopolistic rather than "natu-

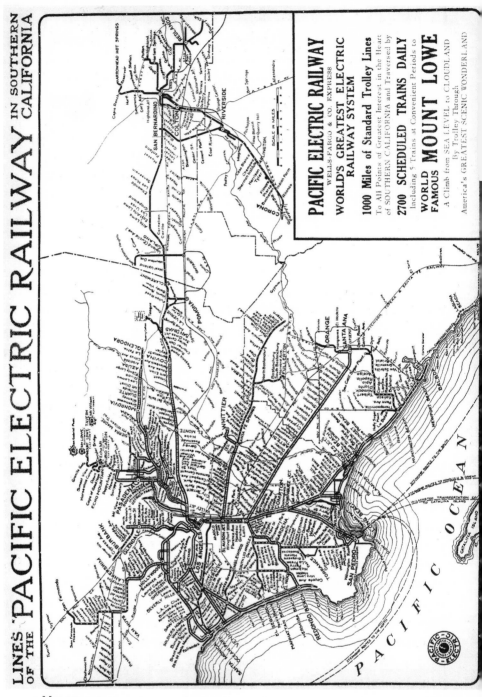

Map 2-1.

ral" free-market economic forces. This conclusion leads to the belief that under "natural" conditions the rail lines would have continued in use, and lends support to those who would resurrect the system today. The alternative view, which historical evidence supports, is that the decline of the Red Cars resulted from the loss of the financial feasibility of operations, a process accelerated as automobiles came to Los Angeles and as rail services became decreasingly suited to meeting the transportation needs of the population. This hypothesis suggests that the Red Cars became naturally extinct as they were displaced by automotive technologies that caused changes in the Los Angeles urban form and provided services better suited to serving its needs.

The Snell Claims

Snell characterized his study as a case of the restructuring of society for corporate ends in which three powerful automobile companies "eliminated competition among themselves, secured control over rural bus and rail industries, and then maximized profits by substituting cars and trucks for trains, streetcars, subways, and buses. In short, it describes how General Motors, Ford, and Chrysler reshaped American ground transportation to serve corporate wants instead of social needs."[34]

The study starts from an anti-auto view, lamenting that "[u]nlike every other industrialized country, we have come to rely exclusively on large, gas-guzzling cars and trucks for the movement of passengers and freight. In the process, we have consumed much of the Nation's supply of oil, fouled our urban air with poisonous exhausts and turned our cities into highways and parking lots."[35] Snell denies that the demise of public transport was a function of a public preference for automobile travel. Instead, he claims, much of the growth in automobile as well as truck travel was due to the decline of rail and bus systems.[36] General Motors, Snell states, saw that (more efficient) public transport could replace large numbers of automobiles and that it was to their advantage to eliminate this competition. "In the course of events, it became committed to the displacement of rail transportation by diesel buses and, ultimately, to their displacement by automobiles. . . . By the mid-1950s, it could lay claim to having played a prominent role in the complete replacement of electric street transportation with diesel buses."[37]

Snell paints a picture of Los Angeles as a Red Car utopia destroyed by the curse of the car. Nowhere, he says, "was the ruin from GM's motorization program more apparent than in Southern California:

> Thirty-five years ago Los Angeles was a beautiful city of lush palm trees, fragrant orange groves and ocean-clean air. It was served then by the world's largest electric railway network. In the late 1930s General Motors and allied highway interests acquired the local transit companies, scrapped their pollution-free electric trains, tore down their power transmission lines, ripped up their tracks, and placed GM buses on already congested Los Angeles streets. The noisy, foul-smelling buses turned earlier patrons of the high-speed rail system away from public transport and, in effect, sold millions of private automobiles. Largely as a result, this city is today an ecological wasteland: The palm trees are dying of petrochemical smog; the orange groves have been paved over by 300 miles of freeways; the air is a septic tank into which 4 million cars, half of them built by General Motors, pump 13,000 tons of pollutants daily.[38]

Snell's description is of violence being done to a living (almost human) organism. "In sum, GM and its auto-industrial allies severed Los Angeles' regional rail links and then motorized its downtown heart."[39] Snell underlines the April 1949 conviction of General Motors by a Chicago Federal jury of criminal conspiracy with Standard Oil of California, Firestone Tire, and others to replace electric transit with gas or diesel buses, and to monopolize the sale of buses and related products throughout the country. This court fined GM $5,000 and its treasurer, H. C. Grossman, $1.[40]

General Motors' Reply

General Motors replied to Snell's report, saying it contained "false accusations, misleading inferences, and erroneous conclusions:" "For those who seek simple explanations for complicated urban problems, " 'American Ground Transport' provides a convenient scapegoat. . . . It also provides a simple cure, namely the forced ouster of General Motors from the mass transportation business, presumably to clear the way for the return of the street car." General Motors (GM) states, however, that "street railways failed for economic and demographic reasons which had nothing to do with any plot by General Motors." GM refers to the violation of antitrust laws as having "nothing at all to do with the replacement of streetcars by buses. There is not one word in either the Government indictment in the criminal case or the complaint in the companion civil case which charges GM with un-lawfully scrapping or eliminating street railway systems."[41]

General Motors states that the opinion of the Seventh Court of Appeals in *U.S. v. National City Lines* in 1951, actually pinpoints the basic flaw in the current accusations that General Motors, for ulterior motives, destroyed healthy streetcar systems: "In 1938, National conceived the idea of purchas-ing transportation systems in cities *where streetcars were no longer practica-ble* and supplanting the latter with passenger buses 186 F. 2d at 565" [emphasis supplied by GM].[42]

General Motors accuses Snell of ignoring the fact that Pacific Electric and its successors had themselves converted from rail to bus service over a period of four decades because of the rail's rundown and unprofitable con-dition and a lack of patronage. The interurban lines were not "severed" in 1940, but abandoned in stages with the approval of the Railroad Commis-sion.[43] As we shall see, GM's claims are supported by the facts.

The Decline of the Red Cars: A Case in Displacement of Technology

George Hilton and John Due characterize the interurbans as a "rare ex-ample of an industry that never enjoyed a period of prolonged prosperity; accordingly, they played out their life cycle in a shorter period than any other important American industry."[44] Demand for transit services began to decline by the mid-1920s, while the financial distress of the industry pre-dates World War I.[45]

Los Angeles was certainly party to this trend. Veysey reports signs of "tightened belts" even before 1920, with labor and maintenance costs rising and patronage declining. "And already the weakest streetcar systems were ceasing operation and selling themselves for scrap." Despite a slow increase in patronage from 1916 to 1918, ridership at the start of 1919 was still less than the peak of January 1914. More significantly, per capita ridership dropped after 1912, reaching a low in 1916 and practically standing still through 1918. "In short, the company was slipping in its hold upon the pub-lic which it served."[46] Substantial increases in ridership in the early 1920s took Pacific Electric ridership from 78.5 million in 1919 to a peak of 115.4 million in 1924. As Veysey reports, ridership then leveled off close to the 110 million mark and, after a surge in 1929, dropped to 106.9 million in 1930.

The revenue peak had been in 1923; revenue then declined steadily until by 1930 it was beneath 1920 levels.[47]

The Automobile Makes Its Mark

In 1915, when Los Angeles County had only 750,000 residents, it had 55,217 private cars, more than any other county in the country.[48] While only slightly more than 100,000 vehicles were owned in Los Angeles County in 1919, by 1923 automobile and truck registration stood at 425,582 vehicles. During this period—one of booming population expansion—per capita ridership on Pacific Electric continued to decline, even while it was peaking in absolute terms. By 1930, there were 842,528 registered vehicles in Los Angeles County.[49] Despite this, unrealistic expectations as to the role of rail abounded during the 1920s. In 1925, for example, an *Electric Traction* editorial stated that "[t]he congestion of automobiles, the many resulting accidents, the limited parking spaces, and the greater expense of operating already is driving more and more people to the street car, the elevated, and the big bus with a regular operating schedule." Even by 1929, Joseph Hallihan, writing in *Electric Railway Journal* declared that, "[u]nquestionably the majority of people prefer to live within reach of public transit facilities. The ownership of an automobile by the family does not fully satisfy the daily transit demands for its different members."[50]

However, signs that the automobile had brought about structural changes in travel behavior were quite apparent by then. Low-density residential development and the early move to automobile travel meant that the expected growth in rail passengers never happened.[51] The car did more than extend the range of settlement radially. It brought accessibility to areas bypassed by rail, distant from stations, or in foothills with steep gradients where rail could not be profitably brought.[52] City maps drawn between 1902 and 1919 show few streets more than five or six blocks from streetcar lines.[53] The automobile, however, allowed development wherever a road could be placed, and thus stimulated the real estate boom of the 1920s. In some cases development actually moved from the areas best served by rail.[54] Of particular significance, the automobile took the focus of travel away from downtown Los Angeles, both bringing the demise of rail and changing the city's structure to create a dispersed autopolis. It provided what Hilton has called a "lateral mobility," and the opportunity to travel directly from anywhere to anywhere, which the electric railway had denied.[55]

Fogelson describes how increasing congestion led drivers to avoid the central business district, not to return to the electric trains. Sixty-eight percent of those living within ten miles entered downtown daily in 1923. This had fallen to 52 percent in 1931, a decline of 24 percent. By the mid-1920s, outlying centers had developed at which suburban dwellers could work and shop. The car not only competed with the electric railway, but also established the urban structure in which that competition took place, charting an automotive future which displaced the patterns the railway had previously inspired.[56]

Increasing congestion slowed down trolleys as well as cars. Meanwhile, during the 1920s the costs of automobile ownership steadily fell.[57] While trolleys had a hard time competing with private cars in terms of speed, convenience, or cost at the best of times, the cash-starved rail companies spent little on capital improvements, leading to service deterioration.[58] Users complained of crowded cars, high fares, and slow service.[59] The automobile competed selectively, attracting weekend and off-peak leisure traffic relative to peak-hour ridership.[60] Equipment previously in use all day now lay idle outside the peaks. This structural change helps account for the fall in Pacific Electric revenues from 1923, even when total numbers of passengers held steady.

During the early 1920s, almost all excursion services were canceled by the Pacific Electric. Its most famous—to Mount Lowe—remained in operation at mid-decade, but saw a drop in patronage from 160,930 passengers in 1921 to 118,404 in 1925. Veysey tells of a new joint enterprise started in 1926 which combined motor coach sightseeing with a rail trip to Mount Lowe. This very fact, he says, illustrates the inability of the railway to reach new attractions away from its fixed tracks but demanded by tourists in an automotive age.[61]

While the depression hit the Pacific Electric hard—by 1934 patronage was one-third below the 107 million passengers carried in 1929—automobile registrations dropped only slightly.[62] The last possession any family would surrender was its car.[63] More to the point, the post depression years saw automobile ownership mount: while there had been 843,536 vehicle registrations in 1931, and 808,640 in 1933, there were 884,521 in 1935 and over one million in 1937.[64] By contrast, while the number of interurban revenue passengers carried did increase from sixty million in 1933 to seventy million in 1936, it dipped again to sixty-three million in 1939. Pacific Electric, furthermore, lagged national business indices for the entire period from 1932 to 1938.[65]

Competition with the Bus

The bus became a potent rival to the Red Cars as early as 1914, long before any claim could be made that General Motors unfairly promoted its cause at the expense of rail. In July of that year Pacific Electric opened its San Bernadino line, but a good paved road allowed bus competition to begin only three months later.[66] Eighteen hundred jitney licenses were also in effect in 1914, carrying 150,000 passengers per day in December.[67] By 1916, Pacific Electric President Paul Shoup was so concerned at the loss of his passengers, that he declared:

> Your Long Beach lines have fallen to the point where they . . . do not even pay the transportation expenses, that is the power to move the cars over the tracks. Some of your tracks have disappeared, and others must go. It is up to the people of Long Beach to decide. If you want the Pacific Electric Railway to continue to maintain a system in this city you should impose the same conditions with regard to franchises, taxes and street regulation that are enforced upon the present railway lines on our present common carrier competitor [the jitney].[68]

On July 12, 1917, Pacific Electric became a bus operator, itself. A competing bus service between San Bernardino, Highland, and Patton was taking away Pacific Electric streetcar passengers. Pacific Electric bought the bus line, starting a trend of buying out the bus competition rather than attempting to beat it by offering superior rail service.[69] Pacific Electric would also establish its own feeder bus services to reach outlying communities well beyond the economic reach of rail. As of January 1927, Pacific Electric operated thirty-two bus routes over two hundred route miles, carrying on average 820,000 passengers per month.[70] In 1930 it bought out Motor Transit, a major competitor. With the Motor Transit service route between Fullerton and Los Angeles more direct, rail service to La Habra, Yorba Linda, Stern, and Fullerton in Orange County ceased.[71]

During the early 1930s, bus design improved significantly, and it became apparent that financial conditions could benefit from substituting buses for unprofitable rail services.[72] The bus showed it was earning favor with the public, and many articles in transit publications boasted of its attractions relative to the streetcar. One article showed how "de luxe" bus service was being developed to lure those who had abandoned the streetcar for the automobile.[73] Another article points to the ability of the bus to provide a kind of direct service not available by streetcar to newly developing low-density suburban areas:

> At first, buses were looked upon merely as potential feeders to existing rail lines, in order that the railway investment might be preserved without loss to security holders. But the patrons of these feeder routes failed to take kindly to the necessity of transferring from [street]car to bus and it soon became apparent that unless through bus routes were operated into the business centers of the community, the traffic would not support the bus service. . . . Then followed the idea of changing the feeder bus lines to through bus routes, the gradual elimination of poorly patronized trolley routes, and the substitution of suitable capacity buses. . . . The extreme flexibility of this easy riding, rubber-tired transportation unit has demonstrated that it has many advantages both to the traveling public and to the transportation merchant. Instead of making extensions with track, buses have been used to develop new territory. Many single-track trolley routes have outgrown their usefulness and the bus is now giving better and more frequent service.[74]

A further article mentioned that buses even coped better with congestion than did streetcars, since they could be rerouted to avoid it.[75] "Express service" could be installed, furthermore, by "taking loaded equipment off congested thoroughfares and expressing them via the fastest possible route."[76] A 1936 article reported on "How the New Gas Buses Have Improved Earnings." "Because the new buses are faster and quieter [than streetcars], and particularly because they have been operated on more frequent headways, they have proved very popular with the riding public." In addition to patronage gains on many lines where substitution took place, "the operating costs have generally been reduced."[77]

Thus historical evidence goes against the claims of modern-day rail proponents that people will prefer trains to buses because they are faster and more "modern," even if a train trip requires a transfer not needed when traveling by bus and operates at a lower frequency. Quite the reverse held true in the 1930s: the more flexible service made possible by buses suited the needs of Los Angeles better, and the same holds true today.

Pacific Electric Shifts to the Bus

As General Motors stated in replying to Snell: "The truth is that both the Pacific Electric and the Los Angeles Railway began to abandon streetcars before GM was even in the bus business and long before National City Lines or any of its affiliated companies were even organized."[78] In 1923, all but four of the streetcar lines in Pasadena had been abandoned, to be replaced by "a complex system of motor coach lines."[79] This was the first large-scale replacement of trolleys by buses. In 1928, replacement of the Upland–Ontario interurban service with buses went ahead, the company arguing that buses would result in labor cost savings. One-operator buses replaced two-operator streetcars, and a faster bus running time allowed the schedule to be operated by one bus instead of two rail cars, providing major savings on two counts.[80]

With existing streetcars becoming obsolete and without capital funds for replacement; with competitive pressures from the automobile mounting; and with the effects of decentralization making the core-focused and inflexible interurban system ever less viable, pressure grew for Pacific Electric to shed its loss-making electric services and replace them with bus operations. Between 1936 and 1940, Pacific Electric officials increasingly favored converting important links in the rail system, convinced that rail could be profitably replaced by bus services. The major rail line between Riverside, San Bernardino, and Redlands was thus converted to road service on July 20, 1936, apparently with little opposition.[81] During that year, the nationwide total miles of motor coach route exceeded total miles of electric railway route for the first time.[82] In 1926, 15 percent of Pacific Electric passenger mileage was accounted for by travelers on its own buses. This had grown to over 35 percent in 1939, the year before it is claimed that General Motors had any role in acquiring part of the system.[83]

The California Railroad Commission examined the condition of the Pacific Electric in an exhaustive survey over 1938–1939, finding excessive debt, aging equipment, and inadequate maintenance.[84] Service was also deteriorating due to increasing road congestion and delays caused by at-grade road crossings. Observing, also, rising costs, and patronage hit by automotive competition, the commission ruled that "[o]ne of the logical sources of reduced costs is substitution of motor coach service for rail lines which were constructed at large investments and designed to carry traffic far in excess of that which now presents itself."[85]

In permitting conversions from rail to bus on the local lines of the Los Angeles Railway in 1941, the commission declared that "New motor coaches of modern design with uniform high rates of acceleration and deceleration, high free running speeds and trackless maneuverability in replacement of obsolete, slow, noisy, rail cars restricted to use of tracks located in the street center, will expedite the freer flow of vehicular traffic and allow a more efficient utilization of the street surface."[86] Railroad abandonment proceeded apace; a particularly major conversion occurred in November 1941, when all rail service on the Pomona and San Bernardino lines east of Covina was terminated, even though rail equipment had been modernized for the service only eighteen months previously. Pacific Electric officials stated that their decision was due to the fact that more people in the intermediate area now lived along or adjacent to the highway than to the rail line.[87]

Postwar Decline

Nationwide transit ridership fell 26 percent from its 1946 peak to 1950, and another 28 percent over the next four years. The greatest losses occurred before the freeway construction boom following creation of the Federal Highway Trust Fund in 1956 and before Congress had authorized 90:10 federal-state matching for interstate construction.[88] Jones cites three reasons for the decline in the fortunes of transit. "Normalization", that is the end of wartime gas and tire rationing; the return to "normal" peacetime employment levels; and the lagged introduction of the five-day workweek, simultaneously increased the feasibility of automobile use and reduced the demand for public transportation. Secondly, declining patronage—further discouraged by increasing fares and decreasing service levels—led to reduced revenues, while postwar wage increases produced higher transit operating costs. Finally, the growth in personal income and automobile ownership not only further reduced the market for transit, but led to a reorganization of urban activities to serve an increasingly "auto-mobile" population. This dispersion of activities placed service needs quite out of sync with the kind of service transit could best provide.

Following the Second World War, during which Pacific Electric patronage had swelled, patronage fell off once more, while the trend toward automobile transportation continued and intensified. Service, especially in terms of frequency of departures, continued to be cut back between 1948 and 1950 in response to rapidly falling patronage.[89] While total passengers carried on the Pacific Electric fell from a peak of 182 million in 1945 to 102.7 million in 1951, motor coach operation significantly expanded between 1945 and 1950.[90] Pacific Electric rail operations had been profitable in 1943 and 1944 for the first time since 1923, but by 1946 the annual loss was $2.2 million and climbed to $3.4 million on a gross revenue of about $10.5 million the next year. As Karr points out, such losses escalated to today's dollars are not only substantial "but simply impossible for private enterprise to bear."[91]

Eli Bail succinctly describes the response. Pacific Electric looked for a more economical solution when it saw there was no quick fix for a run-down rail plant operated by obsolete methods with antiquated equipment. World War II had brought diesel engine and hydraulic transmission technologies to new heights of efficiency and reliability, and along with every major transit operator in the country (except for Chicago, which was attracted to propane), Pacific Electric opted for the obvious. "The fact that General Motors had a better engineered product to sell and had, like a good many enterprising businesses, set up favorable financing arrangements, have somehow been held as "unfair advantages." "Whatever happened to free enterprise, anyhow?"[92]

In 1948, Pacific Electric claimed that three-quarters of its 1947 passenger losses had been due to interurban rail operations and, petitioning the Railroad Commission, stated that "to do anything other than give serious consideration to possible substitutions of motor coaches for rail lines or to abandon service in some measure would be futile."[93] On March 1, 1949, Pacific Electric filed for a sweeping series of rail passenger abandonments, which constituted its complete program for modernization.[94] On May 9, 1950, the commission gave the go-ahead for virtually all that had been requested. It found that Pacific Electric "cannot continue to operate its outmoded and obsolete rail facilities at the losses indicated in this record" and that it was in the public interest to authorize the changes in service as provided in this decision. In so doing, we are taking into account applicant's commitment to provide new, modern motor coaches to replace the rail passenger facilities . . . The record clearly shows that the passenger rail operations of applicant have been conducted at a loss over a long period of time: On the other hand, its motor coach service has been operated at a profit.[95]

A public outcry ensued, and the Los Angeles City Council did get the state Board of Public Utilities and Transportation to temporarily rescind its approval, but the board said it could not force the company to continue operating at a loss and unanimously reaffirmed its original decision, leaving Pacific Electric free to act.[96] The year 1950 saw several rail abandonments. What remained of rail service, along with substantial motor coach operations, were sold to Metropolitan Coach Lines in 1953. The rail services were later entirely dropped, either by Metropolitan or its successor.[97]

The Long Beach line was the last of the Pacific Electrics to go. In 1910, Long Beach service had operated every twenty minutes, and taken forty-one minutes for an end-to-end trip. The following year, service was down to half-hourly, but express trains took only thirty-six minutes end-to-end. By 1926 the trip took fifty-one minutes, due to the effect of at-grade crossings and road congestion. In 1944 it took fifty-four minutes, and in 1954 sixty minutes, with trains usually between ten and thirty minutes late.[98] Frequency, which had risen to quarter-hourly in 1946, was by then back to half-hourly. The last train ran on April 8, 1961, arriving in Long Beach shortly after midnight with a contingent of Red Car fans.[99]

Streetcar service continued for a while on the local lines of the Los Angeles Railway which, like Pacific Electric, had been converting rail to bus operations as fast as possible. In 1958 both of Los Angeles' private operators sold their holdings to the Los Angeles Metropolitan Transit Authority. The authority continued operation of the remaining local streetcar lines through 1963, when the MTA annual report declared that "[o]ne of the highlights of 1963 was the successful conversion of the five remaining local streetcar lines and two trolley coach lines to modern bus operation. This changeover was accomplished smoothly after a concentrated public information campaign to acquaint the public with the added convenience, comfort and efficiency of the new operations in their particular geographic area."

Did General Motors Have an Impact?

That General Motors did have some impact is without doubt. As Jones sees it, General Motors market domination dampened competition in bus manufacturing, and probably retarded innovation in bus design. Few American improvements were made in diesel technology after the 1950s, although innovations did continue in Europe.[100] More importantly, perhaps, GM's dominance in the bus market with a diesel product led to the preemption of routes and markets which might have been more economically served by the electric trolleybus. As a result, Jones says, development of the trolleybus lagged in the United States, leaving operations to a noisy, rumbling, and fume-emitting vehicle. While it may have served its users well, it was unpopular with motorists, pedestrians, cyclists, and residents living on bus routes, and contributed a negative image to the transit industry.

Jones characterizes the National City Lines venture as one in market breaching and market cornering, rather than a conspiracy to remove a threat to the automobile. Its importance, he says, is in its concentration of economic advantage in bus manufacturing, rather than in the supposed destruction of viable street railways. Other developments during the depression years were more influential in steering the transit industry away from rail, although these have been neglected in debating the conspiracy theory of the decline of transit.[101]

Would Public Control Have Saved the Red Cars?

As electric railway service sunk into decline, proposals were advanced for rail solutions for Los Angeles. Support for these was splintered, however, by competition between coalitions representing the interests of downtown and advocates for regionally-dispersed centers away from downtown. This "prevented the emergence of a government agency sufficiently powerful to impose upon the region a downtown-oriented rail transit program, which would have included upgrading the Pacific Electric Railway."[102]

This was a period, furthermore, when public transportation was still generally seen as a private responsibility, while the growing popularity of the car, untainted as yet by associations of environmental damage, was attracting increasing government attention. As early as 1909, local government had begun to assume responsibility for road construction. The state began funding highway construction the following year, and the extensive freeway system, starting with the Arroyo Seco Parkway (later renamed the Pasadena Freeway) completed in 1940, was a state and, later, federal responsibility.[103] While the state commitment to the highway program was clear, there was no coherent plan or any sign of financial support for transit, despite all the publicity of declining service. As Bail says, the increase in car ownership and freeway mileage meanwhile made any plan other than for bus substitution academic.[104]

It can be argued that government actively participated in promoting the highway systems of Los Angeles (even if users contributed to the costs through taxes), while similar support was not provided for the interurban railroads. It should be noted, however, that urban street improvements were slow in coming during the 1920s, whereas by the close of World War II, only eleven miles of freeway had in fact been opened in Los Angeles. By then the Pacific Electric had already been on the decline for well over twenty years.

More to the point, however, while it can be argued that government support might have saved the Red Cars, it is clear that it would have been futile for the government to have done so given auto-driven changes in the spatial organization of Los Angeles County that decentralized employment, shopping, services, and recreation as well as residences. These fundamental changes irreparably took away the market of the Red Cars and left the interurban rail system as no more than a monument to a technological past, one now transcended by the highway, the car and the bus.

The Snell report provides an example of concepts living on even when the technology relating to them has been displaced. Snell thinks of Los Angeles in terms of the Red Car's heyday, and cannot see that the autopian town of ultra-dispersed urban functions and non-centralized transportation demands could not be served by the old technology. In fact, he is dreaming of the old Los Angeles associated with the Red Cars—"a beautiful city of lush palm trees, fragrant orange groves and ocean-clean air"— and will not countenance the reality of the new. The erroneous conspiracy theory, with its roots in the positive imagery of a railborne Los Angeles of yesteryear and its negative imagery of the freeway and the bus, was to play an important role in forming perceptions that rail had a place in the Los Angeles of the twenty-first century.

CHAPTER 3   **Evaluating the Promise and Performance of Rail**

For instance, if rail rapid transit was even a way—much less the way to "cure" the traffic congestion and pollution problem, then why is congestion and pollution so bad in New York, Chicago, Boston and Philadelphia?

— Martin Wohl

"LOOKS LIKE A WINNER SO FAR" read the headline of the January 21, 1991 *Los Angeles Times* editorial praising the downtown Los Angeles–Long Beach Blue Line light rail service after its first six months in operation:

> Nowhere has the challenge for kicking off a mass transit system been greater than in car-crazy Los Angeles.
>
> But now 18,000 commuters daily climb aboard the Blue Line for the run between Los Angeles and Long Beach. Those numbers wildly exceed expectations, climbing to three times the projections of six months ago when Blue Line set off on its inaugural run. Its initial success bodes well for Southern California mass transit . . .
>
> The Blue Line's ridership provides a useful gauge in assessing whether commuting Southlanders might hang up their car keys and give mass transit a try . . .
>
> Freeing the region from the increasing paralysis of gridlock will make commuting easier, the air cleaner and provide wider access to jobs for those in the suburbs. The Blue Line is a splendid beginning.[1]

The evidence of apparently full trains provides tantalizing support for declaring rail successful. It is easy to see why the *Los Angeles Times* took the indicator at its disposal (a misleading one, as we will see in chapter 5: "expectations" were revised sharply downward just prior to system opening) as evidence of both immediate achievement and of hope for the future. Such ridership numbers, while attention grabbing, do not tell the whole story. It is necessary to understand the costs at which such ridership was produced, and not only in terms of implementing service on the Blue Line. Costs were incurred, also, in terms of the threatened viability of the much larger MTA bus system and its ability to attract ridership. When a systematic picture is constructed, the achievements of the Blue Line look less rosy even though, in the fiscal year 2000 (beginning July 1999), ridership was to stand at 61,714 average weekday rides, a result 12.8 percent in excess of the original forecast of 54,702 rides for that year prepared by SCAG.[2] Since this book principally concerns the Blue Line, this chapter will focus on providing an evaluation of the reasonable expectations and actual performance to date of this light rail service. Illustrative examples will also be provided from the experiences of other rail and bus transit systems in various cities, and the Blue Line will be placed in the context of other services operating in Los Angeles County.

Service Description

The Long Beach light rail service operates over twenty-one miles from the Seventh Street/Metro Center Station in downtown Los Angeles, through Watts, Compton, and Carson to downtown Long Beach (see map 3-1). It serves twenty-two stations.

Light rail technology, such as is employed on the Blue Line does not require the grade separation of heavy rail subway systems, and includes street running. Most of the track is on the surface, with six miles of operation in street traffic and numerous grade crossings elsewhere on the route. The line enters downtown Los Angeles in a one-mile subway, and there are several short stretches of elevated structure on the route, provided to cross over congested areas.[3]

Blue Line service commenced on July 14, 1990, together with a complementary bus system designed to provide feeder connections to light rail, mostly through changes to existing bus lines. Other buses provide a grid system of services, operating north-south and east-west (see map 3-2). The twenty-mile Century Freeway Green Line light rail service commenced operation on August 12, 1995. The Green Line operates on an east-west alignment from Norwalk to near Redondo Beach, running close to (but not reaching) the Los Angeles International Airport, and intersecting with the Blue Line at the Rosa Parks station. The Gold Line, originally conceived as an extension of the Blue Line, but now operating as a separate line, began

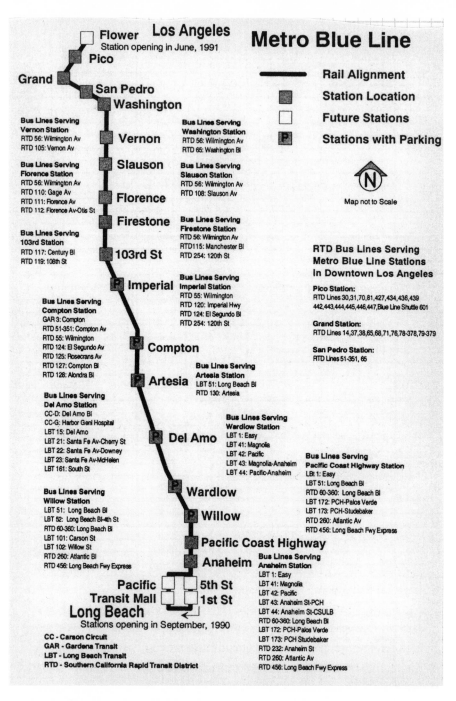

Map 3-1. Blue Line as of opening, July 1990. (Source: Southern California Rapid Transit District)

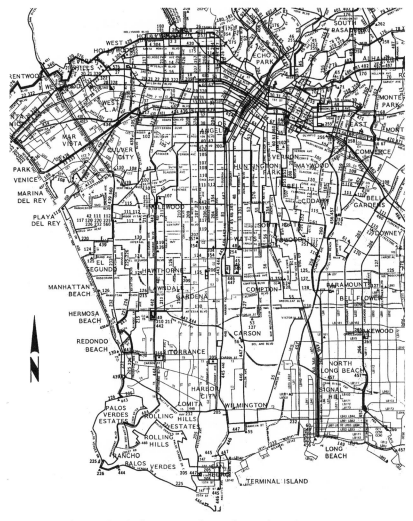

Map 3-2. RTD bus services in the Long Beach corridor as of March 1990.

operations on July 26, 2003. It provides service on a 13.7 mile route between Union Station in downtown Los Angeles and Pasadena.

A heavy rail subway line, the Red Line, opened on January 30, 1993. It initially ran 4.4 miles through downtown Los Angeles and then west along Wilshire Boulevard. It was subsequently extended along Wilshire Boulevard as far as Western Avenue. A second branch, running north from Wilshire Boulevard through Hollywood to a terminal in North Hollywood, was completed on June 24, 2000, taking the Red Line to its current 17.4 miles. A total

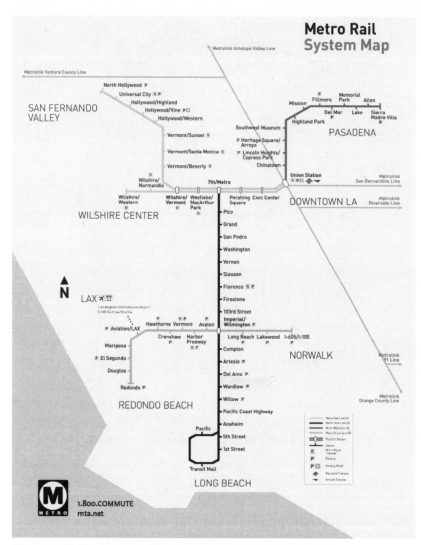

Map 3-3. MTA Rail System as of July 2003. (Source: MTA)

of seventy-three miles of rail route is in operation as of the Gold Line opening. The MTA plans to begin construction of a six-mile extension of the Gold Line from Union Station to East Los Angeles in fiscal year 2004. There is interest, also, in extending the Gold Line a further twenty-two miles from Pasadena to Claremont.[4] Map 3-3 shows currently operating MetroRail services in Los Angeles County. The full Proposition A system was ultimately planned to provide service to the points shown on map 3-4.

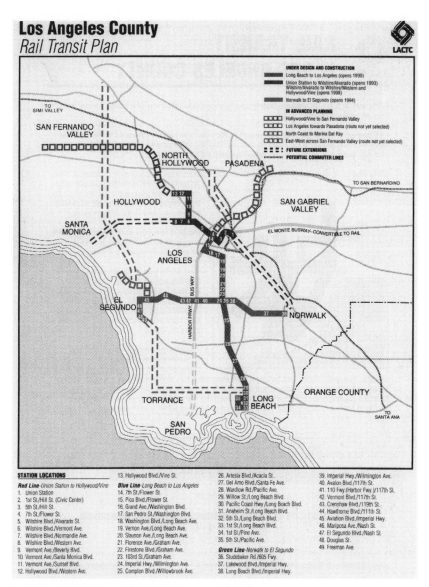

Los Angeles County
Rail Transit Plan

STATION LOCATIONS

Red Line-Union Station to Hollywood/Vine
1. Union Station
2. 1st St./Hill St. (Civic Center)
3. 5th St./Hill St.
4. 7th St./Flower St.
5. Wilshire Blvd./Alvarado St.
6. Wilshire Blvd./Vermont Ave.
7. Wilshire Blvd./Normandie Ave.
8. Wilshire Blvd./Western Ave.
9. Vermont Ave./Beverly Blvd.
10. Vermont Ave./Santa Monica Blvd.
11. Vermont Ave./Sunset Blvd.
12. Hollywood Blvd./Western Blvd.

13. Hollywood Blvd./Vine St.

Blue Line-Long Beach to Los Angeles
14. 7th St./Flower St.
15. Pico Blvd./Flower St.
16. Grand Ave./Washington Blvd.
17. San Pedro St./Washington Blvd.
18. Washington Blvd./Long Beach Ave.
19. Vernon Ave./Long Beach Ave.
20. Stauson Ave./Long Beach Ave.
21. Florence Ave./Graham Ave.
22. Firestone Blvd./Graham Ave.
23. 103rd St./Graham Ave.
24. Imperial Hwy./Wilmington Ave.
25. Compton Blvd./Willowbrook Ave.

26. Artesia Blvd./Acacia St.
27. Del Amo Blvd./Santa Fe Ave.
28. Wardlow Rd./Pacific Ave.
29. Willow St./Long Beach Blvd.
30. Pacific Coast Hwy./Long Beach Blvd.
31. Anaheim St./Long Beach Blvd.
32. 5th St./Long Beach Blvd.
33. 1st St./Long Beach Blvd.
34. 1st St./Pine Ave.
35. 5th St./Pacific Ave.

Green Line-Norwalk to El Segundo
36. Studebaker Rd./605 Fwy.
37. Lakewood Blvd./Imperial Hwy.
38. Long Beach Blvd./Imperial Hwy.

39. Imperial Hwy./Wilmington Ave.
40. Avalon Blvd./117th St.
41. 110 Fwy.(Harbor Fwy.)/117th St.
42. Vermont Blvd./117th St.
43. Crenshaw Blvd./119th St.
44. Hawthorne Blvd./111th St.
45. Aviation Blvd./Imperial Hwy
46. Mariposa Ave./Nash St.
47. El Segundo Blvd./Nash St.
48. Douglas St.
49. Freeman Ave.

Map 3-4. Proposition A rail transit plan. (Source: MTA)

As of July 2001, light rail service between downtown Los Angeles and Long Beach is at peak-hour intervals of five to six minutes, with service every twelve minutes off-peak. Service starts at 5 A.M. from both termini (with earlier trains from some other stations), with the last train north from Long Beach at 10:21 P.M. and south from downtown at 11:45 P.M. The end-to-end trip takes approximately fifty-five minutes, giving an average speed of 23 mph.

Gold Line service operates every ten minutes during peaks, twelve to twenty minutes at other times. A trip takes thirty-six minutes at an average speed of 23 mph. Green Line service operates at seven minute peak and fifteen-minute off-peak intervals, completing a trip in about thirty-six minutes at an average speed of 33 mph. Unlike the Blue Line, the Green Line is grade-separated, enabling the faster operating speed to be achieved. Red Line service is provided every ten minutes during peaks and twelve minutes off-peak on each of the two branches, making for a central area headway of five to six minutes. A journey from North Hollywood to Union Station takes twenty-nine minutes, at an average speed of 30 mph, while the shorter trip from Union Station to Wilshire/Western takes thirteen minutes at an average speed of 23 mph.

Systemwide Metro Rail fares are at the same levels as for local bus service, whatever the distance traveled. The one-way full adult light rail fare was $1.10 from August 1, 1990 to January 31, 1995, when it was increased in line with local bus fares to $1.35. As of January 1, 2004, the local fare for both bus and rail was reduced to $1.25 as part of a restructuring. The fare is substantially lower than equivalent express bus fares (which charge supplements for freeway service). It can be further reduced by purchasing bags of ten tokens for eleven dollars. Fares are paid on an "honor" system: passengers use self-service ticket vending machines at stations or bus transfers, or they purchase passes. Police check for fare payment and also to provide security coverage.

Evaluating Transit Alternatives

"My overall impression of this is that your transportation planners are trying to impose a 19th century technology on a 20th or 21st century city," said John Kain (at the time head of Economics at Harvard University), as he addressed the Executive Committee of Southern California Association of Governments at a special session to which he and Mel Webber (of the University of California, Berkeley) had been invited to review the agency's transportation plans. "I can't understand, on any rational basis at least, this fascination with light rail. . . . The reason is that light rail seems to be nothing more than a slow, low-capacity express bus system that can't run on the city streets. . . . [L]ight rail is incredibly more expensive than a well-designed express bus system." Surmising that the attraction of light rail "has to do with the popularity of Lionel toy electric trains," Kain concluded that "Trains are lots of fun, but if you want to solve the region's transportation system problems, and you want people to use transit, spend your dollars on something where you're going to have some impact . . . and not to make a bunch of hobbyists happy."[5]

We shall see (in chapter 10) that Kain's theory of toy train sets does indeed provide some explanation for the attraction of trains. By the time the conclusion is reached, we will appreciate why a statement of economic facts—which made up most of Kain's presentation—is likely to be inadequate for making a case to political decision makers whose approach to understanding is based on a quite different, symbolic, system. But this chapter will concentrate on providing an assessment of the promise and performance to date of rail transit in Los Angeles.

Kain's frustration typifies that of much of the academic profession as a whole, and of economists in particular. As noted in the introduction to the proceedings of one symposium which concluded that "Rail rapid transit is probably the worst step Los Angeles could take to improve transportation," participants had been unanimous in their analysis and recommendations for transportation policy. "On the one hand, it is novel for 'the experts' to agree so conclusively. On the other hand, this agreement suggests that those who have taken a close and detached look at the transportation problems of this region have something important to communicate to society at large."[6]

Reports by Don Pickrell of the Transportation Systems Center, U.S. Department of Transportation, and by myself, have provided major evaluations of the range of new rail transit systems opened across the United States over the last few decades.[7] Pickrell evaluated the performance of ten new federally funded rail transit projects. He found that only in the case of Washington, D.C. was rail patronage more than half of that forecast and, even in that case, ridership was 28 percent lower than anticipated. The consistent overestimation of future ridership, Pickrell wrote, suggested that the benefits achieved are far below those expected by local decision makers who selected the projects.[8]

Pickrell also found capital cost overruns in nine out of ten cases, while only one out of the ten projects had managed to keep within its operating cost estimates. He concluded that decision makers might have made different decisions, given better data.

My own evaluation of rail transit performance took a different approach. While Pickrell compared forecast ridership and costs to those actually achieved, I was more interested in looking for total system effects: yes, there may be a given number of riders on a rail system, but what sort of impact has the rail innovation had on the patronage of the transit system as a whole? Similarly, what have been the systemic implications of new rail projects on the overall financial viability of transit systems?

The Richmond study evaluated all wholly new rail projects in operation in the United States as of April 1997, together with a light rail reconstruction in Pittsburgh; heavy rail in Miami and Los Angeles; Miami's people mover; and a number of bus innovations. From the evidence of the raw ridership numbers, the results were not encouraging. Only a few of the rail systems had reached initially forecast ridership. The rail system in St. Louis and the San Diego Blue (South) Line had exceeded ridership forecasts by a substantial margin, and the Los Angeles Blue Line, on track to meeting year 2000 forecasts at the time of report writing, has now exceeded them.

The San Diego Orange (East) Line as well as light rail in Buffalo, Pittsburgh, Portland, Sacramento, and San Jose, and heavy rail in Miami, attracted strikingly less ridership than initially forecast. Compared to an initial forecast ridership made in 1971 of 160,000 weekday boardings for the Buffalo light rail system in 1995, and a 1981 pre-opening estimate of 45,400 weekday boardings for 1995, there were only 26,115 weekday boardings that year. Portland, Oregon's 27,000 weekday boardings in 1996 compares to the forecast of 42,500 made in 1978 for the year 1990.

The greatest number of total annual light rail boardings on the new systems in 1995 took place in San Diego, with 15.6 million boardings. St. Louis was in second place, with 12.5 million boardings, and Los Angeles slightly behind that with 12.0 million passengers on the Blue Line (in addition to the 15.9 million boardings on the Red Line heavy rail). In all but two cases—Sacramento and San Diego, where bus boardings were slightly more than double rail—rail ridership was significantly less than that carried by bus in 1995. Taking all systems together, eight times as many passengers boarded buses as trains in 1995.[9]

Table 3-1 shows Los Angeles MTA rail and bus system ridership as of the

Table 3-1. Los Angeles MTA Ridership FY 2000 and FY 2003

Boardings	Rail Lines					Total
	Blue		Green	Red		Total Transit
Rail				Total Bus		
Average weekday						
FY 2000	61,714	29,610	83,230	174,554	1,106,821	1,281,375
FY 2003	69,646	31,935	95,042	196,623	1,126,111	1,322,734
Total annual (millions)						
FY 2000	20.41	9.45	27.96	57.82	359.00	416.82
FY 2003	22.16	9.92	31.46	63.54	365.73	429.27

2000 and 2003 fiscal years (July 1999–June 2000 and July 2002–June 2003). The MTA has experienced ridership data problems in recent years. Alternative methodologies, one based on monthly surveys and the other generating data to meet the National Transit Database (NTD) annual reporting requirements, have produced different results. Average weekday Blue Line ridership, for FY 2000, for example, was shown as 57,484 based on the monthly methodology and at 61,714 in NTD reporting. Red Line ridership that fiscal year, believed by the MTA to be underreported by monthly data, was shown at 62,642 average weekday riders by that methodology compared to 83,230 from NTD data. In FY 2001 and FY 2002, MTA believes the methodology overstated Red Line ridership.

In table 3-1, NTD data, which provides a relatively more favorable view of rail ridership, is displayed for FY 2000, while the FY 2003 ridership data shown is derived from MTA's monthly data methodology. MTA states that for FY 2003, methodology corrections have been made and it expects its annual NTD statistics to be similar to results from its monthly data. The NTD data was not yet available as of July 2003.[10]

Using the NTD results, total Los Angeles rail boardings in the fiscal year 2000 came to 13.9 percent of total MTA transit boardings in that year. By fiscal year 2003, with service extended to North Hollywood, there were 95,042 Red Line average weekday boardings, some 32 percent of the 298,000 boardings forecast for 2000 in the 1989 Supplemental Environmental Impact Statement which defined the system as it was more or less ultimately built.[11] Ridership for all rail lines together made up 14.8 percent of total MTA boardings in the 2003 fiscal year.

If we look at historical trends, rail investments in Miami, Sacramento, and San Diego have been associated with increases in total transit boardings. Between 1985 and 1995 transit boardings increased by 20 percent in Miami and 43 percent in Sacramento. Between 1980 and 1995, total transit boardings increased by 46.1 percent in San Diego. St. Louis saw a 15 percent increase in total transit ridership from 1990 to 1995 during which light rail service began.

In Buffalo, Pittsburgh, and San Jose, the inception of light rail could not stop total transit ridership from falling. In Los Angeles, there had been 497.2 million total transit boardings in 1985 for MTA's predecessor operating agency, RTD, but only 424.6 million boardings in 1988. The fall in ridership reflected bus fare increases and service reductions as resources were diverted to rail. Comparison of Los Angeles ridership immediately following this period is complicated by the transfer of part of RTD bus service to the Foothill Transportation Zone, which began operation in December 1988, because the riders transferred to Foothill services are no longer included in RTD/MTA data. For the fiscal year 1990, RTD boarded 401.1 million transit rides, all of them on bus. For the fiscal year 2000, total combined MTA bus and rail ridership, with the Blue and Green rail lines operational and the Red Line partly complete, was slightly higher, at 416.8 million.[12] The total had increased to 429.3 million in fiscal year 2003, prior to the opening of the Pasadena Gold Line but with the Red Line completed.[13]

All ridership results post-rail should be treated with additional caution because the general trend across the nation has been for the number of transfers to increase with the introduction of rail service. A passenger who previously completed a journey by bus alone would be reported as one boarding, but that same passenger, requiring a bus and then a rail trip to complete the same journey after the inauguration of rail, would be reported as two boardings, making it appear that an increase in transit usage had taken place, rather than merely an increase in transferring between transit modes.

Ridership results must also be understood in the light of fare policies designed to promote usage of new rail systems. Light rail travel was made free in Portland's "Fareless Square" and the length of Buffalo's pedestrian mall, for example. Fare structuring has also often helped the competitive position of new rail systems. In Sacramento, for example, a discount fare applies for downtown trips.

The frequent use of low flat fares has meant that the cost per mile of traveling by train has often been substantially less than the equivalent unit charge by bus. The average trip made by Los Angeles Blue Line in the fiscal year 2000 of 7.2 miles was double the average trip made by MTA bus of 3.6 miles, but the rail fare of $1.35 at the time was equal to the local bus fare.[14] Fares on express bus services, which are designed for the distances typically traveled by Blue Line riders, vary according to the specific mileage on the freeway but are higher, with a minimum of $1.85 as of fiscal year 2000, reduced to $1.75 as of January 1, 2004. Were the now-terminated Long Beach Express bus 456, which made local stops in Long Beach and downtown Los Angeles, and operated nonstop on the Long Beach and Santa Ana Freeways in-between, to still be in operation, its fare would have been $2.85 at fiscal year 2000 rates, $2.25 as of January 1, 2004.

Passengers transferring to the Red Line from buses, which now terminate at the North Hollywood subway station, rather than running express by freeway to downtown Los Angeles, have similarly seen a substantial fare reduction compared to the previous direct-to-downtown bus service, which cost at least $2.35. If rail services were priced equitably compared to bus prices, ridership would be less.

Consideration has been given to charging zone fares on light rail as on express buses. A September 1989 simulation carried out by RTD, however, showed that while zone fares could be expected to produce greater overall revenue than flat fares, this would come at a cost of lost ridership.[15] While the discriminatory fares policy has helped the Blue Line beat forecast ridership for the fiscal year 2000, the Blue Line fare advantage was not modeled in forecasting. Reaching forecast ridership is therefore more a case of artificially pricing the market to produce the forecast than a vindication of the forecast.

It is also critical to understanding ridership results to appreciate that many rail riders were already using bus transit prior to the commencement of rail services and thus do not constitute added transit ridership. The start of rail has generally meant a restructuring of bus services to feed rail stations, and a discontinuation of direct-to-downtown buses. Former bus riders, who generally form the largest component of rail ridership, have little choice but to take the train. Express bus service from Long Beach to downtown Los Angeles was terminated in connection with the start of Blue Line service, while former San Fernando Valley express bus service to downtown Los Angeles now terminates at the North Hollywood Red Line station. Direct crosstown bus service duplicating the route of the Green Line was split into separate bus lines to feed the Green Line and discourage all-bus travel.

While a 1990 survey found that only 24.7 percent of San Diego light rail passengers had previously traveled by bus (36.9 percent had previously driven alone; 13.1 percent had carpooled),[16] a 1994 survey in Baltimore found that half of light rail riders in that city had previously made the journey by bus. Only 22 percent had driven solo (an additional 5 percent had carpooled or had been dropped off).[17] A November 1990 on-board survey (the most recent cited by the MTA in 1996) found that only 21 percent of Los Angeles Blue Line light rail passengers had previously driven, while 63 percent had taken the bus.[18] An additional 6 percent had been driven by someone else—such as in a carpool—while 6 percent had walked and 4 percent were making a trip they would not otherwise have made. Putting aside passengers who had traveled by bus, 37 percent of the ridership constituted new transit passengers. Members of MTA staff estimate that a similar proportion of Red Line subway riders, approximately one-third, are new to transit.

In Denver, 73 percent of light rail riders surveyed were previous bus users. The single greatest reason given for using light rail (32 percent of weekday riders) was that "bus route begins/ends at Light Rail station."[19] In other words, they had no choice but to take the train.

Service Characteristics

This review of differences between the service characteristics of rail and bus options and their impact on generating ridership focuses on evaluating key reasons rail is popularly thought to provide special advantages over bus service developments, reasons we will see used again and again in succeeding chapters in arguments for instituting rail service.

Rail vs. Bus, or Line Haul vs. Access

The basic design report for San Francisco's Bay Area Rapid Transit (BART) conceived an interurban rail system to provide only arterial connections between major urban concentrations. "We are convinced that the interurban traveler, facing the choice between using his private automobile or using mass transportation, will be influenced in his choice more by the speed and frequency of interurban transit service than by the distance he must travel in his own car or by local transit to reach the nearest rapid transit station."[20]

Yet, as early as the 1925 opening of the Red Cars tunnel into downtown Los Angeles, the importance of minimizing the difficulty of access to the service was quite clear. While the tunnel resulted in shorter trip times, surface streetcar lines—traveling in traffic—remained attractive. As the Board of Public Utilities reported, "Objections are raised by patrons to walking the approximately 700 feet and down several ramps into the depot building to use the tunnel cars, claiming that it is quicker and less inconvenient to travel via the surface lines. This has resulted in the surface cars being unduly crowded."[21]

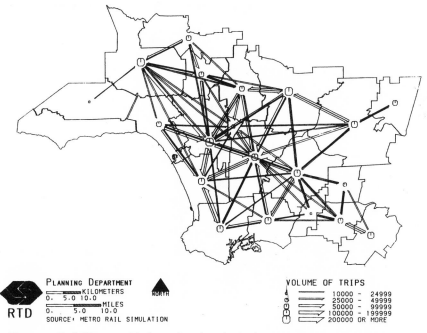

Figure 3-1. Daily Year 2000 PA5 home-based work trips between SCAG Regional Statistical Areas.

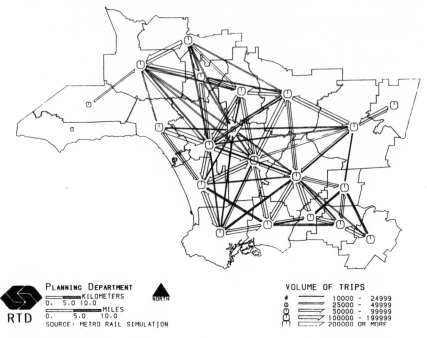

PLANNING DEPARTMENT
━━━━━━ KILOMETERS
0.   5.0 10.0
━━━━━━ MILES
0.   5.0   10.0
RTD
SOURCE: METRO RAIL SIMULATION

NORTH

VOLUME OF TRIPS
━━━━━ 10000 - 24999
━━━━━ 25000 - 49999
━━━━━ 50000 - 99999
━━━━━ 100000 - 199999
━━━━━ 200000 OR MORE

Figure 3-2. Daily Year 2000 PA5 home-based non-work trips between SCAG Regional Statistical Areas.

Figures 3-1 and 3-2 show projected year 2000 trip patterns between re-gional statistical areas conducted for a rail simulation study by RTD: they clearly lack focus. An almost universal criticism of rail in academic circles questions its suitability for providing service between a dispersed set of ori-gins and destinations—such as are encountered in Southern California—precisely because of distances which must be traveled to reach rapid transit. As Mel Webber told the SCAG meeting he addressed:

> [A] transport system has to be able to connect people directly from every-where to everywhere, in effect to offer random access . . . I suggest to you that the perfect model for a transport system is the telephone system, which has a connection directly from everywhere to everywhere. With the tele-phone, it's instantaneous. The closer you can come to instantaneous, direct door-to-door transport, the more likely people would use it. . . .
>
> The reason we failed to eliminate traffic congestion is that the cost of ac-cessing a rail system is high, and I think that's as true here as it was in the Bay Area or more so. The reason it's probably more so is that your land use pat-tern is not linear; you don't match a railroad's geometry.[22]

Reviewing County Supervisor Baxter Ward's "Sunset Coast Line Proposal," then awaiting voter consideration at the ballot (it was to fail to get approval), John Dyckman pointed out that "Mean access time to the rail line is functionally related to the concentration or dispersion of people and activities." With the dispersed and noncentralized population of Los Angeles, the time needed to get to and from rail facilities will be disproportionately high, making rail an unattractive option for most journeys. "The conclusion that one reaches from examining the Sunset proposal is that it has undervalued the costs to the user of waiting, transferring, and interfacing with other 'feeder' modes," Dyckman wrote.[23]

The findings of empirical research show that travelers do not select means of transportation merely on the basis of total journey time but, rather, put extra weight on time they must spend walking and waiting, either to gain access to a mode of transportation or to transfer between modes. In a study in Leeds, England, David Quarmby found that people found time spent walking to be three times as onerous as time spent on the vehicle. Lisco found that Chicago commuters would pay approximately 2.8 times as much to avoid time spent walking as time spent riding. And in New York's Port Authority Bus Terminal, C. Henderson and J. Billheimer found walking time regarded as twice as unpleasant as riding time. Daniel McFadden obtained similar results from surveys of BART passengers.[24]

Allan Nash and Stanley Hille found that "avoidance of a wait of more than five minutes" contributed the greatest perceived difference between the automobile and public transit in Philadelphia and Baltimore. National Analysis reported that the most commonly cited criterion for new system design—expressed by 84 percent of the more than two thousand respondents to a survey in Washington, D.C.—was that the "place to get off [should be] no more than five minutes from the destination." A 1973 survey in Los Angeles suggested adding sedan chairs to the transit system, as Angelenos would not even walk half a mile to a station or bus stop.[25]

The National Analysis study also found that 30 percent of interviewees found any kind of transferring between transportation modes objectionable, while 51 percent objected to specific aspects of transferring. Thus:

[T]he conclusion may be reached that transfer, waiting, and walking time are significant as perceptual choice elements, independent of gross travel time

Figure 3-3. Rail-Bus interchange in Portland, Oregon. (Photo courtesy of author)

and measures, and should be singled out as special elements of public transportation improvement programs.[26]

It is the door-to-door, no wait, no-transfer features of the automobile that, by eliminating access time, make private cars so attractive to commuters—not its top speed. BART offers just the opposite set of features to the commuting motorist, sacrificing just those ones he values most. That was a fundamental mistake.[27]

Buses can provide something much closer to "door-to-door service" than can rail transit. While trains must usually rely upon autos as feeders, or must depend upon bus service and a time consuming mode change as part of a trip, buses can operate on local streets in residential neighborhoods and, after picking up their passengers, enter a freeway or a reserved lane for an "express" trip to downtown at a speed which approximates a rail vehicle.[28]

With a rail system there will be more transferring, waiting, and walking on average than with a bus system. The same bus can provide both local and line-haul service and—lacking attachment to a fixed track—can serve a far wider range of destinations than can be reached by train. The smaller vehicle can be economically used to provide direct service to destinations where rail could never be justified.

Previously, Long Beach and San Fernando Valley express buses entering downtown Los Angeles provided stops every few blocks throughout the downtown area. The Long Beach Blue Line has only one central downtown stop—at Seventh Street/Metro Center, while the Red Line has four downtown stations, including Metro Center, where Blue Line passengers may transfer.

The need for passengers to reach final destinations away from rail stations was shown by experience with the shuttle service linking the initial Blue Line terminal at Pico Street and central downtown prior to the opening of the Metro Center station. According to a September 14, 1990 memorandum from LACTC Executive Director Neil Peterson, the bus shuttle was designed to link the temporary terminal to the permanent one under construction, and not intended to provide circulator service.[29] "There have been requests and inquiries as to the possibility of expanding the Blue Line shuttle to serve more than the immediate Seventh and Flower Street area, perhaps to the First Street and Temple Street vicinity where city, county, and federal employee sites are situated. *The point being that Metro Blue Line riders, for the most part, do not have 7th and Flower Streets in downtown Los Angeles as their ultimate destinations*" (my emphasis). Transfers—some now possible via the Red Line Metro Rail, the rest via buses—are needed to provide distribution to a dispersed set of locations around downtown which cannot be reached directly by the Blue Line alone. This requirement detracts from the advantage of rail as compared to direct bus service.

When BART rail service came into operation, many passengers opted to continue using competing AC Transit bus service over the Bay Bridge to and from San Francisco because it provided for a direct trip, without the need to transfer to BART. More recently, according to MTA management, the opening of the Western Avenue Red Line station on Wilshire Boulevard did not induce most local bus users on Wilshire to transfer to rail to complete their journeys to downtown. They preferred to remain on the slower bus to avoid the transfer. An estimated half of the sixteen thousand daily boardings on local service between Western Avenue and downtown "would be served by the Red Line, but virtually none have diverted." This was the case even though the charge for transfers between bus and rail at Western Avenue was eliminated. "There's a lack of any travel time benefit because the access time into the station and at the other end offsets the travel time saving" [because the train travels at a higher speed than the local bus].

No-transfer trip times from stations on the central part of the Blue Line light rail route to downtown Los Angeles are shorter than by local bus services. Bus route 56 takes approximately one hour to travel between Compton and downtown Los Angeles in local service; the light rail cuts the trip to Seventh Street/Metro Center to about twenty-eight minutes, although passengers needing to reach other parts of downtown must transfer downtown to either the Red Line or a bus, extending trip time. The local bus ride between 103rd Street/Kenneth Hahn station in Watts, and downtown Los Angeles is about forty minutes by bus. This is reduced to about twenty-two minutes to Metro Center by light rail.

Bus trip times for such journeys could be reduced were express service (using the Harbor or Long Beach Freeways) offered. Express freeway bus service between South Central Los Angeles and environs and downtown has never, however, been operated, other than from stations on the Harbor Transitway (exclusive bus and high-occupancy vehicle facility) itself, which is located to the west of the South Central area. A journey between the I-105 Harbor Transitway Station and Union Station in downtown Los Angeles takes about thirty-six minutes by peak-hour express bus. Completing the journey by rail requires taking the Green, then the Blue, then the Red Lines, a forty-five minute trip assuming minimal allowances for connecting time.

A clear advantage of buses is that service may be geared to where it is most needed. Service, such as on the Long Beach line, often uses abandoned former railroad rights-of-way, going where the railroad goes, rather than to locations of greatest demand. Much of the northern part of the Long Beach line lies to the east of the major area population concentrations. As then LACTC Rail Construction Corporation Acting President/CEO, Edward McSpedon stated, "Although these corridors may not be convenient to residential trip origins they often provide a very politically attractive 'path of least resistance' to LRT implementation."[30] Use of such corridors causes relatively less disruption than when an entirely new right-of-way is put into place.

While trip times for midcorridor Blue Line passengers with both origins and destinations near the light rail stations have been shortened relative to those for bus travel, the need to transfer between bus and rail eliminates the advantage for most others. As Anastasia Loukaitou-Sideris and Tridib Banerjee found from interviews with community representatives, "it is a major effort for most people in the inner city to avail themselves of the services of the Blue Line," given a population density the reverse of what would be expected—population density in fact increases with distance from stations—and the inaccessibility of stations from higher-density areas due to a lack of adequate bus connections or station parking facilities at midcorridor stations.[31]

A grid of bus services continues to operate along parallel streets in South Central Los Angeles terminating in downtown. Direct service to downtown Los Angeles is, therefore, provided for a widely dispersed set of corridor origins and, in particular, from the centers of highest population density. To use light rail, instead, passengers not living next to rail stations must take a crosstown bus to the rail station, where such bus service is available, rather than traveling directly. As will be discussed in chapter 11, furthermore, few actual employment locations for South Central Los Angeles residents are actually served by the light rail service. The preponderance of transit trips in South Central Los Angeles are of a local nature, rather than to downtown—an average bus trip is under four miles—and are as widely dispersed as the needs of the community for shopping, health care, and other service and leisure, as well as employment, access. Light rail cannot provide for most trips of this dispersion and short length, and for many of those it can serve, it lengthens journey time by requiring transfers between modes.

Because trains stop at all stations between Long Beach and downtown Los Angeles, the end-to-end travel time is no shorter than was the case on the 456 Long Beach freeway express bus, although this only affects a small proportion of rail service users. A peak-hour trip on the bus took fifty-five minutes, the same as the rail travel time. Passenger protest led to temporary retention of another express bus service, the now-discontinued 457, which operated from the eastern part of Long Beach to downtown Los Angeles. According to one member of RTD staff, "We had an uprising like you wouldn't believe" (because of the plans for termination). Taking the bus to the Blue Line station and transferring to the train added between thirty and forty minutes to the trip, compared to direct service to downtown Los Angeles by bus. "We had huge numbers of people in the area say 'we won't do it, we'll go back to cars,' and the response of the [RTD] Board was, 'we'll keep it.'" The bus fare was $2.30 one-way or $78 for a monthly pass, and these passengers preferred to pay this higher price to go by bus than to travel by Blue Line at a one-way fare which was $1.10 at the time or by purchasing a monthly pass then priced at $42.

The opening of the Red Line extension to North Hollywood has lowered fares for passengers who formerly paid express bus supplements, since the Red Line is priced at a local fare and bus connections are now local, rather than express. For certain journeys, the Red Line has produced a substantial time saving. From the Metro Center station, it now takes only nineteen minutes to reach Universal City, compared to the previous twenty-nine to thirty-four minutes by bus. This comparison does not, however, allow for the time needed to reach or enter and leave the rail system. Not only did the former bus service offer more stops than Metro Rail, lessening the average time needed to reach ultimate downtown destinations, but it takes longer to get access to and from an underground rail platform, compared to walking to a surface bus stop.

The MTA provided spreadsheet examples of before-and-after trip times for locations not directly on the rail network. Ten minutes is saved in the morning and six minutes in the afternoon peak directions on a journey by Rapid Bus (MTA's new limited stop bus service) and Red Line train, compared to the previous trip times of eighty-two to eighty-eight minutes by direct bus between Warner Center in the San Fernando Valley and First and Hill Streets in downtown Los Angeles. For some destinations further out, however, journeys have been extended by terminating buses at Metro Rail and requiring a transfer. From Cal State University, Northridge to down-

Figure 3-4. Light rail street operations in Portland. (Photo courtesy of author)

town Los Angeles, for example, the southbound peak journey takes eleven minutes longer than before with a bus-to-train transfer, and the evening journey takes thirteen minutes longer compared to the previous journey time of between seventy-nine and eighty-five minutes.[32]

Comfort

San Francisco's BART—with its upholstered seats, carpeting, and smooth ride—was designed to be comfortable because it was thought that luxury would lure people who would not otherwise desert their cars.[33] This turned out to be another wrong assumption. As McFadden found from extensive interviewing, travelers were indifferent between riding on BART or on a bus, time and money costs being equal.[34] The time and cost of the trip mattered much more.

This finding has been replicated elsewhere, as demonstrated in literature surveys by Wachs. Gerald Miller and Keith Goodman reported that, while 90 percent of nearly two hundred passengers on special "new feature" buses on the Shirley Highway in Washington, D.C., rated schedule reliability as important to their choice of travel mode, no feature in the comfort or amenity category was rated to be nearly as important.[35] While 71 percent thought air-conditioning important and 62 percent felt that assurance of a seat was significant, other features—including improved leg room, larger windows, carpeting, and absence of advertising—were all significantly less important than fares, travel time, and schedule reliability.

> The significant conclusion to be reached about developing transit improvements is that commuter reactions have consistently shown that it is not necessary to provide luxurious interiors and plush environments in order to attract riders. Meeting basic physiological requirements, providing for a high probability of seat availability, and incorporating temperature control are the most critical aspects of comfort and amenity which should be addressed in vehicle design. To the extent that additional items of amenity, such as space for packages, might be incorporated in the design, the vehicles can provide greater attractiveness. Such features do not seem most critical in attracting patrons out of automobiles.[36]

The results of a survey in Los Angeles are consistent with all of the above findings.[37] Asked what service improvement was most desired, 72 percent of RTD riders and 54 percent of nonriders who responded called for more service—in the form of greater frequencies, new routes, or extended operating hours—while only 7 percent of riders and 16 percent of nonriders who responded said that "cleaner, newer buses" were most desired. The results are in table 3-2.

Table 3-2. Service Improvement Most Desired

	Riders %	Non-Riders %
More Frequent Service	49	1
New Routes	7	18
Extended Hours	12	13
Cleaner, Newer Buses	7	16
Courteous, Helpful Drivers/Employees	14	11
Security	0.5	9
Other Improvements	5	9

Source: Maritz Marketing Research 1989

Finally, we should note that while trains are currently seen as more comfortable than buses, the reverse was happening in the 1930s. The "de luxe" buses of the era were specifically targeted at people who had left the streetcar for the automobile. In an article entitled "Opportunities for Profits in De Luxe Bus Operation," published in 1930, for example, J. R. Stauffer describes the development of bus service for people lost by the streetcar:

> The companies that are utilizing the de luxe bus have almost unanimously had one objective in establishing such service, namely a form of transportation to bridge the gap between the street car and the higher priced means of travel such as is afforded by the private automobile or taxicab. . . . A new rider had to be found and the logical place was in that group of people who had previously left the streetcar for the automobile. It has generally been found that de luxe bus routes equipped with the most modern type of vehicles do appeal to a class willing to pay a higher fare for a service which is fast, comfortable and convenient.[38]

Stauffer discusses the advantages of using smaller vehicles, where "shorter headways could be run with fewer empty seats," and cites examples where "patronage has been built up from those persons who rarely used the regular [rail] service, but were attracted by the new, distinctive vehicles and the advantages of the service they rendered."[39]

One effect to be noted is that the *glamour* of rail service may attract a short-term tourist interest by people who are not going anywhere in particular, but who are simply trying out the train as if it were a Disneyland ride. As an article on the Blue Line by Seth Mydans in the November 21, 1990 *New York Times* stated:

> The verdict among the Sunday riders of Los Angeles's brand new experiment in commuter rail transport: better than the Magic Mountain amusement park. . . .
>
> "You'll like it," said John Skweres, a railroad buff who has ridden the Blue Line for pleasure almost every day since it opened in July. . . .
>
> Minor Cordon, an 18-year-old sailor in uniform, was enjoying the ride with his family, all of whom were turned in their seats to watch the graffiti and barred windows of the Los Angeles suburb of Compton glide by.
>
> "Sometimes we go to the park," he said. "Two Sundays ago we went to Magic Mountain. This is the first time we've ridden on the train. It's nice."[40]

One passenger taking the train for purely pleasure purposes was asked if he would take the train at other times. "Maybe we'll start taking the train more because of the gas prices and all that," said Bryan Valdivius, fourteen years old. "*But it depends on how far away the station is*" (my emphasis). While the train might be a novel transport of delight, whether it is convenient for making a trip counts more than its glamour in deciding whether to use it for routine trips.

Ease of Use

Rail systems may sometimes give the impression of being easier to use than bus systems: trains operate over a limited network of routes and stop at well-identified stations, which are usually equipped with maps and other information systems. Bus stops, in contrast, are often poorly marked, lack schedules, and may cause confusion to the new user if there are a variety of different services stopping at a particular location.

While regular commuters will become acquainted with how to make their particular trip after a few journeys, there is no reason why bus use should not generally be made as "userfriendly" as rail use by the provision of better information. This is a theme of the new "Rapid" bus project in Los Angeles, to be discussed below.

Buses Can Attract Substantial Ridership

With comfort seen by travelers as less important than the ability to provide a quick and convenient trip, buses have demonstrated the capability to attract substantial ridership, including higher-income travelers who can more typically be expected to drive to work. There is no evidence to suggest that because of the supposed lesser comfort of buses potential rail passengers will avoid them.

As Miller and Goodman demonstrate with the results of their survey work, express bus service in the Shirley Highway corridor in Washington, D.C. was chosen not only because bus trip time is competitive with the automobile, but because buses provide an easy-access no-transfer and reliable service. Eighty-two percent of express bus service users had a car available for the trip, while almost half the users of conventionally operated service did not have the option of driving to work. The income of Shirley Highway users was also substantially greater than that of other bus users, and they were more similar to auto commuters than typical bus riders on many socioeconomic and demographic dimensions. Most impressively, 40 percent of those both living and working in areas served by the buses used them in preference to driving to work, demonstrating once more that provision of a fast and convenient trip will attract those who have a choice.[41]

Houston has developed an extensive HOV (High Occupancy Vehicles) program, providing lanes available jointly for express bus and carpool use. By April 1997, 67.4 miles of HOV lanes were operating in five corridors. As of December 1996, 22,990 bus boardings and 55,856 trips by carpool or vanpool were made on the facilities each day.[42] The close relationship between the transit organization and the highway department fostered an attitude of promoting mobility, rather than travel by any particular mode. Carpools were thereby regarded as an environmentally desirable complement to bus ridership.[43]

The El Monte Busway—on the San Bernardino Freeway in Los Angeles—is equipped with bus priority lanes, as on the Shirley Highway. Buses travel express on the freeway, then branch out dendritic fashion to serve many residential neighborhoods with no-transfer service. There is also substantial free parking at the El Monte station. This innovation—which dramatically reduced bus trip times compared to the service previously operated—has also been highly successful. Reporting in 1974, the year following the busway's opening, Crain found that while 80 percent of busway users had annual incomes of over ten thousand dollars, pre-busway transit service included only 46 percent from the income groups above ten thousand dollars. Forty-eight percent of new bus riders on the El Monte Busway, furthermore, had previously used automobiles rather than buses for the same trip.[44]

While the El Monte Busway enables more direct service to operate than would be possible with rail, the facility is also made available to auto drivers who carpool to work, encouraging the more efficient use of automobiles. While the busway—approximately half the length of the Blue Line—was reported by MTA in 1996 to carry 19,366 daily MTA and Foothill Transit bus passengers, it was also used by 20,500 daily automobile carpoolers in that year, commuting in an efficient and environmentally sound way.[45]

Ottawa and Pittsburgh have developed high-capacity bus-only facilities as against the lanes shared with carpool users cited above. Ottawa both expanded existing bus services and built a 20.7 mile busway system. In Ottawa, an additional lane at stations permits buses to overtake, something trains cannot do without far more costly extra track provision. Bonsall characterizes the bus system as operating "just like any other rapid transit facility with vehicles, which in this case are buses, stopping at every station."[46] Ramp access allows for bus lines to collect passengers on surface streets before entering the busway. In addition to all-day busway service with timed-transfer local service available from transitway stations, 78 lines provide direct peak express service which collects passengers throughout the suburbs before entering the busway, while seven additional trunk routes use various parts of the transitway system all day.

Bus service expansion in fact began in 1973, a decade before the busway opened. Substantial new express limited stop and dial-a-bus services were added from that year onward. Bus lanes were provided on regular arterial roads and parkways, as well, with 14.6 miles of them in place by 1974. Priority was provided for buses at several traffic signals. Ridership increased dramatically following the initial service expansion. Urban bus mileage increased 3.5 times between 1971 and 1983, while the number of annual boardings increased from 44.3 million to 109.1 million over the same period.

By 1995, total transit ridership on Ottawa's upgraded all-bus system had declined to 101.2 million, but nonetheless almost equaled the bus and rail ridership in Baltimore that year (103.4 million), although Ottawa had one-third the population. Total transit ridership that year in light rail cities such as Buffalo (29.0 million), Sacramento (23.1 million), San Diego (50.4 million), St. Louis (51.2 million), and Portland, Oregon (64.0 million), was dwarfed by Ottawa's result.[47]

Pittsburgh has constructed two busways in addition to completing a thorough rebuilding of its light rail system. The 6.8 mile East Busway is the more significant of the two, with seven stations and six ramps for bus access and egress, permitting twenty-five express routes to collect and distribute passengers in the suburbs while using the busway for high-speed operation. There are also two routes specifically to provide service to busway stations, one running local, the other express. The variety of operations made possible by this implementation is far greater than on light rail, where trains are limited to the availability of track and generally provide only all-stations local service. "The EBA [local] route operates much like a light rail or rail or other fixed guideway service. The diverted routes, however, combine the service flexibility of regular bus routes with the efficiency of fixed guideway service and eliminate the need for a transfer."[48]

East Busway operating speed is 29 mph, compared to 16 mph on light rail outside downtown. The average busway-bus speed including busway and downtown route portions is 20 mph, compared to 15 mph overall for light rail. As Crain reports, "passengers on the EBA, the major new route, have reduced their travel times by about 21 to 24 minutes, a reduction of 40 to 45 percent." As of May 1996, 4,100 passengers were carried in the peak direction during the East Busway's peak hour of operation. The peak volume on the light rail system was also 4,100. In 1996, the light rail system carried a total ridership slightly less than the East Busway on a route system three times its length.[49]

While the above examples illustrate the benefits to be achieved from investing in bus capital facilities, buses offer the flexibility to provide improved transit service through simpler, lower-cost measures. Indeed, it is significant that in Ottawa the vast increase in system ridership took place before the expensive busway opened, reflecting a response to expansion in service quantity and quality, with higher speeds made possible by regular street bus lanes and traffic priority measures and the implementation of an expanded route network. In Houston, the impacts of real fare reductions (14 percent between 1980 and 1990) and general increases in service (an 80 percent increase in bus miles over the same period) are likely responsible for the largest parts of the increased ridership which went from 39.9 to 88.0 million annual boardings over this period.[50] In 1995, non-transitway buses carried almost twelve times the volume of HOV lane passengers.

In Portland, the period from 1969 to 1980 saw policy focused on increasing the level of existing bus service and lowering fares. This led to an annual ridership increase of 150 percent, while service hours grew by 110 percent. Transit rides per capita doubled over this period. "Tri-Met [Portland's transit operator] became a national model of a successful public transportation system." Notable daily bus ridership increases included those over 1975–76 (22,300) and 1979–80 (26,500). Each of these increases compares favorably to the twelve thousand total gain in transit ridership Tri-Met says is attributable to light rail through 1996 (the remainder of the twenty-seven thousand total light rail passengers that year being people who would have used bus transit anyway).[51]

Los Angeles has recently taken steps to implement an improved model of bus service delivery, based on incremental levels of costs and improvements. The Metro Rapid Bus Program, initiated by the MTA Board of Directors in March 1999, sought to emulate attributes of the bus rapid transit system in Curitiba, Brazil. Two demonstration lines, the 720, extending along the high-demand corridor from Santa Monica via Wilshire Boulevard through downtown Los Angeles and to the Eastside, and the 750 Ventura, providing service connecting points in the San Fernando Valley to the Red Line Metro Rail began operation on June 24, 2000.

To improve operating speeds, the Rapids have fewer stops than local buses, can turn traffic signals green at certain key intersections, and use a low-floor design to speed boarding and alighting, as well as to improve comfort for elderly or other customers who find boarding a challenge. The buses are actively managed from the MTA Bus Operations Control Center. Overall speed improvements of 29 percent were achieved on the Wilshire line, 23 percent on the Ventura line, with average system speeds of 14–30 mph, depending on time of day and service direction.[52] In addition to providing attractive new buses in a distinctive livery, Rapid bus stops have shelters with service information and lighting, aiming at amenities more in common with those provided for light rail passengers than the simple bench and bus stop pole installed at most regular MTA bus stops. Installation of indicators at stops which countdown the minutes to the next bus arrival has begun. Each shelter carries an advertisement that generates income to maintain the facility.

Following Rapid service installation, bus ridership in the Wilshire corridor increased from 63,500 pre-Rapid daily riders to 90,000 as of August 2001, a net increase of 26,500 or 41.7 percent. 41,000 daily riders out of this total were using the new Rapid 720 line, with remaining passengers riding local buses serving a larger number of stops. Ventura corridor ridership increased by 4,200 or 38.8 percent over the same period, from 10,800 riders to 15,000 daily riders, with 10,000 rides out of the total on the new Rapid Buses.[53]

On-board surveys three months into the new service operations demonstrated that "Metro Rapid riders perceive a quantum leap in service performance and quality. Changes of this magnitude in performance ratings are rare, particularly over a relatively short time frame (90 days)." Of particular interest, survey data showed that one-third of the overall ridership increase came from people who previously never used public transit, while one-third was from existing transit users who have been induced to ride more often. The remaining one-third of passengers have come from users of other MTA services, who have switched to the new Rapids. Over 13 percent of Rapid riders had incomes over $50,000 compared to 6 percent for local buses, indicating the Rapids' attractiveness to those who have alternative choices. Improving service for longer-distance travelers was a major objective of the program. The average distance traveled by Rapid is approximately double that traveled on a typical MTA local bus service. This "is not inconsistent with expectations from a similar light rail service."[54]

Given the success of the program, MTA plans to both expand the Rapid Bus system and to introduce additional improved service attributes, including exclusive bus lanes (including lanes in existing arterials and on separate rights-of-way); to introduce higher-capacity buses with multiple door boarding and fare prepayment; and to provide more elaborate station-type structures for bus stops. As MTA concludes, "[T]he Metro Rapid program has demonstrated two critical elements: (1) customers perceive Metro Rapid as clearly superior to MTA's existing bus services; and (2) Metro Rapid's ability to increase transit's market share among discretionary travelers."[55]

The lessons of these examples seem clear. First of all, bus systems already play significantly greater roles than rail in all new light rail cities. Secondly, even modest improvements to bus services can generate ridership increases which are competitive with those achieved on many light rail systems. Thirdly, innovative and well-designed bus systems that replace slow local buses with buses operating on their own rights-of-way with greater frequency and reliability can not only provide at least the capacity of light rail systems, but can prove highly attractive to a wide range of potential passengers.

Environmental Factors

Congestion Relief

While passengers make decisions on how to travel primarily based on trip time and cost—not on levels of comfort or luxury—and there is no reason to expect that a rail service could attract more travelers than a bus service of comparable service quality, caution must be used when making claims that either mode could contribute to a meaningful reduction in freeway congestion in an urban environment such as Los Angeles.

Only 131,400 weekday passengers were using BART in June 1976, compared to the 258,000 forecast.[56] But this was not the only problem:

> Although the Composite Report expected that 61 percent of riders would be diverted from private automobiles, in fact only 35 per cent formerly made the trip by car. . . . Half of BART's transbay riders come from buses which BART has replaced—at very high cost. . . .
>
> BART initially reduced the number of cars on the streets and highways by 14,000 (not 48,000, as forecast). In turn, in accordance with the Law of Traffic Congestion (which holds that traffic expands to fill the available highway space until just tolerable levels of congestion are reached), other people began driving their cars on trips they would not otherwise make.[57]

The existence of congestion "signifies the existence of a great deal of latent highway travel demand, ready to be expressed as and when additional highway capacity becomes available," according to Alan Altshuler.[58] If cars are taken off the road, the "cost" in terms of automobile trip times on the highway is reduced, and additional automobiles attracted. Congestion, Altshuler says, also helps to maintain transit ridership: if road speeds were to actually increase as a result of transit improvements, transit would ironically become a less competitive option compared to the automobile.

There was, in fact, a reduction in auto travel throughout the San Francisco region during the 1974–75 period. But the Bay Bridge, which runs parallel to BART, experienced a smaller proportional reduction than the other bridges across San Francisco Bay. "Traffic congestion even on the Oakland Bay Bridge and its approaches did not go away."[59]

As Hilton states, BART had the advantage of paralleling a highway crossing of a water barrier, with a toll acting as a disincentive to driving, but had only been able to reduce Bay Bridge peak-hour vehicle counts 2–4 percent. "There is no reason to believe that a rail system in Los Angeles, which is without water barriers or toll crossings, would be able to do that well."[60]

In another example, Hilton comments that the Quincy branch of Boston's MBTA (Massachusetts Bay Transportation Authority) Red Line is believed to have only taken nine hundred to one thousand vehicles per day off the Southeast Expressway. "The Southeast Expressway, however, carries between 80,000 and 120,000 vehicles per day. Thus the diversion cannot be perceived, relative to the daily variance."[61] The Southeast Expressway experience is consonant with Lave's "Law of Large Proportions" which states that "The biggest components matter most." Making a small improvement to a large part of the total makes more difference than a major improvement to a minor component.[62] With such a small proportion of Los Angeles commuters using public transportation, even a large increase in transit patronage would be unlikely to have a noticeable effect on congestion, especially since any slight initial improvement in speeds would simply attract more vehicles to the facility, returning it to its former equilibrium level of congestion.

This is not to say that public transportation cannot increase mobility—more capacity is certainly available when transit services are laid on, if only used by a small percentage of travelers. But if congestion is to be reduced, action must be taken to change the ways in which highways are operated, actions such as charging tolls, controlling access through the use of ramp meters, and increasing parking charges. Such measures may be unpopular, of course, and—for that reason—politically tough to implement, but providing transit of any sort is no substitute for going through with them. While overall congestion reduction cannot be expected from implementing improvements in transit, by whatever mode, exclusive bus facilities—if open to carpools—do provide for the avoidance of congestion by drivers who share their ride, something not possible with rail.

SCAG conducted an analysis of likely traffic impacts in the Long Beach corridor of the operation of light rail service, assuming attainment of the full ridership they predicted.[63] For the route eventually selected, only 1,385 Long Beach corridor passengers who would otherwise have traveled by car were forecast to use transit because of the presence of light rail. The other light rail users, according to SCAG, were people who would otherwise have traveled by bus, but who would now be attracted to light rail because it would supposedly enable them to complete their trip faster.

Because some automobiles carry more than one passenger, the actual daily reduction in vehicle trips on the road system—given adoption of the light rail system which was chosen—was only 1,030, according to SCAG, 0.3 percent of the 270,288 daily vehicle trips projected for the corridor. In no alternative studied was the reduction found to be greater than 0.5 percent.[64] SCAG concluded that:

> From a county-level or even a corridor-level, the LB/LA LRT project has only a very minor positive impact on traffic. Within the LACBD, it appears that east-west traffic might experience a small increase in congestion levels and only an insignificant decrease in north-south traffic. The southern portion of the Midcorridor and Long Beach, however, show significant decreases in traffic volumes especially with the LA River Route alignment [which was not, however, adopted]. The northern portion of the Midcorridor shows almost no traffic volume impacts—positive or negative—from the project.[65]

As we have seen, only 37 percent of Blue Line passengers were found to be new to transit, a result comparable to that achieved with the implementation of Rapid Bus service. Neither of these systems, however, can be expected to have a noticeable effect on congestion reduction. The El Monte Busway on the San Bernardino Freeway achieved a higher proportion of new transit riders, while also accommodating carpools. This joint approach has paid particular dividends in Houston, where implementation of the HOV lanes was found to have reduced areawide congestion levels by about 4 percent.[66]

Does Removing Buses in Itself Reduce Congestion?

Providing new rail service may be seen to reduce congestion merely by taking buses off the street. The evidence for such an assertion is, however, slight. According to the *Highway Capacity Manual*, for buses in uninterrupted motion, the bus equivalency factor is approximately 1.5 passenger cars.[67] In other words, while a peak-hour bus on a freeway may carry fifty passengers, its effect on road space is no worse than that caused by 1.5 automobiles. Given the small number of buses in operation relative to automobiles, their negative effect on traffic when operating in continuous motion is therefore unnoticeable.

Looked at another way, buses increase the capacity of the road system by taking up less room than the automobiles they displace. One example in the *Highway Capacity Manual*:

> assumes a maximum freeway capacity of 1,800 vph [vehicles per hour] without buses, a bus-passenger car equivalency of 1.5, and occupancies of 1.5 and 50 for cars and buses respectively. As the number of buses on the freeway increase to 300, the total person-capacity increases from 2,700 to nearly 17,000, while the vehicle capacity drops from 1,800 to 1,620.[68]

In city streets, where buses make regular stops, the matter is different. If stops are made in a lane:

> not used by moving traffic (for example in a curb parking lane), the time loss to other vehicles is approximately 3 to 4 sec. per bus . . .
>
> Where buses stop in a normal traffic lane, the time loss involves the dwell time for buses plus a time loss for stopping and starting, and the associated queuing effects on other traffic.[69]

The queuing effects are typically six to eight seconds, according to the *Manual*. The *Manual* also translates buses making city street stops at signalized intersections into automobile equivalents. If a bus stops for twenty seconds where traffic signals are green 50 percent of the time, the bus is equivalent to seven automobiles in the traffic stream. Where it stops for forty-five seconds, where there is 60 percent green time, the bus is equivalent to fifteen cars.

It is clear that buses are a part of congestion on city streets, although, even in the last example given, it may be argued that a peak-hour bus will typically take more than fifteen cars off the road. When measures are taken such as the provision of separate curb lanes at bus stops, which keep buses at rest out of the regular traffic flow, the impact on traffic is minimal.

A case may also be made for providing bus-only lanes on city streets—providing the equivalent of the exclusive right-of-way enjoyed by trains. In Seattle, a bus tunnel project provides a facility similar to a rail subway: of course, buses can be run on city streets at lesser cost, but if a fair comparison is to be made between buses and trains, it could be argued that they should be evaluated operating under equivalent conditions.

Where light rail lines operate in city streets, they also occupy space which could be taken by other traffic. At-grade crossings also impose delays on road traffic when it is stopped to allow trolleys to pass. As one *New York Times* article put it, Blue Line riders "gazed in wonder at the lines of traffic backed up at the crossings."[70]

If long-haul buses are replaced with rail services in urban environments such as Los Angeles, furthermore, buses are not necessarily eliminated, as city street feeders are still required to provide distribution between rail stations and final destinations.

Reducing Energy Consumption

Lave has made the point that the difference between the operating efficiency of the average car and a modern rail transit system is much less than the energy efficiency difference between the average imported car and a standard-size American car. "That is, people who switch car sizes can save more energy than those who switch to trains."[71]

Because of the "Law of Large Proportions," Lave declares that transit's "potential contribution to solving the energy problem was always insignificant. . . . That is, cars use most of the energy, and we ought to concentrate on improving their efficiency."[72] Alternatively stated, if we increase the fuel efficiency of the average car from 15.0 mpg to 15.2 mpg, we save more energy than we do by doubling transit patronage. Public transportation cannot make any significant contribution to energy conservation in the near term, and even in the long term its contribution is likely to be insignificant.

SCAG did find that the light rail project would result in energy savings, compared to a no-project alternative. Annualized regional transportation energy requirements were calculated. With the Broadway/Spring light rail routing, 822,800 billions of BTUs were found required, as against 823,426 billions of BTUs with the no project alternative. This represents an annualized saving of 1,338 billions of BTUs, or 0.16 percent of the total estimated no project requirement. On the other hand, the "All Bus Alternative" was found to save almost as much energy: 1,120 billions of BTUs, or 0.13 percent of the total estimated no project alternative.[73] In other words the bus alternative to light rail (which as will be seen in the next chapter was designed extremely poorly) was found to perform comparably to light rail in terms of saving energy.

Reducing Pollution

While trains may operate on electricity, the power to operate them has to be generated, and this process may cause pollution elsewhere. Nonetheless, the advantages of localized reductions of pollution in city environments which might result from the introduction of light rail could be thought valuable.

Buses, nonetheless, also reduce pollution: a bus produces far less emissions than would result from the cars it keeps off the road. On the other hand, diesel buses do produce harmful particulates, although the proportion of these contributed by buses as compared to other diesel vehicles such as trucks is low. The answer to this problem, however, might be the increased development and implementation of buses using "clean" fuels such as methanol. The Los Angeles MTA has in fact been a leader in demonstrating such technology.

For the same reason that transit improvements cannot be expected to have a significant impact on reducing energy consumption, however, they are unlikely to have a noticeable effect on air quality in a city such as Los Angeles. SCAG concluded that, comparing a case study taken of the:

> Broadway/Spring and No Project alternatives shows a decrease of only 43,000 VMT as a result of the project. The average speeds [on highways] during the A.M. and P.M. peak periods would increase only marginally. *Neither of these two factors contribute to any meaningful change in the level of emissions or in the ambient air quality in the project study area* [my emphasis] . . . [In addition,] it is found that none of the light rail alternatives would have a significant localized air quality impact compared to expected conditions without the project.[74]

Development Impacts

Rail is often seen as a valuable tool for encouraging more concentrated development patterns. Toronto is frequently held to be a city where a new rail transit system precipitated central-area development. Warren Heenan, for example, claimed that $10 billion out of a $15 billion increase in assessed property valuation over the ten years following construction of the first segment of the rail system was directly attributable to the transit system. However, as Robert Knight and Lisa Trygg point out, Heenan failed to consider the effects of other factors which encouraged downtown office development in Toronto or to compare Toronto's results with those of other cities. "Many policymakers have been misled by this widely-publicized overstatement." Carol Kovach found that other large Canadian cities with no plans for rail transit systems had similar or greater growth rates over the same period.[75]

Many commentators have said that rapid transit shaped and intensified development in Toronto.[76] But, claim Knight and Trygg, "the transit system alone did not cause these impacts."[77] The city enacted several pro-development policies, including the payment of density bonuses around stations, encouragement of coordinated station development with developers to provide direct access from office, retail, or apartment buildings, and city zoning changes to permit much higher intensity development, especially near stations.[78] "Toronto planners and developers interviewed in our check of these results agreed that these incentives were essential in inducing the extent and type of development which occurred."[79]

"BART was intended to do far more than bring commuters into San Francisco; it was conceived from the start as a regional system that would foster the growth of the entire Bay Area."[80] As Webber points out:

> BART officials like to claim credit for the spectacular change in San Francisco's skyline; they say it is a direct result of improved commuter access from the metropolitan region. However, they also argued from the outset that BART was primarily needed because forecasts of impending downtown office employment raised the specter of the ultimate traffic jam. . . .
>
> Large-scale office construction in other Western and Midwestern metropolitan centers suggests that the building boom might have happened anyway, for many of them have had similar booms, although none except Chicago had anything like BART in sight. During the 12 years following BART's successful bond election, San Francisco's high-rise office buildings were expanded by 4,200 square feet for every 1,000 people in the metropoli-

tan region. By contrast, Houston, the automobile city par excellence, added 5,500 square feet per 1000 population, Chicago 4,500, and Dallas 3,500.[81]

Knight and Trygg add that downtown San Francisco was experiencing increasing construction activity even before bonds were approved for BART and cite Douglass Lee and D. F. Wiech, who estimated that much if not most of the new development may be due to conditions independent of BART.[82] As in Toronto, certain density bonuses were made available for buildings at or near downtown San Francisco BART stations.[83] As in Toronto, however, many other influences encouraged concentrated office development there, while there was little evidence of any substantial land use impacts outside the CBD station areas.[84]

While two modest office buildings did appear in Berkeley and Walnut Creek in response to the arrival of BART, suburban stations for the most part "stand in virtual isolation from urban-development activity, seemingly ignored by all except commuters who park their cars in BART's extensive lots." Development near the Fremont BART station occurred at a slow pace relative to that of Fremont as a whole. Looking back at the development impacts of two decades of BART operations, John Landis and Robert Cervero found that apart from in the city of San Francisco, "population has grown faster away from BART than near it," with an actual fall of about 11 percent in residential units within a quarter mile of BART stations over a period when the region added 25 percent to its housing stock.[85]

A similar profile held for employment growth. With the exception of San Francisco once more, "From 1970 to 1990, job growth mostly occurred away from BART. Employment grew 84.5 percent in non-BART superdistricts compared to 39.5 percent in the BART-served ones, mirroring the trend of job decentralization that was occurring throughout the U.S." As rents grew in San Francisco, where public policy aimed at office concentration in downtown to avoid encroachment on residential neighborhoods, tenants increasingly looked elsewhere. "These tenants found cities with excess highway capacity, plentiful supplies of developable land, relatively liberal zoning and land use policies, and a yen to become a suburban office center. . . . Except in downtown Walnut Creek—and even there, not until the mid-1980s—BART was not a significant inducement to office developers." BART's lack of influence on development "stands in marked contrast to the effect of freeway interchanges."[86]

Knight and Trygg conclude that "mere accessibility and land availability are not enough," while Garrison provides a warning against an uncritical assumption of particular desirable outcomes: "The problems to be solved are complex; land development is not a simple function of transportation."[87] Webber comments that BART added only a small increment of accessibility to existing levels, "scarcely enough to have significantly affected the location plans of many households and firms."[88] As he said in his address to the SCAG Executive Committee:

> The Bay Area, like the Los Angeles area, like all Western cities, has a very high concentration of roadways . . . If you did a contour map of the accessibility, it would be a flat plane almost . . . some ridges along the freeways, of course, but very high levels all over the place. BART came along and created a few points of high accessibility at its station stops, which were spaced at 2 1/2 miles, but that was a little peak on top of this big, high plateau, and it did not seem to have made enough difference to developers who were putting in offices or housing, etc . . . We've got too much accessibility already with the automobile highway system, and we can't make enough difference with a rail system to change the land use pattern.[89]

More recently, Portland, Oregon has been cited as a successful example of desirable urban development tied to a new light rail system,[90] although this view has been challenged by local critics.[91] Peter Gordon and Harry Richardson point to the financial failure of The Round at Beaverton, conceived as "an entire community being built around a light rail urban transit station in the greater Portland, Oregon metropolitan area."[92] "Beaverton and Tri-Met officials hailed the project as the kind of development that light rail could inspire—anti-sprawl, anti-strip-mall and foot-traffic-friendly," reported *The Oregonian*, but "Although land-use planners love transit-oriented developments, the projects are far from a proven proposition to the financiers."[93] Despite imagery of an urban utopia served by bright red trolley cars, the developer, BCB Group, was unable to find financing for the project, sited atop a former sewage treatment plant and surrounded by a run-down commercial area. Following unsuccessful attempts by the City of Beaverton to bail out the company, title to the development was handed to the city by the U.S. Bankruptcy Court.[94]

Where development close to light rail has occurred in Portland, the major question is whether light rail, with its tiny incremental addition to mobility relative to the highway system, of itself stimulated the development patterns, or whether the zoning and other incentive measures instituted would have achieved the same result in the absence of light rail. Robert Cervero stresses the need for pro-development actions to encourage development, and says that for some cities "the current auto-highway system seems so firmly rooted that any major structural changes in urban form seem unlikely."[95]

Systems adopting bus-based improvements have also claimed land-use advantages related to their transit projects. Wohlwill found that fifty-four new developments occurred along or near Pittsburgh's East Busway since its 1983 opening, most located near busway stations. Management at OC Transpo, Ottawa, describe $1 billion Canadian worth of new development around the stations of their busway system. The Los Angeles MTA, furthermore, has plans to promote transit oriented development in Rapid Bus as well as rail corridors.[96]

Developers seek accessibility. When the Red Cars provided it, they catalyzed the development of Los Angeles. A new light rail or rapid bus system today, however, adds negligible accessibility compared to the extensive highway system of Los Angeles, and cannot be expected to attract significant development relative to locations with existing substantial highway access. Such development as may occur is likely to require active incentive programs. There is no evidence to suggest, furthermore, that rail systems have advantages over bus systems with similar quality and quantity of service provision in attracting development. Richmond provides further discussion on this issue, while the special case of the potential of rail to bring about renewal in depressed communities such as Watts and Compton is discussed in chapter 11.[97]

Finally, we should question whether centralized patterns of development, said to result in connection with rail system construction, are even desirable. As Altshuler points out, transit mode share increases with urban density, but so does the absolute level of highway traffic demand per unit of road capacity. If new developments are induced, a certain proportion of trips will be served by the new transit system, but there will also be new trips generated by people who choose to gain access to the development by automobile, resulting in a worsening of highway congestion. "By comparison, land use dispersal has the opposite effects; transit shares decline, but so does the absolute level of highway traffic demand per unit of road capacity."[98]

Capital Costs

Capital costs for rail service projects are almost invariably greater than for similar bus-based improvements, although rail advocates often claim such extra expense is justified by the higher levels of ridership and lower levels of operating costs they say rail systems can achieve. As we have seen, there is no justification for the claim that comparable-quality rail service will attract more passengers than buses. As we shall see shortly, potential operating cost savings for rail as against bus services in low-density cities are also highly questionable.

While many of the new light rail transit systems were initially said to be "low-cost" alternatives, subsequent cost escalation has seriously affected many of the projects. San Diego's Blue (South) Line—below budget—was an exception. The 15.9-mile system cost $116.6 million as of completion of double tracking and other improvements in 1983.[99] While the total cost of the San Diego Orange (East) Line was estimated at $164.8 million in 1983 dollars, however, the actual cost was $250.5 million in current dollars.[100]

The Buffalo light rail system, estimated in 1971 to cost $241 million, and in 1978 to cost $450 million for a smaller system, actually cost $552 million as of 1987. Pickrell, whose study found that all four heavy rail systems and three out of four light rail systems he examined were over budget, estimated the Buffalo system was 51 percent above budget in inflation-adjusted dollars. Ottawa's 20.7-mile busway system, joins the list of projects which were over budget: Bonsall reports actual costs of Canadian $420 million compared to the budgeted Canadian $277 million.[101]

Inflation has played a significant role in capital cost escalation. There have, however, been many other elements contributing to higher-than-expected costs. There has been a frequent failure to reflect the complexities of construction requirements in estimates, and political needs which have emerged in the process of policy implementation (putting an extensive section of light rail line in tunnel rather than at-grade in Buffalo, for example, to appease local interests) have added to costs. In some cases, the scope of the project has changed considerably between the completion of initial planning and the date of service opening (a route extension to the main airport terminal in St. Louis, for example). In other cases, extra money has been spent after project opening to improve the system (double-tracking in Sacramento, for example).

Figure 3-5. The San Diego trolley. (Photo courtesy of author)

Estimates of capital costs—like those of ridership—have frequently be-come more realistic following the decision to proceed. Further work has often uncovered potential new costs and provided for more accurate pro-jections. Tri-Met, unhappy with the critical treatment of their MAX light rail line in Pickrell's report, declared in their response: "FACT: MAX was built on time and under budget." Construction costs "were $7.5 million under the budget, established in the 1982 Full Funding Agreement, *agreed to and signed by UMTA officials*" (original emphasis). According to Tri-Met staff member Bob Post, furthermore, "The 1978 estimates used in the proj-ect DEIS were not presented as the estimated construction costs. They were estimates in 1978 dollars for purposes of comparing alternatives."[102]

Going back to those estimates, conveyed in the staff recommendations to the board, however, there is a heading: "Capital costs are based on firm estimates."[103] The staff recommendations do acknowledge that "several modifications currently under consideration could affect the project costs, though not to such an extent as to change the conclusions of this report." Several of the changes which were to in fact happen—and increase costs— are then listed.

The 1982 Full Funding Agreement with the federal government was for $328.5 million, and Tri-Met reports the actual cost as $321 million as of "experience 1986." If the original projected capital cost estimate of $161 million is escalated to 1986 dollars according to Tri-Met's methodology, the result is $259.2 million, implying a 24 percent real cost increase. Pickrell uses a somewhat different methodology, using a different early projection (of $188 million, made in 1980) and escalation system, and concludes that there was a 55 percent increase in real costs.[104]

While operators such as Portland's Tri-Met have protested at the comparison of their initial estimates with actual results on the basis that they did in fact keep to their ultimate funding agreement to the federal government, the fact remains that the initial estimates have generally been used in major decision-making, after which alternative options (which might have seemed more attractive in comparison with more realistic projections for the chosen project) are eliminated and the focus shifted to one of achieving a funding structure and project implementation for light rail.

The Long Beach–downtown Los Angeles Blue Line light rail system line was to be modeled on the San Diego experience, and was originally heralded as a "low-cost" answer to the corridor's transportation problems.[105] Its cost increased from an initial estimate of $146.6 million in 1981 to $890 million in current dollars, making the system actually implemented far more expensive than decision makers had anticipated when they put aside alternative uses of the money.[106] Table 3-3 shows the escalation.

Table 3-3. History of Capital Costs, Los Angeles Blue Line Light Rail

Estimate Year	$ million
1981	146.6
1982	254–280
1983	350–400
1984	393–561
1985	595
1990	887
Actual	
1996	890

A California Department of Transportation (Caltrans) feasibility study produced the first estimate of $146.6 million. Parsons Brinkerhoff, under contract to the LACTC, estimated "baseline" costs of $194 million in 1982 dollars, which translated into a range of $254–280 million when escalated to account for inflation. This higher estimate was said to account for certain costs (such as right-of-way acquisition) which the consultants said Caltrans had omitted. This was the estimate associated with LACTC decision-making and which enabled Los Angeles County Supervisor Deane Dana to compare the project favorably with the "highly expensive" Wilshire subway and claim that it would "make maximum use of limited dollars" at the commission's March 24, 1982 meeting. The April 15, 1982 *Long Beach Press-Telegram* announced that the commission had given the "go-ahead for the proposed $194 million train last month."[107]

As Rich Connell reported in the October 20, 1985 *Los Angeles Times*, "Predictions that the Long Beach line could be built quickly for about $200 million faded soon after the line was selected. Transportation Commission officials said they found that a workable and politically acceptable system required double tracks, a downtown subway, street improvements in downtown Long Beach and other costly additions not initially anticipated." As of October 1983, the project was said to cost between $350 million and $400 million. By the May 1984 *Draft Environmental Impact Report*, the capital cost estimate was in the range $393–561 million for a variety of options. The November 11, 1984 *Los Angeles Times* quoted then LACTC Executive Director Rick Richmond as discounting the possibility that the project would reach $1 billion in capital costs, although he admitted that "things beyond our control" could drive costs up to somewhere between $500 million and $600 million.[108] The June 1985 issue of LACTC's publication *The Rail Way* announced that the estimated cost of the project was $595 million in 1985 dollars.

Costs continued to climb as described in an article by LACTC's Edward McSpedon. The relocation of pre-existing Southern Pacific railroad tracks led to "numerous and substantial" complications:

> As might be well expected in a case where the owner (SPTC) does not bear the burden of the construction costs, the railroad has been extremely stringent in the application of its construction approval authority, resulting in change orders to contracts and pending contractor claims totaling hundreds of thousands of dollars.[109]

LACTC was also required to provide a $50 million railroad protective liability insurance policy.

> The physical configuration of the SPTC trackage has also added to the cost of the LRT project . . . At heavy traffic locations it has been necessary to grade-separate the LRT to avoid crossing conflicts with the railroad . . . Because the alignment of the railroad is being shifted, it has meant physical changes to each of the 37 railroad grade crossings in the midcorridor. . . . The cost of relocating and replacing freight railroad facilities is estimated to exceed $40 million.[110]

Perhaps the most greatly underestimated difficulty, McSpedon says, concerns dealing with other right-of-way users: it was necessary to relocate, replace, remove, or protect twenty-three hundred individual utility lines.[111]

Additional extra costs were imposed by demands from municipalities for grade-crossing improvements, upgrades of adjacent streets and sidewalks, installation of new street lighting, computerized traffic signals and signage, addition of landscaping, and construction of new fences and retaining walls. Due to demands by the City of Compton, four miles of Southern Pacific track running through that city was removed and relocated to a corridor to the east at an additional cost of $67 million, $57 million paid by LACTC, and $10 million by the City of Compton (largely through a long-term zero-interest loan from LACTC).[112]

In a letter to the author dated July 3, 1990, Edward McSpedon declared that "Forecast total cost is $887 million." Following opening, a succession of change orders added to cost. For safety reasons, for example, fencing replacement at intersections was deemed necessary at a cost of $40,000 per intersection. In February 1996, the MTA reported total construction cost of the Long Beach Blue Line at $890 million, some $42.4 million per mile.[113] Subsequent further work has included extensive station platform work to accommodate longer trains; MTA was not, however, able to supply updated total capital cost numbers.

As McSpedon concludes, while the use of existing rail corridors "will always be high on the list of least-undesirable alternatives" for "the construction of a new rail transit facility in a mature, densely developed urban area with the objective of minimizing construction costs through maximum use of at-grade construction," the use of such corridors "will probably be much more costly, time-consuming, and complex than might be presumed initially."[114]

Red Line costs for 17.4 miles of subway route came to $4,318 million as of December 2000, as compared to the $3,024 million, in 1985 dollars, projected in the 1989 environmental report, a smaller proportionate escalation of actual costs, given the impacts of inflation, than occurred on the Blue Line, if greater in terms of total number of dollars.[115] The subway cost per mile comes out at $248 million, almost six times the cost per mile for the Blue Line. The recently completed Pasadena Gold Line cost $859 million as of July 2003, or $62.7 million per mile, not including the costs of improvements around local stations paid for by the municipalities served.[116] The 6-mile Eastside extension of the Gold Line, which includes 1.7 miles in tunnel, is estimated to cost $880 million.[117]

Capital costs of bus alternatives are generally less than for their rail equivalents. In the East Corridor from San Diego to El Cajon, for example, the most favored of the bus alternatives considered was estimated to cost $99.2 million in 1983 dollars, as against $145.6 million for the rail alternative eventually selected.[118] In Ottawa, a LRT alternative was estimated to cost 50 percent more than the Canadian $277 million estimated cost of the 31 km bus transitway selected.[119] The Los Angeles County MTA is currently considering busway or light rail alternatives for the twelve-mile Exposition Corridor which runs to the south of Wilshire from the downtown area to Santa Monica. The busway alternative is estimated to require capital costs of $188 million ($15.7 million per mile), compared to $431–589 million ($35.9–49.1 million per mile), depending on the degree of grade separation, estimated for light rail.[120]

Pittsburgh's rehabilitated light rail system, with a capital cost of $537 million as of 1987[121] was constructed close to or even under budget in terms of inflation-adjusted dollars according to Pickrell. This compares to the Pittsburgh East Busway capital costs of $113 million as of 1983. Biehler compared the busway and rail project capital costs in 1989 dollars.[122] The light rail came out at $49.8 million per mile, compared to $20.3 million per mile for the East Busway. Because the downtown Pittsburgh rail subway construction had been costly, Biehler calculated an equivalent light rail capital cost assuming the subway had not been built: $45.7 million per mile, two and a quarter times the busway unit cost. This difference is particularly significant since, as of 1996, the light rail system, with three times the route mileage, carried slightly less passengers than the East Busway.

An "All-Bus Alternative" to the Long Beach Blue Line light rail was considered in demand modeling (see next chapter). This option was not routed via any segregated rights-of-way, but on local streets. According to one calculation, $168 million would need to be spent to provide the buses and operating yard facilities needed to operate a bus service equivalent to the Long Beach light rail.[123]

One of the principal objectives of the Metro Rapid Bus demonstration program implemented in Los Angeles in June 2000 "is to provide high quality rail emulation service with significantly lower capital investment." Costs were incurred for station development, bus signal priority, and bus purchases. The total cost for the program was $12.5 million ($7.8 million for Wilshire, $4.8 million for Ventura), representing a cost per mile of $295,000 over 42.5 miles of route.[124] The 30,700 extra weekday Wilshire and Ventura corridor riders generated by the program by August 2001 amounts to half of the 61,714 Blue Line weekday ridership for the fiscal year 2000 for a capital cost less than one-seventeenth the cost of the Blue Line.

On July 26, 2001, the MTA Board approved the San Fernando Valley Bus Rapid Transit Project, a fourteen-mile busway connecting with the Red Line North Hollywood station to include thirteen stations, five of them with park-and-ride lots. Journey time is projected to be cut from the current fifty-five minutes to thirty minutes at a capital cost of $285 million, $20.4 million per mile.[125] While this represents less than half the cost per mile of the Blue Line light rail, it is clearly a far higher level of expenditure than for the initial phase of the Rapid Bus experiment. The point, however, is that the bus approach offers the flexibility of alternative levels of implementation. Simple upgrading has demonstrably achieved outstanding results at very low capital cost. Not only may more elaborate, higher-capacity/higher-speed extensions be added later, but the same buses can operate on routes combining sections of high-speed busway with improved regular road operation. In contrast, the cost of providing new right-of-way must be incurred for all sections of light rail before operations may commence.

Operating and Maintenance Costs

"One of the arguments made most often for the rail line is that it will be cheaper to operate because a single driver on a train can carry up to five times as many passengers as a bus," wrote the *Los Angeles Times*, reporting on the Blue Line light rail project. Tri-Met's staff recommendation to proceed with light rail in Portland echoed this view, stating that "Light rail costs less to operate largely because fewer drivers are necessary . . . In the context of Tri-Met's entire system, light rail transit will reduce annual operating costs several million dollars compared to the other alternatives." Even greater optimism was to be found in Sacramento: "Light rail is a bargain. It will allow RT to operate more cost-effective transit service. With a train, an RT employee can move ten times more passengers than with a bus."[126]

In six cities studied in Richmond, Baltimore, Buffalo, Dallas, Miami, Los Angeles, and Pittsburgh, the financial performance of rail transit is clearly inferior to the average of bus system performance in the respective cities.[127] In Baltimore, for example, fares were expected to cover 69 percent of the costs of rail operations "after the first or second year of operation."[128] In 1995, which followed two years of light rail operation, rail's farebox recovery (the proportion of costs paid by fares) was 26.6 percent according to local data (for buses, the same source, which does not account for all overhead elements, reports a farebox recovery of 57.2 percent). In Buffalo, project consultants Voorhees had estimated an 84 percent systemwide operating and maintenance costs farebox recovery ratio (rail and bus) for 1995, given construction of the light rail system now in place.[129] The actual systemwide farebox recovery ratio in that fiscal year was 32.6 percent.[130] Except for a few months, the farebox recovery ratio has been consistently worse for rail than bus service.[131]

As of 1995, Denver's overall farebox recovery ratio had slightly improved since the recent introduction of light rail service, while light rail was operating at a cost per passenger of $1.39, compared to $2.24 for the bus system.[132] San Diego has achieved the top financial performance, however, with a 92 percent farebox recovery for its Blue (South) Line in the 1996 fiscal year. Financial results have been disappointing on San Diego's Orange (East) Line, however, which has shown inferior financial performance to the average for San Diego Transit's bus system.

The Portland, St. Louis, Sacramento, and San Jose light rail systems can also be seen to have lower costs or higher farebox recovery ratios than the average of existing bus systems. Such comparisons of average performance can be misleading, however, both because they fail to pit rail performance against equivalent bus lines and since they mask the effects of structural changes to the bus system made to coordinate with rail. In St. Louis, for example, the overall farebox recovery ratio declined from 24.4 percent in 1993 (reflecting bus services only) to 23.0 percent in 1996 (including the new light rail service), while the bus farebox ratio fell from 24.4 percent to 20.9 percent.[133] Bus system costs increased relative to revenues with the route system reconfiguration aimed at focusing on having buses feed light rail.

St. Louis fares were also restructured to reduce fares for longer journeys. While a former bus passenger, who now uses two modes (bus and rail) to complete a journey, is counted twice for purposes of ridership data, they pay their fare only once. That fare would have previously been credited to the bus system. It must now be shared between bus and light rail, and therefore contributes a lower share of total system costs.

The operating performance of new rail systems is invariably reported in comparison with data representing the average of bus system performance. Light rail generally runs in high-demand corridors, however, where trunk bus lines typically perform better than in the system as a whole.[134] A private bus company started a new bus line to replace discontinued bus service competing with the San Diego Blue Line light rail and within a year was breaking even on avoidable costs.[135] Subsequently incorporated into San Diego's MTDB system, the farebox recovery of this route stood at 81.4 percent in 1996, much higher than the 47.3 percent for San Diego Transit systemwide bus services, and above the 65.7 percent attained by the light rail system as a whole, if below the 91.7 percent recovery for the Blue Line light rail service in that fiscal year.[136]

While bus lines with comparable markets to rail show above average performance, feeder bus services established to support new rail systems exhibit very poor financial performance, although transit management do not allocate their costs to rail. In Portland, Oregon, for example, the operating cost per passenger on light rail of $1.83 in 1996 compared to a systemwide average bus cost of $1.87 (and the rail's farebox recovery ratio of 27.8 percent compared to that for systemwide buses of 23.8 percent). The cost per passenger of providing rail feeder bus services, however, came to

$3.12 in 1996. By comparison, the average bus route radiating from downtown Portland had a cost per passenger of $1.77 in that fiscal year, while certain high-volume bus lines had costs per passenger well below those for light rail ($1.43 per passenger on line 4; $1.17 on line 72).

Since passengers tend to travel further by rail than by bus, the cost per passenger mile will be relatively lower for rail than bus as compared to a cost per passenger measure. On this statistic, the average radial bus line (at forty-seven cents per passenger mile) costs more than light rail (thirty-eight cents), but line 4 (thirty-nine cents) is roughly equivalent and line 72 costs less (thirty-two cents). Since light rail is located in the financially better performing corridors, it makes sense to evaluate its results in comparison with high-achieving bus services, rather than with the average for all buses, which not only includes a variety of lower volume local bus services, which will necessarily operate at higher cost, but very high cost rail feeder buses only present through the need to support light rail.

Light rail's lower projected operating subsidy requirements were critical to making it the preferred option for the Pittsburgh South Hills Corridor. While the estimated capital costs of a busway system were lower than for light rail, operating costs for a busway system were estimated at $39.1 million, compared with only $22.6 million for light rail, both for the year 2000. Annualized 1985 total (capital and operating) cost requirements showed a clear advantage to the light rail alternative, with rail coming in at less than two-thirds of the subsidy and debt service requirements of the busway, half the subsidy per passenger mile and less than half the subsidy per passenger.[137]

The Port Authority of Allegheny County has produced some unusual and provocative statistics on the costs of their East Busway operation, and these are presented in table 3-4.

Table 3-4. Light Rail and Bus Operating Costs, Pittsburgh 1995

Cost per ($)	Light Rail	Busway Bus	Non-Busway Bus
Passenger	3.22	0.95	2.55
Passenger Mile	0.46	0.14	0.64

Note: passenger mile data based on an average trip length reported for both busway and light rail of seven miles.

Busway buses perform substantially better than light rail. The analysis is controversial, however, because costs allocated in calculating busway journey costs are based only on the time buses are on the busway and in downtown, not on the time many routes require to complete journeys from suburban origins beyond the end of the busway. By removing the feeder portions of journeys, busway performance is shown in an unduly positive light, although no differently than the way light rail is seen when the costs of (bus) feeder systems are excluded. When challenged on the reasonableness of this procedure, one Pittsburgh manager replied: "They don't include the feeders for rail, so why do it for bus?"

For the Pittsburgh bus system as a whole, farebox recovery was 40.6 percent in 1995, so it would be reasonable to assume that for busway trips the result was somewhat better. By comparison, the farebox ratio for light rail in that fiscal year was 28.2 percent, making it fair to conclude that the assumption of better rail operating performance critical to giving rail the edge in the consultant's evaluation was wrong.

Table 3-5 shows the estimates and actual results for Los Angeles Blue Line operating financials.

Table 3-5. History of Financial Performance, Los Angeles Blue Line Light Rail[138]

Recovery Forecast	Operating Cost ($ million)	Revenue ($ million)	Farebox %
1985 for 2000	12.5	8.4	67.0
1990a for 1991	27.7		
1990b for 1991	33.6		
Actual Fiscal Year			
1991	38.6	3.9	10.1
1995	41.9	5.5	13.1
2000	39.3	9.2	23.5

The first estimate, from the Blue Line Final Environmental Impact Statement, projected a 67.0 percent farebox recovery for the Blue Line. Just as LACTC reduced ridership estimates close to opening date, LACTC increased its estimate of the costs of the first year of operation (1990a in table). The RTD budget set the 1991 fiscal year costs yet higher (1990b in table).[139] For the first fiscal year of operation, 1991, operating costs came to

$5 million more than this high estimate, while fare revenues only covered 10.1 percent of operating costs. By the year 2000, the farebox recovery had increased to 23.5 percent. In the same year, bus operations covered 33 percent of operating costs with fares.

Driver costs do make up a lower proportion of light rail costs than bus costs, but maintenance costs tend to be substantially greater for the rail system. For the fiscal year 1995, light rail operators accounted for 9 percent of Blue Line light rail operating costs, while bus drivers accounted for 40 percent of bus operating costs.[140] This makes for a driver cost per passenger of $1.47 on the bus as against thirty-one cents by light rail and of eighteen cents per passenger mile by bus as against four cents per passenger mile by light rail.

Security costs budgeted for the Blue Line in the fiscal year 1995 came to $10 million as against $19.7 million for the MTA bus system. Given 344 million fiscal year 1995 bus passengers as against 12 million fiscal year 1995 Blue Line passengers, this means that while $0.83 was spent on providing security for the average Blue Line passenger, only six cents was spent per average bus passenger during that fiscal year.

Facilities maintenance costs—for stations and other buildings, tracks, catenary and fare collection machines—accounted for 22.8 percent of light rail operating costs in 1995. Facilities maintenance for the less elaborate requirements of bus operation constituted only 4 percent of bus operating costs. To put it another way, bus facilities maintenance cost was three times the cost of rail facilities maintenance, but MTA buses carried twenty-nine times as many passengers as light rail and thirteen times the passenger miles. While this may not be unacceptable per se, it is nonetheless inappropriate to make simple comparisons based on obvious characteristics such as the number of drivers required to provide a service.

The busway and light rail alternatives currently under consideration for the Los Angeles Exposition right-of-way are projected to have the same operating costs of twenty-six cents per passenger mile, but when operating and capital costs are combined, MTA's consultant found that the incremental cost per incremental passenger in the year 2020 would be fourteen dollars for the bus project as compared to between eighteen and twenty-one dollars for light rail as compared to the alternative of taking no action to improve services.[141]

Table 3-6 provides a general overview of performance characteristics for fiscal year 2000.

Table 3-6. Performance Indicators, MTA Bus and Blue Line Light Rail Services, FY 2000

(Operating Expense, Passenger Revenue, Unlinked Passenger Trips, Unlinked Passenger Miles are in Millions)

| MTA service | Bus | Rail Services | | |
		Blue	Green	Red
Total Operating Expense $	668.02	39.31	22.07	46.55
Passenger Revenue $	216.50	9.24	4.31	6.83
Subsidy $	451.52	30.07	17.76	39.72
Unlinked Passenger Trips	359.00	20.41	9.45	27.96
Average Trip Length-Miles	3.6	7.2	6.5	2.7
Unlinked Passenger Miles	1237.21	146.98	61.77	74.73
Cost per Boarding $	1.86	1.93	2.34	1.66
Cost per Passenger Mile $	0.54	0.27	0.36	0.62
Revenue per Boarding $	0.60	0.45	0.46	0.24
Revenue per Passenger Mile $	0.17	0.06	0.07	0.09
Subsidy per Boarding $	1.26	1.47	1.88	1.42
Subsidy per Passenger Mile $	0.37	0.20	0.29	0.53
Farebox Recovery %	32.41	23.51	19.54	14.68

Source: MTA[142]

Analysis of this data is complicated by the relationship between fares and ridership. Fare revenue is lower on the rail modes than for bus service, particularly in terms of revenue per passenger mile. In the case of the Blue Line, with its long average trips, the revenue per passenger mile is little more than a third the equivalent revenue by bus. A fares policy which has priced rail low relative to bus has clearly stimulated ridership relative to a policy which would have charged equivalent rates, and resulted in lower relative costs per passenger and per passenger mile. Even so, the cost per boarding is higher than bus for the Blue and Green Lines, if lower than bus on the Red Line. The cost per passenger mile is substantially lower on the Blue Line than by bus, given the long average Blue Line trip length.

The operating subsidy cost per boarding is greater on all the rail modes than by bus, but it is lower in terms of passenger miles for the Blue and Green Lines, once more reflecting high average trip lengths. For the Blue

Line, the subsidy per passenger mile is only 54 percent of the subsidy per passenger mile on the bus system.

While the subsidy per passenger mile is lower on the Blue and Green Lines than on the bus, low rail fares have helped lead to fares covering a lower proportion of costs for all of the rail modes than on the bus system. With the Red Line operational to North Hollywood effective July 2000, Red Line financial performance improved, with a 22.0 percent farebox recovery through March 2001 for the 2001 fiscal year, a performance closer to that achieved by the two light rail lines, if still behind the bus system as a whole.

While the above comparison pits relatively high-volume rail lines against the average for all bus services, including costly low-volume bus services, the subsidy per passenger on the 720 Wilshire Rapid Bus as of August 2001 came to sixty-three cents, well under half the $1.47 per ride subsidy on the Blue Line for the fiscal year 2000 (the lower passenger volume 750 Ventura route experienced a subsidy per ride of one dollar, making the

Figure 3-6. Buses at the transit plaza at Los Angeles Union Station. Interchange is available between buses and AMTRAK and MTA Red Line services. (Photo courtesy of author)

average for the two Rapid lines seventy cents). With similar average trip lengths (6.0 miles on the Wilshire Rapid, 7.2 miles on the Blue Line), the subsidy per passenger mile on the 720 came to eleven cents, compared to twenty cents for the Blue Line, demonstrating the lower subsidy cost for a bus operation similar in many ways to the Blue Line light rail.[143]

Opportunity Costs

It is too easy to forget the capital costs of expensive projects once they are built, but those costs represent opportunities forgone for making improvements in less capital-intensive ways, at least in cases such as the Los Angeles Blue Line light rail which was constructed with purely local funds. Modest improvements to basic bus services combined with an attractive fare policy have shown they can secure substantially greater ridership increases than capital-intensive projects involving either light rail or busway construction. The examples from pre-busway or HOV service upgrades in Ottawa and Houston; ordinary bus improvements in Portland which stimulated greater ridership increases than the subsequent light rail program; and the success of the new Metro Rapid Bus installation in Los Angeles bear this out.

Fare reductions present one promising path to gaining increases in ridership. As Wachs reports, "Dozens of transit fare changes have been monitored, and we know with reasonable certainty that changes in transit fare have elasticities in the range of –0.3 to –0.4."[144] A 10 percent reduction in fares, for example, can be expected to lead to a 3–4 percent increase in ridership.

The 1982 bus fare reduction from eighty-five cents to fifty cents funded by initial disbursements from the new half-cent local sales tax authorized by Proposition A of 1980 took bus ridership out of a trough at 359.5 million annual passenger bus trips and to a peak of 497.2 million passengers in 1985, an increase of 137.7 million passengers, or 38 percent over three years.[145] Some of these new passengers were people making trips they would not otherwise have made; the rest were people who formerly drove, carpooled, or walked. All these new trips represented increased use of transit.

When the fare reduction ended as of July 1, 1985, and the Proposition A funds formerly used to support it went to light rail construction instead, bus patronage fell dramatically: down 46.8 million annual passengers in just one year—more than double the annual ridership of the Blue Line in the fiscal year 2003. RTD regression analysis shows that 84 percent of the ridership change is explained by the fare changes.

By 1990, RTD reported that reductions in subsidies to bus services and bus fare increases had been principally responsible for the loss of eighty-six million rides (17 percent of ridership) since the ridership peak five years previously, and warned that the diversion of operating funds to support the rail services would further weaken the viability of bus operations and ridership.[146] Seemingly without seeing the inconsistency between the two parts of the sentence, the memorandum from senior RTD management states: "While no one doubts that rail is an absolute necessity for urban mobility in Los Angeles, funding the construction and operations of rail out of bus funds could prove to be extremely counter-productive."[147]

In fiscal year 2003, total rail ridership on the Red, Green, and Blue Lines came to 63.7 million. The total ridership gains to rail to date are less than the losses in bus riders occasioned by the diversion of funds from bus operations and fare subsidization to the rail program. Decreased funding to the bus system disproportionately hurt the urban poor who depend on those services the most, furthermore, as alleged in a civil rights lawsuit responding to the deterioration in bus services occasioned by the start-up of rail and the anger of those who had seen diminished service and higher fares.[148] It is only through this lawsuit that a court-implemented consent decree now requires the MTA to provide new infusions of money to its run-down bus system.[149]

Rail Has Not Been Good Value for Los Angeles County

Superficially, the Blue Line has done well by beating its forecast 54,700 average weekday boardings for the year 2000, with actual results of an average 61,714 weekday boardings during that fiscal year, and 69,646 during fiscal year 2003. The Red and Green Lines in Los Angeles County, along with the majority of other new rail projects around the country are far short of target ridership levels, but when one looks at the whole story of the Blue Line's impact on the total transportation system, it has hardly done any better.

Ridership, fares, and system financial performance are clearly all related.

Figure 3-7. The Green Line operates in the middle of the Century Freeway. (Photo courtesy of author)

The policy in Los Angeles, as in many other cities, has been to price rail low to boost ridership. With revenue levels for rail service far lower than on the MTA's bus system, it is not surprising that farebox recovery is also lower for rail than for bus—and at 23.5 percent in fiscal year 2000, is well below the 67 percent recovery originally projected for that year. The Blue Line cost and subsidy per boarding is slightly above that for bus service as a whole, although the cost and subsidy per passenger mile is substantially lower, given the considerably longer distances Blue Line passengers travel.

Such comparisons against the aggregate bus system are not helpful, however, since the bus system includes a variety of low volume and low productivity local services as well as higher-volume trunk routes. It makes more sense to evaluate light rail services in the light of comparable bus lines, and the new Wilshire Rapid service, which provides limited stop high-volume service in a dense corridor, and attracts passengers traveling the same average distance as on the Blue Line, is the obvious candidate. Developed for $5 million, the service boosted Wilshire corridor ridership by one-third—20,666 daily passengers—in its first ninety days while requiring an operating subsidy per rider and per passenger mile 20 percent lower than for Blue Line operations.

The initial capital costs of instituting Metro Rapid service were a bargain, yet first-generation service—in mixed traffic—will be improved by further measures to increase speed and reliability, and this will require higher levels of expenditures. Dedicated busway, for example, is expensive to build, but can nonetheless be provided at less than half the cost per mile of light rail. Most importantly, a staged approach can be taken with bus improvements, with low-cost measures followed by larger-scale capital projects. Parts of bus lines can operate over dedicated rights-of-way, while other sections run in mixed traffic, offering a choice of approaches to system configuration. Buses can offer a variety of no-transfer services diverging from a single fixed guideway, furthermore, while rail travel requires a far higher degree of transferring, an attribute found highly unattractive to passengers. While it has not been practical to trace the total financial impacts of overall system changes in the complex environment of Los Angeles transit, bus systems in other cities which have adopted rail have shown

Figure 3-8. Loading a bike onto a Rapid Bus in Santa Monica. (Photo courtesy of author)

reduced productivity and increased cost when reconfigured from direct-route to rail-feeder service.

While rail has shortened trip times for passengers beginning and ending their journeys near rail stations and, in the case of the Red Line, for some passengers served by new Metro Rapid Bus connections, passengers not located close to rail have in some cases experienced longer journeys when their bus alternative was discontinued. Where grid bus service remains in place in South Central Los Angeles, a direct trip on the bus system remains faster than the Blue Line for those located away from rail stations. With most area trips of a short-distance local nature, getting to and from stations makes up a relatively high proportion of total trip time, making light rail unattractive for such journeys.

About a third of Metro Rapid passengers are new to transit, about the same proportion as on the Blue and Red Lines, indicating that bus service can be just as attractive as rail in pulling people from their cars. While comfort has been shown to be of much less importance in attracting passengers than journey time and fare, rail service has traditionally been provided with far more amenities by way of station facilities, information services, and attractive vehicles than bus service. The Metro Rapid has brought a fundamental change in Los Angeles by aiming to offer equivalent levels of amenities to light rail. Its attractive lighted shelters, information panels, and smart low-floor buses provide fundamental improvements to passenger comfort, compared to the sparse bus stops and basic vehicles traditionally provided. If bus travel appears less comfortable to some people than rail travel, Metro Rapid has demonstrated that it does not have to be.

Capital costs are often, strangely, forgotten once spent, yet they represent real opportunities foregone, especially as, in the case of the Blue Line, construction was paid for entirely out of local funds. Not only did Blue Line construction costs escalate dramatically, but the removal of funds programmed for bus fare subsidization to make building of the Blue Line possible caused a sharp downturn in areawide bus travel. The five years following increases in bus fares and reductions in bus services saw a loss of 86 million riders. By the fiscal year 2003, the Red, Green, and Blue lines together were carrying 63.6 million passengers. More recently, bus ridership has been increasing as funds have been devoted to bus service augmentation, but only as the result of a consent decree following a civil rights lawsuit which alleged that the diversion of funds from the bus system to build rail had hurt minorities and the poor, both through increased fares and a marked deterioration in service quality. Far more passengers would be transported by transit if Proposition A money had continued to subsidize bus fares and/or paid for expanded bus service, rather than going to construct light rail, especially if innovations such as the Metro Rapid had been introduced at an earlier date.

In the above discussion, we have made a critical assumption: that money should be spent on some sort of transit improvement, whether rail or bus. Yes, if we are to expand our inquiry from its immediate assumptions, as Churchman would have us do, we need to question whether this is the right question: perhaps there are other needs more urgent than transportation on which resources could be spent.[150] It turns out—as we shall see progressing through this study—that transportation metaphorically represents all sorts of other benefits, whether in terms of community development, the enhancement of job prospects, and in general bettering the lot of the poor. We will question such assumptions in chapter 11.

For now, we conclude that rail provides a service unmatched to the majority of the dispersed travel needs of Los Angeles County, and at a far higher overall cost per new transit passenger than would be incurred by reducing bus fares and/or developing bus services instead. The rail program, in fact, imposed a long-lasting decline on overall transit usage in Los Angeles County, which is only now being reversed with increased funding to the bus system.

A recent consultant study for the Exposition Corridor, which takes a southerly alignment between the downtown Los Angeles and Santa Monica areas, found that putting buses rather than trains on the right-of-way would be more cost-effective, yet there is still considerable advocacy to put rail there. We are left with a mystery. Why, given its extraordinary expense not only in actual costs but in opportunities foregone, and its unsuitability to the dispersed Los Angeles autopolis, has rail construction taken place. In the next two chapters, we will examine the ridership forecasting process for the Long Beach line. In chapter 6 we begin to unravel the mystery.

CHAPTER 4

Inventing Demand for the Long Beach Line

> Everything is estimated optimistically. If you start with a pessimistic
> estimate, it never gets built.
>
> —SCAG staff member

FORECASTING IS AS OLD as the human desire to know tomorrow's history today: it responds to a fear of the unknown. The Ancient Greeks formalized forecasting in the institution of The Oracle—at Delphi, for example—which "helped reduce uncertainty, often by relying on the principle of self-fulfilling prophesy."[1]

Today's attempts to picture the future perform a similarly reassuring and stabilizing role in society, even if they are the result of technical analysis and involve complex computer programs. How could we formulate economic policy without a forecast of where the economy is going? How can we decide on the number of weapons we will need without predicting what the other side will do? "Transit systems, power plants, hospitals and airports are constructed only after forecasts have demonstrated that a 'need' exists for their services, and that their costs are justified by expected benefits."[2]

Computer modeling has become so accepted and commonplace in both the social sciences and government that "many educated people treat computers and the ensuing recommendations as objective fact."[3] The desirability of quasi-scientific approaches is intensified by their apparent ability to provide simple and clear-cut answers to complex problems. Given our basic intolerance for uncertainty, it is reassuring to have "hard numbers"—which we may presume were "rigorously" obtained—to tell us in which direction to go. But, as Donald A. Schön remarked, "formal modelling has become increasingly divergent from the real-world problems of practice."[4]

While the apparent complexity of high-powered techniques lends them authority, all quantitative models—however complicated—must simplify. As Richard Chorley and Peter Haggett state in their classic book, "The most fundamental feature of models is that their construction has involved a highly *selective* attitude to information, wherein not only noise but less important signals have been eliminated to enable one to see something at the heart of things."[5]

To simplify, "rules" are needed to decide both what is relevant information and how the chosen information is to be processed. These rules—or assumptions—are selected subjectively, not determined "objectively," but color the whole analysis of which they form the fabric. Models are therefore a compilation of judgments, not science-based "value neutral" tools.

A mathematical statement has no social content of itself: it is correct or not to the extent that it follows the rules of mathematics. But mathematical statements, though empty of themselves, may powerfully organize information; they do so according to the assumptions under which they operate. Danger lies when, according to Ida Hoos, "in the absence of clearly specified limits and conditions, the assumptions and biases of the analyst are taken as representative of the real system under study. This often leads to oversimplification, neglect of vital facets, and inappropriate or unwarranted recommendations and conclusions."[6]

The problems introduced by models exist on a number of different levels, but they center on the need to adopt a series of assumptions given the lack of necessary validity of any assumptions in particular. The biggest assumption is that the model-in-use is simulating relationships central to the problem at hand. In many cases, however, the model is merely simulating the interactions of variables which are easy to represent. Model relationships act as surrogates for the real problems—such as of the environment, mobility, and equity—which are often thornier, not directly amenable to quantification, and which may, therefore, remain unaddressed.

On a basic level, there is scope for political agendas to influence which assumptions get chosen: if leading politicians wish a particular project to proceed, it is tempting to go for optimistic assumptions which will make that project appear attractive. In some cases this manifests itself in outright dishonesty, in the deliberate misapplication of techniques.[7] In many other instances, however, there is no such simple black-and-white case of unethical behavior. As Martin Wachs explains, "[F]orecasts often require so many assumptions that there is leeway to allow the forecaster to satisfy both organizational goals and technical criteria. Indeed, if he or she has become a 'team player' and has internalized the goals of the agency, there may not even appear to be conflict between the two loyalties."[8]

On a number of deeper levels, the very nature of computer modeling—with its basis in technical expertise and ability to process information only in predetermined standardized formats—arouses questions about its desirability for social decision-making, even given the best of honest intentions on the part of those doing the modeling work.

To start with, public officials, concerned with policy questions, generally know little about the technical methods by which forecasts are obtained, and are therefore in a weak position to critically evaluate them. Those preparing the forecasts, in contrast,

> are usually drawn from the ranks of social scientists, engineers, and planners whose education and professional identity are based primarily upon technical methodological skills. They are likely to believe and promote the belief that forecasting is impossible without the use of computers, mathematical methods, and complex data sets.
>
> Sophistication in the technique of forecasting is more apparent, however, than real. Computers are used because there is often a great deal of data: many variables, many units of analysis for each, several time periods. These conditions lead to the requirement for training in mathematics, statistics, data manipulation, and computer programming. But together, such skills ensure no special perspective on the future, and there is relatively little theory derivable from the social sciences to help one arrive at reasonable core assumptions.[9]

Given a lack of theory with which to choose assumptions, the door is left open for arbitrariness in their selection. On a micro scale, there are frequently dozens of detailed assumptions which must be made for any particular forecast. At what rate will the population grow? How will the geographical distribution of housing and employment develop? How many times will people be prepared to change vehicles during a trip? Given several mathematical functions which "fit" a given data set, which should be chosen? And so on.

Perhaps one of the most egregious practices is the unfounded use of scientific analogues for social processes. Despite a lack of proof that scientific theory maps from the domain of the physical to the world of the social, work upon which it is based can look impressive—and it does appear to provide an answer. Brewer demonstrates this phenomenon at work in his account of modeling for the Community Renewal Program of the City of San Francisco, which uses analogies from chemical kinetics and physics. "The assumptions, built into the rent pressure relationship," for example, "are offensive to sense, common and otherwise," he says.[10] But "If a model-builder has never been sensitized to the details of a specific empirical context, one should not find fault with his great inferential leaps, from decaying isotopes to decaying houses or from expanding and collapsing magnetic fields to expanding and collapsing rentals."[11] We shall see a similar problem in the use of a Newtonian gravity model for transportation forecasting below.

In the San Francisco case, it was not simply that a bad job had been done, but that the city planners had wanted to ask detailed questions which the model could not address. But, says Garry Brewer, "Even though the model can't answer 'those kinds of questions,' it was decided to build in so much detail that those questions nonetheless appear to be asked."[12] As we shall see below, the model used for Long Beach forecasting is incapable of picking out flows on individual transit lines, such as the Long Beach light rail line, as against representing an aggregate of parallel transit lines. Even so, it appears capable of providing such detail, and was used to do so even though the results had no statistical significance.

Ascher terms the reliance on old core assumptions "assumption drag," and it is a problem endemic in forecasting.[13] Assumptions made during model formulation may quickly become invalidated; yet, the model may live on, especially if costly work—in conducting surveys, for example—was required for model development. A model established at a time of gasoline plenty, for example, might well be invalid at a time of shortage, but the model might continue in use.

Users can easily remain unaware of problems with outdated or otherwise invalid assumptions when running a model. It can be difficult to fathom the complexity of a model, even if an attempt is made to do so. Technicians who make forecasts often run data through already prepared packages. Rarely is every step displayed let alone understood in such a process, and it is much easier to simply use the model as supplied than to attempt to dismember it and put its assumptions to the critical test.

Even sophisticated users, who make attempts to check behind the scenes of the models they use, can easily be foiled in their attempts to understand the assumptions upon which those models are based. Not only is it often difficult to verify matters such as the validity of sample surveys upon which estimation of the model formulae may be based, but many assumptions for input to one model—on population growth and distribution to be used in a transportation model, for example—may be the product of other models, with their own webs of assumptions. It becomes impossible to trace everything to the source, and the result is that unreasonable and sometimes absurd assumptions can be unknowingly allowed to influence forecast outcomes.

A more general problem than the use of old assumptions (to which a technician's answer may be made that the modeler is at fault for not using newer ones) is that the modeling process inherently uses "old data" since we have no evidence of what the future will hold other than our experience up until today. Although past data must be relied upon, it may not be an indication of the future, especially as actions taken as a result of the forecast may change that future. Patterns of the past, furthermore, might not form a desirable program for the future.

We come, via this path, to realize that our model may encapsulate a view of the future we have not knowingly sanctioned. The model takes on a political life of its own, one which might be quite independent of the intentions of either its creator or of the technician feeding in the data and extracting the results according to "correct" technical procedure.

Jacques Ellul conceived of the villainous "la technique" (the soul of forecasting) in anthropomorphic terms: it is something created by humans but which, through its tacit but powerful assumptions, itself becomes "the judge of what is moral."[14] Joseph Weizenbaum illustrates this phenomenon in remarks on modern warfare in which "it is common for the soldier, say the bomber pilot, to operate at an enormous psychological distance from his victims. He is not responsible for burned children because he never sees their village, his bombs, and certainly not the flaming children themselves."[15] Pilots follow correct technical procedure and—parts of a machine themselves—do not need to be aware of what they are doing.

Computer systems can similarly divorce their users from the realities of the real world, and therefore from responsibility for the actions taken as a result of those computer operations. In Vietnam, Weizenbaum writes, "computers operated by officers who had not the slightest idea of what went on inside their machines effectively chose which hamlets were to be bombed and what zones had a sufficient density of Viet Cong to be 'legitimately' declared free-fire zones, that is, large geographical areas in which pilots had the 'right' to kill every living thing."[16] The computer provides the authority for action, and responsibility appears to rest with the computer rather than with the person executing its instructions.

Edward Leamer's study of the use of econometric models to determine the efficacy of the death penalty in reducing murder provides another example which illustrates both the extent to which prior assumptions influence predicted outcomes and how a model can itself tacitly encapsulate a moral outlook independent of the user's intentions.[17] Leamer finds that a regression of murder rate on variables thought to influence murder "leads to the conclusion that each additional execution deters thirteen murders with a standard error of seven. That seems like such a healthy rate of return that we might want just to randomly draft executees from the population at large."[18]

But the conclusion changes when the set of variables thought relevant to the model is altered. A result which looked convincing under one set of assumptions loses credibility when those assumptions are changed. "Individuals with different experiences and different training will find different subsets of the variables to be candidates for omission from the equation."[19] So a right winger will look to the punishment variables and regard others as doubtful, while "an individual with the bleeding heart prior sees murder as a result of economic impoverishment."[20] The conservative thereby "finds" that execution has a strong deterrent effect upon murder, while the liberal "finds" that execution actually encourages further murder.

The death penalty case, "Perhaps the single most important legal use of multiple regression thus far," according to Franklin Fisher, presents a twofold problem.[21] Firstly, the outcome is most heavily influenced by the prior beliefs inculcated into the assumptions, rather than by the data they purport to analyze. Secondly, and on a deeper level, not only are the assumptions employed in the procedure subject to "bias," but the procedure itself reflects a point of view, the implicit belief that the death penalty should be used if it will deter murder, which might be rejected were it to be brought to the surface and directly subjected to critical attention.

The uncritical use of statistical analysis therefore has the potential to distract us from deciding whether society should—as a matter of principle—have the right to kill someone, a debate which is embarrassing because it exposes the roots of our ethical values, lays them open to criticism, and leaves us uneasy since there is no "sure" solution. It is tempting for those on both sides of the death penalty debate to stand behind the illusion of science provided by the apparent precision of econometric technique. But when opponents become entangled in technical arguments over the alleged deterrent effect of capital punishment, their case is weakened because the "right to kill" is tacitly (if unintentionally) presupposed by the calculus employed.[22]

Quantitative techniques, then, are not simply subject to abuse; their use for honest purposes may imply a set of beliefs which their users might reject were they aware of them. "The quantitative approach tends to divert our attention away from the evaluation of the concepts and variables themselves" says Young.[23] "We can thereby be drawn unwittingly into an uncritical acceptance of the overall framework of theories and approaches to nature and society."[24]

In summary, there are objections to forecasting-as-practiced on a number of levels, some related to how users conduct forecasts; others related to flaws inherent in the forecasting tools themselves; others at the boundary of these two cases. Users may knowingly misuse models, but they may also be placed in a situation where, given choices of assumptions all of which are technically justifiable, it seems quite professional as well as ethical to choose those assumptions which are most politically convenient. They may, furthermore, use complex models without checking and—in many cases—without being able to check the full implications of the assumptions upon which their forecasts are based.

On a deeper level, "la technique" has a political life of its own, representing particular questions asked, and asking them from a series of value positions implied by the model's structure, rather than with the user's knowing assent. The ready virtuosity of the model produces results of apparent solidity and reliability, and attention is not only diverted from the more fundamental questions which the model does not ask, but from the values the model presupposes, values which we might not want reflected in our policy choices.

Chapter 5 examines the larger ethical implications of modeling, considering issues of responsibility on the part of clients purchasing modeling, practitioners performing modeling, and the nontechnical decision makers who consume its results. This chapter provides an in-depth critique of the attempt by SCAG to model future ridership demand on the downtown Los Angeles–Long Beach Blue Line light rail service.

Demand Forecasting and Alternatives Analysis

Peter Stopher and Arnim Meyburg describe the "transportation-planning process" as a succession of seven technical steps, at the core of which are a series of forecasts to estimate the future demand for transportation and determine the relative attractiveness of alternative means of travel.[25] The quantitative models they describe play a key role in the federal system of transit project evaluation known as "Alternatives Analysis," which is mandatory for federal capital grant applicants seeking to build or extend transit systems. The results are used to rate the merit of different projects competing for federal funds. "The rating system has been designed to provide a rational approach to the allocation of Federal funds in a setting where the demand for Federal assistance far exceeds available resources."[26]

This federal approach applies the techniques of systems analysis to quantify the costs and benefits of alternative projects within particular cities and to establish whether they fall within a threshold of cost-effectiveness which might make them eligible for federal funding. Quantification is central to its operation, and an emphasis on precision in translating phenomena to be evaluated into quantitative terms is seen as central to ensuring the "rationality" of the approach. "The primary emphasis here is on transportation service and the mobility it provides. Several other considerations, ranging from economic development to pollutant reductions to energy conservation, are secondary, but are so closely related to improvements in mobility that they are implicitly included in the Federal objectives and the evaluation system."[27]

Ridership demand modeling plays a large part in evaluation since a significant relationship is seen to hold between the number of riders a project will carry and the total benefits a project will bring. Peter Stopher and Arnim Meyburg, for example, outline the "transportation planning process" as follows. An inventory is to first be taken of existing travel and

land use, socioeconomic population characteristics and existing transportation facilities. A series of forecasts follow "of land uses that should
occur in the forecast period, and then of the demand that may be anticipated and the way this will occur throughout the region."[28] Four models
are used: (1) to gauge total demand; (2) to allocate it between origins and
destinations; (3) to allocate it between competing modes of transportation;
and (4) to allocate it amongst the set of available network paths. Finally, alternative strategies for providing transportation are evaluated in light of
the above, and policy choices for planning are made.

> The conduct of these transportation studies and their general structure is
> based on the premise that the demand for travel is repetitive and predictable,
> and that future transportation systems should be designed to meet a specific,
> predicted travel demand. This demand is itself based on an analysis and ex-
> trapolation of current travel, and an investigation of this relationship to the
> patterns of population, employment and socioeconomic activity.[29]

Planning as defined here is to take place by following a predefined procedure. Planning is a technical problem; there is a "correct" way of doing it;
and there is a "solution" to the transportation "problem." The systems
analysis the authors prescribe "provides a framework for a *systematic* approach to solving *complex problems* under conditions of *uncertainty*. The
technique provides an approach in which *objectives* are defined and *alternatives* are assessed against these objectives. The entire process is carried
out in the framework of identified *systems* and *subsystems*" (original emphasis).[30]

Since the LACTC planned to build the Long Beach light rail with the
proceeds from an extra half-cent sales tax enacted by county Proposition A
of 1980 and without the use of federal funds, the agency was not required
to perform a federal alternatives analysis. As part of the environmental impact process, however, an analysis was conducted with many features common to the federal process. In particular, several alternative projects were
considered, including an "all-bus alternative" and one which involved no
change in transit service levels beyond those already provided in the SCAG
Regional Transportation Plan. Ridership forecasts were prepared for each
alternative considered, although costs were only estimated for the light rail
alternatives.

Despite the appearance of rigor, the modeling of expected human behavior in response to change in transportation and other systems requires the adoption of an array of assumptions about an uncertain future which must be subjectively chosen. There are so many ways in which error may enter this process that claims to reliability can be at best weak. As noted in chapter 3, modeling has typically overestimated demand for new rail projects. This has not been the case for the Los Angeles light rail line. As we will see below, this has been a result of happy circumstance, and not because of the quality of modeling work.

Initial Demand Assessments

SCAG's forecasts were not the first to be produced. In October 1981, Caltrans published the results of a "feasibility study" into light rail transit from Long Beach to Los Angeles. It was based on the use of existing Southern Pacific right-of-way and city streets for the trackage, and construction cost was estimated at $150 million.

A simple method was used to estimate patronage: "An assessment of the northbound morning peak-period (6–9 A.M.) patronage on ten relevant RTD routes now operating within the Long Beach line corridor served as a basis for estimating peak-period light rail ridership demand."[31] Bus occupancy data for 1979/80 indicated a morning peak-hour patronage of 5,100. With a Los Angeles DOT cordon count that indicated that about half of peak patronage was between 7:30 and 8:30 A.M., a morning peak light rail patronage of 2,500 was generated, with weekday patronage estimated at 15,000 and 7,500 on the weekend.

Of the ten bus lines included in the study, those closest to the light rail line right-of-way "were statistically weighted at or near 100% of the cumulative occupancy counts, whereas lines located at or near the corridor's outer limits were trimmed back in some cases by as much as 40 percent."[32]

The following year Parsons Brinckerhoff built upon the Caltrans estimates.[33] In addition to using data from existing bus operations, demographic and land use characteristics of the service area, the projected capacity deficiency in the road network along the light rail corridor, existing transit operations and usage in the corridor, and connectivity of the proposed light rail line with a regional transit system were considered.[34] Prior bus users, prior car users, and new transit trips induced by the presence of the project were considered as sources of potential ridership. As a reasonableness check, estimates were compared with results obtained on the then recently opened San Diego South (Blue) light rail line.

It was assumed that only parallel bus lines within one mile of the proposed light rail alignment would be affected by the rail service. It was estimated that 17,900 passengers would be diverted from these lines. But,

> because the Los Angeles–Long Beach Corridor is not projected to experience significant highway congestion in the near future the proposed LRT line would have less potential for auto trip diversion than in other corridors of the region. Therefore, it was estimated that 15% of the initial ridership for the baseline LRT project would be from prior auto users and other non-bus sources (induced trip making, pedestrians and others).

In all, 21,000 daily passengers were estimated to use the new service. While this was based on the assumption of a 1982 service, escalation factors were provided to produce daily patronage of 30,600 (with 2 percent annual growth) or 53,000 (with 5 percent annual growth) in 2000.

During August–October 1982, Caltrans ran light rail patronage estimates on the LARTS system to be later used by SCAG, forecasting 15,013 daily home-to-work trips (27,802 total daily trips) for the defined "Baseline" system, a substantially lower estimate than SCAG was to obtain later.

Review of SCAG Long Beach Corridor Demand Modeling

"I know the model overestimates . . . You can quote me on that. It overestimates." So said a member of the SCAG staff of the modeling technique used to project that in the year 2000 the Los Angeles–Long Beach light rail line—with the particular route which was eventually selected—would each day transport 54,702 passengers, a forecast adopted by the LACTC as a result of their environmental impact process, and a number constantly cited in the media as the decision to construct the Long Beach line moved forward.

As part of environmental impact assessment for the Long Beach light rail project, SCAG produced a report on "Patronage Estimates and Impacts,"[35] under contract to LACTC: SCAG was retained in the role of a private consultant and paid $200,000 for its work. Ridership forecasts were prepared for seven alternative light rail routings, for an "all-bus alternative," and for the adopted Regional Transportation Plan (RTP) system with planned light rail lines removed (in other words, a "no-build" alternative).

All of these forecasts were for the year 2000. A forecast was also prepared for the 1980 base year, without light rail. This was provided for purposes of "validation," to see—in other words—if the model could properly replicate already existing conditions.

In addition to the development of ridership forecasts, SCAG provided an analysis of the likely effects the Long Beach light rail project would have on road congestion, pollution, and other areas of environmental concern. As was illustrated in chapter 3, the findings here were less optimistic than the high ridership forecasts would suggest: the light rail project, SCAG said "has only a very minor impact on traffic" because most users would be people who formerly took the bus, rather than automobile drivers. Nor would the light rail project "contribute to any meaningful change in the level of emissions or in the ambient air quality in the study area."

SCAG, in short, had produced an impressive-looking forecast of ridership on the light rail line, but rated the system poorly in terms of the benefits it would bring, as compared to the continued operation of transit systems equipped solely with buses. These negative findings received little publicity, compared to the spotlight shone on the ridership forecast, which was repeated in countless media accounts.

Given the role of SCAG's ridership forecast in legitimizing the project, it is important to understand the basis and evaluate the reliability of the analysis which produced it. This demands an approach on a number of levels: "did the way the model was used produce reliable results?" is a different question from "can the model produce reliable results at all?" Both must be asked.

The Modeling Process

SCAG conducts travel demand forecasting in the context of the "urban transportation planning process" developed by the U.S. Department of Transportation, and which,

> has evolved to a generally accepted and widely applied practice. In this process, planners develop information about the benefits and impacts of implementing alternative transportation improvements. This information is used to help decision-makers (elected officials or their representatives) in their selection of transportation policies and programs to implement. Thus, the planning process *leads* [my emphasis] to development and adoption of the Regional Development Guide Plan, the Regional Transportation Plan,

the Transportation Improvement Program, and the Air Quality Management Plan.[36]

A "complex set of interrelated models" is used to forecast travel patterns. "As it would be carried out for a complete forecast of regional travel, SCAG's Regional Transportation Modeling System has many applications involving more than 100 separate executions of computer programs."[37]

The steps SCAG takes in modeling follow the standard process described by Stopher and Meyburg, which will be outlined below. Having assembled a base of socioeconomic data and land-use forecasts, a four-step model is applied. In "trip generation," the total trips made by a particular market segment are estimated for each zone of the region under study. "Trip distribution" then forecasts how the total trips originating at each zone will be distributed among possible destinations. "Modal split" forecasts what proportion of travelers will use transit and which proportion automobile. "Network assignment," finally, assigns forecast trips to paths through the transportation network.

The modeling process will now be examined in detail. As will be seen, the approach taken by SCAG involves the use of a substantial number of questionable assumptions about what is essentially an unknown future. The compounding of likely error from piling one fragile assumption upon another serves to render the results SCAG reaches meaningless. As we shall see from study of the results of SCAG's "validation," the model is in fact incapable of predicting the demand for travel on a particular transit line, such as the Long Beach Blue Line, since it cannot differentiate between the demands for the use of parallel lines packed close to one another.

Inventory

Compilation of an inventory constitutes the first of the seven steps Stopher and Meyburg outline in their description of the "transportation planning process." A database is developed for evaluating existing travel demand and existing transportation performance, and to provide a basis for predicting demand and future system requirements.[38]

An inventory is taken of the distribution of activities in the area under study. Information is then collected on how people currently travel, how often, and on which facilities. Demographic data is compiled, reflecting both the size of the total population and its socioeconomic characteristics.

Finally, the capacity of existing transportation systems and their level of service characteristics are determined. There are two basic functions of the data which is assembled: it is needed to establish the relationships described by the models in use (in technical terms, to "estimate" the coefficients which act on the variables under consideration in the model); and to provide inputs to be processed by the models themselves.

A number of difficulties emerge at the inventory stage. Most basically, data inventories represent how things are now. This might not make for a desirable model of how we would like things to be in the future; yet, this data is used in constructing a model to predict the future. Data used by SCAG in setting up model relationships mostly came from a 1 percent sample survey of travel behavior conducted in 1967. While factors such as changes in behavior resulting from increased population and congestion, shifts in employment patterns, or other factors can be estimated in data assembled for input to the model, the mathematical relationships of the model itself—estimated as they mostly are on a 1967 basis—do not allow for them. As one SCAG staff member said,

> Most of the travel surveys were funded by the FHWA late '60s and then the money dried up, and no one's conducted any surveys since. We did one, a short one here, a follow-up in '76 and they adjusted some of the interception points—they didn't change the trip rates. In fact, some of the trip rates do change, but we don't have any new information to change them. And so we have to go with the assumption that nothing's changed since '67.

Secondly, there are problems in how current conditions are represented in the inventory, especially descriptions of existing transportation networks. These problems will be further discussed under "network assignment."

Land-Use Forecasts

The next stage of the urban transportation process, as described by Stopher and Meyburg and by SCAG, requires the forecasting of a variety of data to itself be used as inputs to the demand forecasting methodology:

> Forecasting future urban systems that the transportation system is to be designed to serve requires an estimation of the intensities and spatial distributions of population, employment, social activity and land use. Socio-economic data for a future year is derived from the region's growth forecast policy, a product of SCAG's ongoing Development Guide Pro-

gram . . . The current growth forecast policy, called the "SCAG-82 Growth Forecast Policy" was adopted by the SCAG Executive Committee in October, 1982. It is the culmination of extensive analysis and discussion by SCAG committees, local government, and other affected/interested parties and agencies.[39]

Each of these inputs results from separate forecasting efforts. The reliability of transportation forecasts—conducted with their own host of assumptions—is therefore contingent on the reliability of the plethora of assumptions employed by a separate group of planners working with their own sets of unknowns. It is common for those conducting transportation forecasts to be unaware of the details of how each of the inputs to the models they are using were arrived at. This became clear during interviewing at SCAG, for example.

Factors such as population and the distribution of employment can be highly significant in influencing the number of riders a potential new means of transportation might attract. The number the new system might draw, given "today's" population and employment base might be extremely modest, suggesting only a modest diversion from pre-existing alternative transportation modes. The effects of population or employment growth forecasts, "both notoriously prone to error," can multiply the estimated potential ridership manyfold beyond the effect of introducing that new mode alone.[40] That growth—and the associated growth in ridership—is, however, far from assured.

As Andrew Hamer comments:

> The projection of future urban growth has proved to be beyond the capacity of the techniques now available to urban planners. Whether the techniques used are judgmental, as in San Francisco, Los Angeles, and St. Louis, or "statistical," as in Atlanta and Washington, the results depend heavily on the whims of local planning agencies. The lack of substance of such projections is readily apparent when naive technicians create confident estimates of what will be built around each rail rapid transit station, or of how many persons will travel between any two points at any given hour of the average weekday in the year 2000. Even [in] those areas, like Los Angeles, which have experienced repeated revisions of future forecasts, the transit planners seem incapable of comprehending the simple fact that such projections are meaningless.[41]

At SCAG, the "adopted growth forecast policy reflects regional and local growth policies, and is intended to represent the best judgment of association membership in terms of a likely and viable direction for the region."[42] This "best judgment" reflects a consensual political agreement on how SCAG representatives would like growth to be shaped in the region. Thus, the politically determined assumptions of one forecast form the building blocks for the construction of another forecast. The computerization of value judgments lends them the appearance of truth, helping to obscure the political process by which they were determined.

In addition to employment projected by the adopted SCAG-82 forecasts, "the LACTC consultants identified some planned commercial developments within the Los Angeles–Long Beach corridor that would attract tripmakers who might use the light rail line."[43] The employment generated by these developments was added to that assumed under SCAG-82, and the extra transportation movements implied by this additional employment were then calculated. This is problematical, since SCAG-82 forecasts already included a certain level of projected employment increase for the areas where these developments were to be located. This implies a degree of double counting. There is also a biasing in favor of light rail implied by specifically including those new developments which might attract light rail ridership, as against those which might be better served by other means of transportation.

Finally, there are problems in the way the number of trips to be generated by these developments is calculated. When only the acreage of a development is known, SCAG says, the number of trips is generated by a regression equation.[44] For a planned fourteen-acre shopping center at the intersection of Imperial Highway and I-105, for example, the regression equation was used to estimate 9,670 daily trips by all modes. Of these, five hundred were said to be work trips (implying an employment base of 250 if each person makes one trip to work and one going home each day), even though such shopping centers' "employment is usually under 200."[45] A member of SCAG staff was unable to explain the discrepancy. "I don't know. I don't know if it's a typo," he said.

A comment from SCAG on a draft of this chapter asserted that the 500 work trip estimate was reasonable, since people made other work-related trips apart from travel to and from home.

Trip Generation

"Trip generation" and "trip distribution" involve building a picture of the total transportation market by all modes of travel. Total travel by all modes is then divided among competing transportation modes in a "modal split" model. The problem is analogous to baking and serving cake. First, the size of the cake to be baked must be decided. Later, a decision has to be made on how to divide the cake. A quarter of a big cake will be more than a quarter of a small cake. So, the bigger the cake out of which a potential new mode gets a proportionate slice, the more its predicted patronage will be. Errors both in estimating the total market and in dividing it up will affect forecast validity.

Trip generation, as defined by Stopher and Meyburg, consists of "the estimation of the total number of trips entering or leaving a parcel of land as a function of the socioeconomic, locational, and land-use characteristics of the parcel." The trip generation model SCAG used "estimates the number of person trips generated by the residents of each analysis zone on an average weekday. It does not consider their other characteristics such as direction, length, or duration."[46]

Predictions of distributions of population, employment, and other activities and land use for the forecast year are taken as inputs. Base-year relationships between these distributions and the trip-making between the zones which represent the different parts of the city are extrapolated to produce estimates of how many trips will be made to and from each of the zones in the year for which the forecast is to be prepared, based on the population, employment and so on estimated to then prevail:

> The underlying rationale for trip generation can be seen as comprising a number of factors. First, travel is an aspect of derived demand. The frequency and distribution of travel is a function of the distribution of activity and land use in an urban area. Second, it is assumed that the intensity of travel to or from a given zone is a function of the activities and land uses that are contained within that zone. It is then assumed that the intensity of travel can be estimated independent of the transportation service provided and independent of the set of opportunities available. This particular assumption is one that perhaps is most suspect of all those used in the trip-generation modeling process. Next, it is assumed that relationships between trip rates and zone characteristics may be assumed to remain stable over time. Finally, it is assumed that trip making and activity may be related by the specifica-

tion of trip purpose. There is a heavy focus on trips to and from home, at least in part because these trips form a major part of total trip making in the urban area.[47]

While factors such as car ownership, family size, income, number of persons sixteen years and older, number of persons sixteen years old and over who drive, and distance from the CBD can be used in trip generation, at SCAG, "[a] basic assumption of the model is that the best indicator of trip generation is the number of vehicles owned."[48] The number of households owning zero, one, or two or more vehicles is estimated, based on the ratio of single housing units to total units, the ratio of population to total units, and median household income. "Trip generation rates . . . developed from survey data by cross-classification techniques"[49] provide the basis for estimating the trips generated by the total number of units of each housing type. While all trips are estimated on this basis, those trips which do not start or finish at home are reallocated on a basis of population and employment.

A key problem with the method of trip generation is that the trip volumes being generated for a study of the potential of a new transit system are said to be a function of private vehicle ownership alone. While this may seem reasonable in autopian Los Angeles, there are unsettling normative implications to this procedure. A relationship is presupposed between the ability to purchase cars and the propensity to travel. Areas with a high proportion of households with no cars or one car are said to produce fewer trips proportionately than areas with a high proportion of households with two or more cars, and these volumes are ultimately used as a source of estimating the demand for transit.

We would, however, expect areas with lower car ownership to have a proportionately higher demand for transit, all other things being equal, because of the lack of alternatives. One would, in fact, expect a negative income effect: the demand for transit would be expected to decrease, relatively, as increasing income provides increasing opportunities to travel by car.

Secondly, if more opportunities to travel are provided than are currently available, residents of poorer districts might travel more. We might want to supply areas with low car ownership with a relatively higher level of transit service, simply in order to increase transportation opportunities to those for whom these are currently denied. The formulation used, however, implicitly makes a normative statement favoring those areas which already have a high degree of transportation opportunity by showing that transit will have a higher relative demand in those areas than in parts of town which are, in fact, more heavily "transit dependent." The "latent" demand of the poorer areas is, of course, much harder to gauge, for it is itself related to social policies and the values upon which those depend, variables not subject to quantification.

Trip Distribution

Trip generation tells us how many trips will be made, not where they will be made. A trip distribution model provides this information. This is most frequently accomplished by use of a gravity model: "This model is based on the concept of Isaac Newton's Universal Law of Gravitation, which states that the attractive forces between two bodies is proportional to the product of their masses and inversely proportional to the square of the distance between them."[50]

Newton's Law of Gravitation has the following form:

$$f_{12} = \frac{GM_1M_2}{d_{12}^2}$$

where f_{12} = Force of attraction between bodies 1 and 2
M_1 = Mass of body 1
M_2 = Mass of body 2
d_{12} = Distance between the bodies
and G = The gravitational constant.

In the transportation application, the body masses are replaced by the total volume of trips sent out from or attracted to each zone on the network. This volume, calculated in the previous "trip generation" model, is a function of factors such as population and employment which, therefore, indirectly constitute the "masses" under study. The trip distribution model distributes this total volume between the different origins and destinations. As in the Newtonian model, distance (generally specified primarily in terms of travel time) acts as a form of "friction," which constrains attraction.[51]

We end up with:

$$T_{ij} = \frac{P_i A_j (F_{ij})}{\sum_j A_j (F_{ij})}$$

where T_{ij} = Number of trips generated by zone i and attracted to
                       zone j

         P_i = Total number of trips produced (generated) by zone i

         A_j = Total number of trips attracted to zone j

and F_{ij} = A measure of the spatial separation of zones i and j,
                       generally an inverse function of travel time.

The gravity model provides a metaphorical representation of a complex problem of social science in vivid, easily understandable terms. Such simplification and clarification is a hallmark of metaphor.[52] This borrowing from physics also gives an impression of scientific rigor, for which planners yearn. Yet the mapping from physical bodies interacting in space to human bodies commuting across cities lacks theoretical grounding. People do not interact across space in the same way as objects. A large employment site, for example, may not generate a large number of trips to and from a residential area if the jobs available are not suitable for the residential population. Nontraditional cities with their dispersed and specialized activities provide a particular obstacle to modeling by this method.

In the case of the SCAG application, a number of problems emerge. Most significant is the failure of the gravity model approach to match residential population demographically with the types of employment appropriate to the educational and skill characteristics of residents of the neighborhood in question. Thus, for example, the presence of financial sector jobs in one area and of an unskilled workforce in another area, will lead to the distribution of trips between those areas even though such trips would not occur in practice. This has a particular impact for the Watts and Compton communities with their largely unskilled workforces, for whom work opportunities are mostly scattered in an arc to the east and west, rather than to the north and south (the dispersion of work trips is indicated, for example, by census data prepared for the *Los Angeles Times*.)[53] The model therefore leads to the prediction of movements which would not in practice materialize. In SCAG's review of a draft of this chapter, this serious problem was confirmed: "In the discussion of the SCAG application of gravity model to trip distribution, he has a reasonable argument for work trip but not for other trip types."

According to one SCAG staff member who was interviewed, furthermore:

> The computer absolutely does not know that jobs in Watts do not pay enough to attract people that work in Long Beach—they just know that there is a job there. They know that the people in Long Beach work because they have a certain household income. And there is not a model in the country that has income at the employment area in order to have a rational trip distribution. So the model is just a simulation. It works fine for overall travel on a regional basis. On a strict corridor—without income at the destination or without salaries at the destination—the model cannot connect people coming from high-income areas to places where there are jobs that would pay enough. People get off at an average of seven miles, which would get them from Long Beach up to Watts or from downtown to Watts. It doesn't take into account the socio-economic factors that might prevent the lily-white people from Long Beach from riding up to downtown.

Said Paul Taylor, then deputy Executive Director of the LACTC, "What you're trying to do is get an aggregate picture of behavior. And you can't get down to such an atomized approach as to say, will this job here be held by a white person from here, a green person from there or a blue person from there?"

The claim that the model might perform better on a regional level does not lessen the problems it has in gauging travel which might be drawn to the Long Beach light rail service in particular; and, it is the potential demand for this service which is the subject of the study. The gravity model, in short, cannot represent travel over a complex, dispersed and multifocused urban landscape. The patterns of travel it produces for the Long Beach light rail show that its formulation bears no relation to actual travel patterns in the corridor.

A last point worth making is that travel times used for trip distribution are based on the highway network alone. The transit network—and specifically the new light rail service—has no role in influencing how the model distributes trips.

Modal Split

Modal split consists of slicing the "cake" sized up by the previous trip generation and trip distribution models.

> The basic rationale of modal split is the assumption that travelers, either individually or in groups, make rational choices between the available modes, based in part upon characteristics of those modes, and in part upon characteristics of the travelers. It is assumed that this evaluation is performed within the constraints imposed by the purpose and destination of their trip, and their attitudes towards alternative transportation system characteristics.[54]

There are a number of ways modal split may be accomplished, but all deal with predicting how consumers will behave when faced with alternative choices with different characteristics. The most often considered characteristics are the time taken to complete a trip, the cost of a trip, and the frequency of departures by a particular mode of transportation. These and other factors can be seen as barriers to an ideal trip, which would take no time at all and be free of charge. The negative effects of time, cost, and the need to wait for a departure can be seen as a form of "disutility," and the objective of the "rational" consumer to minimize the disutility incurred. Regression analysis can be used to construct a demand function based on utility analysis.

The LARTS model SCAG employed uses a binary mode split model. The model does not allocate travel to each individual mode of transportation, but splits it only between automobile and transit use. While more advanced models do include separate algorithms for rail and bus modes, the binary model lumps them together as if they had the same characteristics. Actual allocation to a particular type of transit only occurs in the network assignment stage of modeling, which comes next.

The LARTS mode choice model is "a demand function based on time expressed in terms of marginal utility. Marginal utility is defined as the difference in disutility between transit and other modes."[55]

The function is specified as:

$$U = (Tr - Ar) + 2.5\,(Tx - Ax) + (Tc - Ao)/(0.25I)$$

where U = marginal utility

Tr = transit running time

Ar = auto running time

Tx = transit excess (access and waiting) time

Ax = auto excess (access and terminal) time

Tc = transit fare cost

Ao = auto operating cost

and   I = zonal median household income.[56]

Time spent reaching transit and waiting for it is weighted by a factor of 2.5 to allow for the fact that travelers find this time less pleasant (and therefore of greater disutility) than time spent on the vehicle itself. This factor is usually, however, only associated with transit—not automobile—usage, since no transfers between vehicles are required when traveling by car, and since the one vehicle takes the traveler directly from origin to destination. Some studies may apply this factor to time needed for auto parking and walking in central business districts.

SCAG does not specify how or why this factor was also applied to parts of auto trips, and staff were unable to supply this information. When asked how "auto excess time" was arrived at, one staff member replied, "I have no idea. It's something that we inherited when we brought the model over from Caltrans . . . I'm not quite sure what the access time is . . . there's a table in the computer, somewhere."

As will be discussed in reviewing validation, below, the mode split formulation produces a consistent overestimation of transit usage. A last point to be noted here is the lack of current empirical data employed in the estimation: mode choice is a function of a number of behavioral characteristics observed on an aggregate basis; such information as actual current transit usage is not used as part of the model, although it is employed during validation for comparison with predicted data.

Network Assignment

Some of the greatest problems in the modeling occur during the final stage in the travel forecasting process, which involves the assignment of trips to alternative routes in the highway and transit networks. It is important to note that only at this stage is a choice made between rail and bus transit modes.

Before investigating how assignment to the network occurs, we need to understand the nature of the network, itself. As Stopher and Meyburg point out, both because surveying work results in a "sample survey and because the amount of data for an urban area is so vast, conventional transportation studies utilize some sort of spatial aggregation over the entire region for which the study is being made."[57]

Trips are said to originate and terminate at the "centroids" of a honeycomb of "analysis zones" into which the region is divided, and for each of which data is aggregated. "The actual zone structure that is arrived at exerts considerable influence on the accuracy and validity of later analyses. This is particularly so when those analyses employ zonal mean values as representative of the entire zonal population, and utilize a point centroid of a zone as an areal representation of that zone."[58]

As part of the aggregation process, a coded network is constructed: "This is effectively an abstraction of the existing network of highways or transit routes into a graph of links and nodes. It is *not* usual that the coded network specifies every single link in the real-world system. It *is* usual, however, that the network will include all major links that are used by through traffic and all rail lines used, as well as bus routes."[59]

For SCAG's Long Beach study, the highway system is "represented by a set of links that describe the road segments in the network," including supply attributes for "total travel time and distance" and "total travel cost, including out-of-pocket vehicle operating cost and parking costs at the terminal end."[60]

The transit system is:

> represented by a set of transit line descriptions, i.e., the list of links (route segments) that constitute the route, the headway of each route or line, and a set of all the links in the network. Headways and average bus speeds are generally obtained from operator schedules. Each link is assigned a length in tenths of miles and an average speed crossing the link, or a computed running time in tenths of minutes. In addition to transit links, there are links that provide walk or auto access, walk egress, and transfers between adjacent lines if there is no common station.[61]

The transit network provides these attributes of transit supply: "Total travel time, broken down into waiting time for first boarding, waiting time for subsequent transfers, and in-vehicle time," and "total travel cost, including transit fare, transit fees, and parking charges."[62]

Parts of the network providing access to major links would not normally be coded for a regional study. Instead, abstract "load links" are created to represent that part of the trip taken getting to and from major transportation facilities. These links represent an average of access facilities to each zone centroid (the point which represents the zone as a whole).

The use of zones and representation of whole zones by one point introduces an element of coarseness that can make the model poor at making comparisons between alternative transit modes operating within the same zones. The use of abstract links for representing access to major transportation facilities requires the making of assumptions about how access is to be "averaged." There is a large measure of subjectivity in making such assumptions, and they may not be representative of large parts of the zone which do not conform to the "average." Where rail stations have parking lots, "auto connector links" may be added to represent the possibility of driving to a station. In other cases "walk links" may be put in place. In all cases judgment is required and there is no one "correct" answer.

There are a number of serious problems with this specification. Firstly, while the "centroid" represents the origin/destination for all points in a zone, it is not necessarily true that all points in a zone are equally accessible to a particular transportation system: some parts of a zone may be nearer a light rail station; other points more convenient to a bus stop. Moreover, because the model deals only with "centroid" trips, the effect of having more than one transit stop in a particular zone cannot be represented: the stop most convenient to the centroid for a given trip will be selected by the model, others ignored.

Clearly, however, the more stops on a system, the more accessible it will be, compared to a system with fewer stops. Recognizing that having more than one station in a zone "not only influences the ridership forecast, but also distorts the station volumes,"[63] certain zones containing more than one light rail station were split by the SCAG modelers. The same was not, however, done with respect to bus services, and the advantageous effect of having bus lines crossing a zone with stops every few blocks—as compared to a single light rail station—is not represented in modeling. "Bus stops are every 2 or 3 blocks. And you don't have that in this model," said one member of the SCAG staff.

While the details of walk-time assumptions are not specified in SCAG's report, a SCAG staff member provided the information that within the Long Beach corridor this was mostly in the range of between two and five minutes. "This is an average walk time within the zone . . . If the lines are ¼ mile apart, then the average walk distance will be ⅛ mile . . . We're saying that all the trips around here have the same access time." For the purpose of defining "lines," bus and rail lines are put together. Thus the fact that the single light rail line will on average be a longer walk from residences than the choice of multiple parallel bus lines is not allowed for.

The basis on which a path for a given trip is selected is a "minimum impedance criterion based on the assumption that the trip-maker always takes the shortest route available. Impedance is simply the total time in minutes that it takes to traverse a given path from the origin zone to the destination zone; it includes the times transit passengers spent waiting, walking, and riding the various modes."[64]

A key problem is a failure to weight access and waiting time more heavily than on-vehicle time, as is done in the mode-split part of the model. Thus, while the inconvenience of transferring may be represented in the choice between transit and automobile, it is not specified in the choice between transit modes. Each transit transfer (subsequent to initial boarding) is said to take one-half the headway (the length of time between departures) of the line the passenger is transferring to, and up to four transfers are allowed.

The model is sensitive to the nearest tenth of a minute. It will therefore allocate all the traffic for a given trip to a route involving four transfers, but involving as little as one-tenth of a minute less travel time than a route with no transfers or one, two, or three transfers. Not only is this unrealistic: it biases forecasting in favor of rail. It does this because, with only one north-south rail line—as compared to several parallel bus lines—rail trips generally require more transfers than do bus trips, and the inconvenience of these transfers is not represented. Commented Paul Taylor, then of LACTC, on hearing that up to four transfers were allowed: "That's ridiculous." But that was included in the model, he was told. "I know. But it all comes out in the wash, I submit."

While the modal split part of the modeling—which predicts what proportion of travelers will opt for transit (including both bus and rail modes) and which proportion for automobile—imposes a penalty to reflect the fact that people find time spent transferring from vehicle to vehicle more inconvenient than time spent on the vehicle itself, no such penalty is applied in modeling the choice between rail and bus modes, which occurs during the network assignment stage of modeling. With a light rail departure every six minutes, only three minutes is allowed by the model for waiting for a train. For very small assumed time savings over an alternative direct bus link, passengers are said to instead use buses to get to and from light rail stations, and then transfer to light rail service. According to a SCAG staff member, "most of the trips that are using the light rail line are coming off buses." Especially given the slow speed of light rail service, it is unlikely that such transfers would actually take place except where the direct bus service is terminated.

A member of the SCAG staff was asked if the failure to reflect the fact that people prefer not to have to transfer between vehicles properly in modeling was adding a lot of people to the light rail who probably would not use it. He replied, "And the Metro Rail [Red Line], yeah, sure. That's part of the problem, I think . . . People don't really behave that way. They don't select their paths that way. But the model doesn't know that."

The problems with this modeling assumption were brought home especially with the forecasting results for the northern part of the light rail line, where bus lines operating adjacent to the light rail line run at high frequencies. In initial runs, a SCAG staff member said, the passenger loadings on the light rail links in downtown Los Angeles were zero:

> The lines would get unloaded before they got to Spring St. What happened was you come up here, and then you turn up Washington. In here, passengers were all jumping off because there was a bus line that came up one of these streets—Central Avenue—and went into downtown and it was much faster—remember, they had on the surface only 8 mph [light rail] operation. People will not change—in the peak hours these buses, they're all jammed to capacity; they're standing . . . *I had to change the speeds on those bus lines so they would not get off [the light rail] and go this way . . . It was a fictitious path. I reduced those speeds so they would stay on the LRT* [my emphasis].

But, while such special adjustments were made to eliminate these unrealistic transfers, similar adjustments were not made to eliminate unrealistic transfers from buses to light rail. The source of the unrealistic predictions of people's willingness to transfer is in the inappropriate model specification, but the "solution" was to knowingly use false data on trip times to smooth out selected errors, rather than to change the model. A SCAG staff member commented, "We made some adjustments *in* the model, not the model itself—you cannot adjust the model because it's based on regression analysis, unless you go and collect new data and perform a new regression analysis."

One problem resulting from the "all or nothing" assignment of trips to the shortest route, is that insertion of an "auto connector" link to a planned park-and-ride lot on the light rail route has the effect of attracting an unrealistic flow of rail passengers. This happens because an auto connector link is essentially like a very fast bus connection, and makes access to the light rail seem extremely quick. The assumption made that everyone for whom the auto connector would be faster would use it presupposes that all such people would have access to or would wish to use cars, which is clearly wrong, especially in low-income areas. The failure to represent the inconvenience of transfers adds further bias here, for an unrealistic preference is shown to be given for slightly faster trips involving drives to a light rail station as compared to direct trips by bus. Commented one SCAG member of staff: "there's park-and-ride auto connector links in there which automatically—right there is an overestimate."

This problem is, in fact, even more severe for the East-West Century Freeway Green Line light rail line forecasts. All Century Freeway light rail stations have park-and-ride lots, and so were assigned auto connectors which effectively made it appear that all transit users produced by the mode split part of the model within a five-mile radius of a station would drive to the station and transfer rather than take a direct bus, according to one member of the SCAG staff. In reality, many—especially those without access to cars—would continue to use existing bus services.

The model was also tried out without the assumption of auto connectors, this staff member said, resulting in substantially lower forecast ridership for the Century Freeway Green Line light rail service. These lower projections were not included in the results that SCAG provided the LACTC, and the staff member refused to provide information on exactly how much lower that projected ridership was. Actual experience during the first year of Green Line operation demonstrated the bias introduced by the auto connectors. According to a July 1, 1996 memorandum from CEO Joe Drew to the MTA Board of Directors, only 14 percent of Green Line riders were using the park-and-ride facilities.

As a potential added problem, SCAG's methodology is incapable of estimating the effects of differences in fares between rail and bus modes on light rail ridership, since cost is not a factor taken into account by the model when allocating passengers to rail or bus. Early plans called for a zone fare system, similar to that used for pricing express bus services. A review drew attention to the substantial price increase this would mean for pre-existing local bus users who might potentially use rail instead,[65] and a decision was made to initially price light rail at the local bus flat fare. This is an instance where a policy which would boost light rail ridership was unaccounted for in modeling, and represents a bias against rail.

The forecast results produced by SCAG's modeling reflect home-to-work trips only. To generate full day ridership, these results were divided by .54. According to a member of SCAG staff, "SCAG did a survey of light rail properties around the country, and the average [percentage of trips made up by home to work trips] appeared to be 54 percent . . . And the .54 is used throughout this." The particular relationships of peak to off-peak ridership, or its responsiveness to lower planned off-peak operating frequencies of the Long Beach rail line, compared to those of peak-hour service, were not specifically considered.

Bus Service Assumptions

LACTC supplied SCAG with assumptions for a "complementary bus network" designed to feed light rail stations. Modeling work to estimate demand for light rail was conducted with the assumption that this network would be in operation. LACTC also specified the elimination of certain competing express bus services from the transit network to be modeled, notably Long Beach–downtown Los Angeles freeway express 456. The deleted bus service would, in fact, have been faster than light rail for trips from one end of the line to the other since, while the light rail service makes all stops, the bus traveled directly on the freeway. "And, of course, the reason they took it out is to force those riders to use the LRT," said one member of SCAG staff.

The critical issue here is not only that bus riders would have no public transport alternative to light rail, but that the removal of the bus link assigns riders to a light rail service which the model would have shown as unattractive compared to the bus. Since assignment to transit links is on the basis of minimum time paths, leaving the bus service in would have resulted in all trips originating in Long Beach and destined to downtown Los Angeles being assigned to the bus by the ridership model, and none to the light rail.

All-Bus Alternative

In addition to forecasting ridership for the light rail service, an "All-Bus Alternative" to light rail was selected for modeling by LACTC on the basis of work by Parsons Brinckerhoff/Kaiser Engineers. While the chosen light rail alternative was forecast to have 54,702 daily boardings in the year 2000, the "All-Bus Alternative" was forecast to achieve only 21,983. For this latter forecast, pre-existing express bus lines were kept in the system and a new route paralleling the light rail line was added. To reduce competition with this new bus line, the frequency of two pre-existing local bus services was reduced.

The "alternative" was set to operate at low speeds in local traffic, taking a total of eighty-six minutes to complete its end-to-end journey from Long Beach to downtown Los Angeles. While buses on this slow and uncompetitive service were assumed to run at six-minute intervals, a comparable frequency was not provided for existing express bus lines: to have done so would have been to make such lines more competitive than the light rail service.

In the "All-Bus Alternative," there was an overall reduction in service with respect to the full RTP system because frequency on competing lines in the corridor was reduced. In addition, obvious improvements which might be made under an alternative strategy of improving bus service—such as implementing express service from mid-corridor points to downtown Los Angeles—were not incorporated. Nor were frequency increases applied to local bus lines—an obvious alternative use of light rail funds. The "alternative" was, therefore, little more than a "straw man," and not representative of the types of changes which could lead to real improvements in bus service. As one SCAG staff member observed, "No, [the way] it [was designed] doesn't make any sense . . . it was more political." Even so, while the eighty-six-minute route is itself seen to carry less than half the passengers of the light rail service, total corridor transit home-to-work trips with the "alternative" are forecast at 78,021 daily trips, only 2.6 percent less than the 80,163 daily corridor trips forecast for the light rail alternative which was eventually selected.

The poor design of the "All-Bus Alternative" is particularly brought home when it is considered that forecast corridor home-to-work transit trips without either light rail or the supplementary alternative bus route were forecast at 78,778 trips, slightly more than with the "All-Bus Alternative" in place. The alternative was therefore forecasted to do less well than the service already planned under the RTP.

A comparison between the chosen light rail service and the pre-existing RTP bus network, however, yields some particularly significant observations: total corridor daily transit home-to-work trips with light rail service in place come to 80,163, or only 1.7 percent more than under the pre-existing RTP bus network.[66] Transit usage for the year 2000, either with or without LRT, is forecast to be substantially greater than 1980 base year corridor home-to-work daily transit trips of 53,200, an indication that the increase in ridership is attributable to forecast population increases, rather than to improvements in corridor transit services. As SCAG reports, the increase in travel demand from 1980 to 2000 is almost entirely due to population growth, because trip generation and distribution are largely independent of the transportation system.[67]

The results from modeling of the "All Bus Alternative," in conclusion, do not show that light rail would perform better than alternative improvements in bus service, but only that the bus system "improvements" considered were poorly designed. Comparison of light rail ridership performance with the already planned bus system shows, furthermore, that putting in the light rail system does not result in significantly higher total corridor transit ridership.

Validation Tests

SCAG compared the results of its work (using the "Broadway-Spring Couplet" light rail option, for which SCAG had produced a forecast of 29,401 daily home-to-work trips and 27,802 daily trips) to those of the forecasts Caltrans had prepared two years earlier (15,013 daily home-to-work trips; 27,802 total daily trips).

> This difference is attributed to differences in both socio-economic data and in the networks. First, the Baseline LRT was run on SCAG-82A, whereas the

Broadway-Spring Couplet LRT was run on SCAG-82 socio-economic data; when the Broadway-Spring H-W transit trip table was loaded onto the Baseline LRT network, the LRT carried 19,859 boardings, an increase of 4,846 over the 15,013, demonstrating the effect of the new growth forecast. Some of this increase is attributed to the effects of splitting zones and adding special attractors. In another test, SCAG loaded the Broadway-Spring H-W transit trip table on to the 2000-RTP (i.e., the without-LRT) system, which resulted in 27,470 boardings; this indicated that 1931 trips on the Broadway-Spring LRT resulted from network changes: higher frequency bus service to the LRT, and elimination of the Long Beach Freeway Express and Park-Ride buses, in the Complementary Bus network. The rest of the higher boardings on the Broadway-Spring LRT can be attributed to the 6.0 minute headway; the Baseline LRT had been assigned a 13.0-minute headway.[68]

In other words, 5,680 out of the total 12,457 difference in home-to-work trips (10,519 of the 23,069 difference in total daily trips) is attributable to differences in frequency assumptions alone. Meanwhile, there was no guarantee that any particular frequency would be operated. SCAG's comparison illustrates, furthermore, the peril in choosing which socioeconomic data to use: such decisions are necessarily subjective, but they have significant impacts upon the results that are produced.

As a test of model performance, SCAG also performed a "validation" test. The idea was to see how well the model did at predicting demand for a previous year for which actual performance data were available for purposes of comparison. The degree to which the model replicates the past can be used as an indication of how well it might represent the future. Accordingly, SCAG used 1980 census socioeconomic data, and assumed the 1980 highway and 1980 Sector Improvement Plan transit networks were in place, and ran the model through its four stages. The modal split/trip distribution models predicted that 55,590 daily home-to-work trips were made by transit in the Long Beach corridor in 1980, more than double the 27,579 trips actually reported in the 1980 census. In trip assignment, "the model predicted 80 percent more boardings (174,023 vs. 97,248) on RTD bus lines within the Los Angeles–Long Beach corridor than were observed."[69]

The overprediction was largely attributed to the fact that the planned service frequencies upon which the model was based were not in effect in 1980. The headways of five lines were subsequently adjusted to those actually in effect in 1983. The effect of headway changes was dramatic: a change from initial model headway of 5.0 minutes (a bus departure every five minutes, in other words) for line 53, for example, to actual 1983 headway of 7.9 minutes, reduced predicted ridership from 30,652 to 22,683 daily passengers. Total transit ridership with the new headways was forecast at 110,824, still 14 percent higher than actually observed. This remaining bias toward transit is therefore still reflected in final forecasts produced for buses and light rail.

While the 14 percent overall overestimate might not appear to be drastic, errors in forecasting ridership on particular transit lines is substantially more significant (see table 4-1). For some lines, ridership was underpredicted—by as much as 92 percent, as compared to 1980 actual ridership—and for others it was overpredicted by as much as 138 percent. For lines for which a direct comparison was possible with 1980 actual ridership data, the average error in prediction was 59 percent.

While SCAG did not report the results of any statistical tests to compare the goodness of fit of "expected" (or estimated) data with "observed" (actual) data, a Chi-Squared test was conducted by this author for these twelve lines, with the null hypothesis, H_0, that the actual observed distribution of ridership on these lines is given by the model distribution. Since Chi-Squared tests require the sums of the expected and observed data sets to be equal, expected data (which indicated a total overestimation, as compared to observed data) was factored down so that its sum equaled the sum of observed data. The test, then, is valid, for testing whether the model is distributing patronage between the lines in the same proportions as are implied by observed data, not for examining the degree of overprediction.

The function of Chi-Squared used in this instance was:

$$\text{Chi-Squared} = \Sigma \frac{(O - fE)^2}{fE}$$

where   O = Observed data

             E = Expected data

and   f = A factor to equate
ΣO and ΣE, in this
case .91413.

The data used are shown in table 4-1.

Table 4-1. Data for Chi-Squared Test

	Observed	Expected (before factoring)
	6526	5349
	16062	4950
	14967	22683
	6150	6855
	22134	30878
	3402	939
	3403	9395
	3404	6683
	3405	81
	3406	1355
	3407	459
	1066	2538
TOTAL	84243	92156

The value of Chi-Squared comes out at 57572 which, with eleven degrees of freedom, required rejection of the null hypothesis at the 99 percent confidence level, with the implication that the modeled distribution of passengers across the twelve transit lines bears no relationship to the actual observed distribution. There is a small problem in the application of the Chi-Squared test that the data are not entirely independent—one traveler may use more than one transit line. The value of Chi-Squared is so large, however, that its use for inference in this instance is valid.

Commenting on the failure of the model to predict ridership on individual lines, one SCAG staff member said, "you have to look at it in the aggregate, not the disaggregate . . . You find that you cannot compare the boardings line by line." If these errors in predicting ridership line by line occur in validation, however, the same degree of error is to be expected in actual forecasts, including those for light rail service, which is represented as just another transit path, no different from a bus route path. This leads to the most damaging defect of the forecast since it was conducted precisely to produce a disaggregate result: in particular, to estimate ridership for the light rail service. Since the model cannot satisfactorily separate out and predict the flows on particular transit lines—cannot differentiate between particular bus lines or between bus lines and light rail line ridership—the expected error rate for the results of SCAG's light rail ridership forecast is so great as to render the forecast meaningless.

Presentation of Results

In any exercise where error exists, it makes sense to present ranges within which it may be reasonably expected that results may lie, rather than giving single numbers, subject to substantial error.[70] SCAG presented only single numbers, however, for each of the alternatives considered. "Going through this, we've seen so many places at which error can creep in. How can we have confidence in that 54,000. Wouldn't it be better to have a range?" a SCAG staff member was asked during an interview. This was the reply:

> Well, I have to give the numbers that come out of the model, because we're comparing alternatives, and some of the numbers are so close together that if I gave you a range it would become meaningless. In fact, they looked at it and they said, [then LACTC Long Beach project manager] Dan Caufield said, "this means to me that they're the same." And I said "yuh, for all practical purposes, but I'm just giving you specific numbers that come out of the model for identification purposes."
>
> But, what I'm saying is, if it's an overestimate, if the patronage estimate is too high or too low, that's something for the decision makers—they have to evaluate the model.

It could be argued that it is not the ranges which would be meaningless, however, but the actual results produced, since by the SCAG staff member's own admission there was no practical difference between the performance of the alternatives.

Report Production

The SCAG report—for which LACTC paid $200,000—is difficult to decipher because it is, in fact, a compilation of a number of smaller reports and memoranda stapled together, not a final work product. Many tables are not typed out, but reproduced in indecipherable handwriting. Several sections lack summaries or conclusions, making them hard to understand. There is no overall conclusion to the report. According to one member of SCAG staff "My boss told me to write this in three hours, and we were running out of money. I told him I couldn't do a half-assed job. So I spent a lot of time putting this together. And you'll notice it's pretty rough. But it's there."

The SCAG report was not widely distributed; its results were, instead, put into the *Environmental Impact Report*.[71] But "if you read the EIR, there's nothing in there about what the modeling—they just took the numbers. So we had to put something together," a staff member said.

Meaningless Model Paints Optimistic Picture

The model—or perhaps more accurately the system of models—we have just examined does not come close to representing real-world transportation-choice behavior. The problems inherent in the model's structure, compounded by the impediments introduced during its implementation, take it ever further from the possibility of realistic representation. If modeling is complex, it is also error-prone, and the errors introduced at every stage of analysis contribute to a final result that is quite meaningless.

We start with the initial model calibration and its representation of conditions in 1967 (with some updating in 1976), and failure not only to reflect changes since then, but its implicit normative assumption that the patterns it does represent are necessarily desirable as planning objectives, when projected into the future.

At the next stage—land-use forecasts—we see the uncertain assumptions of a variety of other modeling efforts cascading into the transportation model in use. Model users are obliged to accept assumptions on population and development established elsewhere in what is essentially a political process. These external speculative assumptions, which have frequently changed in the past according to the political winds of the day, turn out to have a major impact on the size of the projected ridership since increased population and development indicates a greater demand for a new rail passenger service. If the projected population and development increases fail to materialize, then neither will the ridership.

The estimation of trip generation reflects the tendency to concentrate on what Godet referred to as "better lit" aspects of the problem.[72] Information on car ownership was readily available, and was used to generate the total number of trips expected to begin and end in each zone of the network. This implicitly (and not necessarily deliberately) biases results against poorer areas—with lower car ownership—since it presumes that these areas will have a lower demand for transit services, given their lower current consumption of transportation as indicated by car ownership.

Trip distribution provides a prime example of the dangers of relying on physical analogues, alluring in their simplicity but which—in reality—bear little relationship to the problem at hand. A Newtonian gravity model provides an attractive metaphorical understanding of how people interact across space, a conception, moreover, that is simple and concrete. It also looks "scientific" and rigorous. But people don't behave like objects under the spell of physical forces. The failure to take demographic patterns into account—the fact that a particular employment site may be ill-matched to the skills of a given residential neighborhood—is problematic under the best of circumstances, and a fatal flaw when dealing with the complex and diverse urban morphology of Los Angeles County. Another layer of not just error but misrepresentation is piled onto that already accumulated.

At this stage, as well, there is no evidence of any sort of deliberate misrepresentation: just difficulties arising from taking an attractive and easy representation, whether or not it matches the situation at hand.

While modal split estimation provides a substantial overestimate of the market for transit, the network assignment stage of the model drives the final nail into the analysis. Access to competing modes is not properly represented. The "minimum impedance criterion" (which assumes people will always take the fastest route even if it involves the need to transfer vehicles up to four times on a trip), and the failure to represent the inconvenience of getting access to a mode or transferring between modes at this stage of the model causes routes to be chosen which people would in reality reject.

The need to make a myriad of further assumptions causes more problems: these assumptions must be chosen subjectively, but have a massive influence on results. How does one choose frequency when the assumption of high frequency will guarantee a high rail ridership forecast but when actual start-up demand or operating budget may subsequently warrant a lower frequency which the model would have predicted to produce fewer riders? Bus service assumptions had to be provided, which might or might not materialize. And so on.

In most cases the assumptions chosen—given a choice—were optimistic with respect to rail. In modeling an "All-Bus Alternative" the assumptions chosen were unfavorable to the bus. Many assumptions were dictated by the LACTC, not selected by SCAG.

The validation part of the exercise showed up the failure of the model, but this was not properly acknowledged. In addition to overall overprediction for transit, the model proved incapable of distinguishing between loads on competing individual links in the network, with no statistically significant relationship showing between the loads the model predicted and those actually observed (according to a test by this author). Since the model cannot pick out flows on particular transit lines, it cannot distinguish the demand for the light rail line from the demand for parallel bus lines. It was therefore of no value for forecasting light rail demand. The result for light rail was nonetheless taken, and given out in absolute terms (rather than as a part of a range). Thus we have a case of the model being used to address questions which it could not answer, and of providing detail in the face of massive error.

We have seen two basic sorts of error here: "errors of convenience," and "errors of optimism." The Newtonian gravity model provides a convenient representation of interaction, for example, even if it in reality bears no relationship to actual flows of travel in a complex urban environment. The use of automobile ownership to gauge travel demand as a whole is also an "error of convenience." There is surely no intention to impose a bias against people in areas of low auto ownership, but the assumption used nonetheless does so. "Errors of optimism" include a whole range of assumptions such as of transit service frequency, willingness of passengers to transfer, and development sites which will generate transit ridership. There are also "errors of pessimism" relating to the performance of a possible alternative bus service. In Kain's study of ridership forecasting in Dallas, he referred to such errors as "deception."[73] Altogether, a model quite disconnected with real world travel patterns combines with a set of over optimistic assumptions to produce light rail ridership forecasts which are both inflated and of no statistical significance: which are meaningless, in other words.

A basic question to ask at this point is: "Was the model used properly?" The modeling work SCAG conducted could certainly have been done to higher standards. Different sets of assumptions could have been used, for example, to reflect different scenarios. A fifteen-minute frequency could have been modeled along with a six minute one; a restriction could have been placed on the number of transfers people would make; a more favorable bus alternative could have been considered. Different population growth and development assumptions could have been tried out. The ranges produced, however, would have been so wide as to be meaningless.

The pretense at precision produced by the exact presentation of singular results is, of course, also meaningless, but exudes far greater confidence.

Perhaps yet more significantly, using more reasonable assumptions could not have rescued the modeling, since it was structurally flawed. Its gravity model approach failed to properly represent demographic information crucial to shaping transportation demand. And the methodology was in any case incapable of distinguishing between the demand for close together lines, so could not pick out the potential demand for the light rail line in particular. The problem, then, is not primarily one of the need to use the modeling system better but of whether it should have been used at all: clearly, it should not have been.

All of this critique might seem overshadowed by the fact that Blue Line light rail ridership in the fiscal year 2000 has in fact exceeded the forecasts. Trains operating at peak-hour frequencies between five and six minutes (twelve minutes apart during the off-peak) carried an average of 61,714 weekday passengers over the period.[74] It might have seemed more convincing to have deconstructed the modeling in the majority of new rail cases where ridership has remained far short of forecasts.[75] That is not, however, the point. The fact that forecast ridership has been achieved does not demonstrate the "accuracy" of the modeling since, as we have seen, it was incapable of properly representing the relationships under study. Rather, the result was a happy coincidence, its chances of achievement boosted both by the decision to price rail fares substantially below those for comparable express bus services and by the sheer decline in level and quality of alternative bus services as subsidy spending was diverted from buses to rail. This diversion lost far more passengers from bus service than it took to the rails, a result too often overlooked when the ridership of rail is presented, but one which is critical to a proper evaluation of investing in rail.

More important than a finding of whether forecasts were actually achieved is the more subtle point that the need to make such a large number of subjectively chosen assumptions defeats any pretense at "accuracy" a model such as this might possess. This is the lesson both for Los Angeles and other cities where modeling is used to demonstrate high potential rail ridership.

Earlier studies of the Long Beach corridor, especially the initial Caltrans study, have one particular advantage over the SCAG approach. While Caltrans adopted an essentially back-of-the-envelope approach, it was simple and its assumptions visible. It used empirical data, especially on existing bus ridership, and employed a small number of assumptions, to get an order of magnitude idea of the potential for ridership of a new service. SCAG, in contrast, but in common with most other studies conducted for alternatives analysis, employed a complex multistage model using generalized—not corridor-specific—behavioral information mostly based on a 1967 survey.

The Caltrans approach has a semblance of reality: data on today's actual travel is being used, rather than some product of a metaphor from physics, combined with speculative assumptions about future population and employment growth. The lack of complexity enables the user to understand what is being done. The back-of-the-envelope approach promotes an awareness of the fragility of the results—no more fragile, in fact, than those of the supermodels—but more humble and less liable to provoke false confidence. As we will see in chapter 5, however, humility is not in demand from political systems. Complex models which paint optimistic pictures which may be taken for real are much more the order of the day to prove to politicians that their preferred projects are desirable.

CHAPTER 5

Questions of Responsibility in Forecasting

LACTC Commissioner: [asked the role of technical analysis]
"I think it's to protect our asses."

THE GREAT SWISS MATHEMATICIAN EULER got into an argument about the existence of God while staying at the court of Catherine II of Russia. He asked for a blackboard, and wrote:

$$(x + y)^2 = x^2 + 2xy + y^2$$

Therefore God exists.

Euler's argument was accepted because the court's literati could neither dispute the relevance of a formula they did not understand, nor confess their ignorance.[1] The apparent power of quantitative methods and, in particular, the alleged omniscience of the computer, continues to dazzle us today. This is especially so in the case of forecasting which—as we saw in the previous chapter—can endow biased results with seemingly "scientific" validity.

With large amounts of public money spent on such approaches, profound ethical questions demand attention. What is the responsibility of the client, the modeler, and the politicians who are the ultimate consumers? What ends does modeling further, and to what extent is such work political? To what extent does it produce information with which to make decisions, as against legitimating decisions which have already been made?

The Responsibility of the Modeler

Modelers and their clients may define their ethical and professional responsibilities in narrow ways; the result can be a loss of overall responsibility, leading to forecasting which is deficient or misleading. Asked about the ethical considerations of doing forecasting, one SCAG modeler replied, "My ethical considerations were to leave the numbers relatively untouched," in other words to use the model as far as possible according to specification and to report the findings. The responsibility was to use the procedure "properly," not to ask whether the procedure was either asking or answering appropriate questions.

Asked about problems with the modeling system just reviewed in chapter 4, one SCAG staff member said "it was the only one available." And, regarding the use of possibly unreasonable assumptions supplied by his client in his modeling work, he said "The system was defined for me. They [LACTC] told me what the Complementary Bus System would be. So I didn't invent any of this myself. They told me." And, even though parts he was "told," such as the "All-Bus Alternative" did not "make any sense," he felt his job was to use the information specified, not to challenge it.

Peter Stopher, co-author of *Urban Transportation Modeling and Planning*, then working at Schimpeler-Corradino on projects for the former RTD, was asked if he would refuse to do work should he not be happy with the conditions specified, and replied:

> If I refused to do it, basically, I'm going to end up out of business. I'm not sure whether I've really examined in depth the ethics one way or the other, but if I am asked by a client to do any particular job, whether it's patronage estimation or something else, I know that because generally the client that is directing what I do is not skilled in the same technical areas that I am, doesn't necessarily understand the ramifications of that, that the client may specify something the client thinks they want that I don't believe is what they should be getting from a technical and professional standpoint. To the extent I can, I will persuade the client to change their specifications of what they want. To the extent I cannot do that, then I will generally live with what I am directed to do, do the best job that I can as a professional.

Stopher, who was not at all involved in work on the Long Beach project, was, however, critical of the relationship between LACTC and SCAG on that project, especially of the way in which LACTC specified assumptions SCAG was to use, assumptions which affect how well light rail would be shown to perform:

> In my experience, clients that I have worked for do not specify assumptions to that level of detail. That's a very unusual situation. I think it has to do

somewhat with the characteristics of the two agencies in this case, particularly the fact that neither one of them is really very experienced in the business of patronage estimation. It also seems to me to be a little bit questionable in general to have a public agency, SCAG, operate like a consultant.

Such problems in the relationship were reviewed with a member of SCAG staff, who replied:

> I told you before, you have to learn to live with the system. You have to learn not to let the system beat you down. I wrote this report to say, hey, I did an honest job. You gave me the model, you gave me the system, I used my professional judgment to make some adjustments, so I don't get ridiculous numbers; but, here it is folks, I did an honest job, and I've got all the fucking tabs here, and tapes down at USC to back me up. I'm going to archive all this shit, that's L.A.–Long Beach.

This member of staff described a meeting at LACTC during which

> we admitted that the model overestimates. And some of the other problems we were concerned about were things like the year 2000 socio-economic projections; who the hell knows, you know, whether it's going to take place? They're reasonable, we think they're reasonable based on trends . . .
>
> Actually, the truth is, to do a thorough professional job, you look at several states of nature. You look at several probable states of nature; not just one. But, who models that? We weren't asked to do that and, besides which, we generate SCAG '82 and SCAG essentially says—I disagree with my own agency?
>
> I'm an engineer. I'm trained as an engineer . . . And there's only one forecast that we work to. That's what we use.

Asked if he was happy with that forecast, he replied, "that's not for me to say." He insisted, however that:

> I did an impartial analysis. We technicians did an impartial analysis . . . Like I said before, it's up to the decision makers to evaluate the model and the modelers, like a good submarine commander. A good submarine commander evaluates his sonar man and his communications man, otherwise he doesn't really know what's going on. But they don't do that. They take the numbers and use it for their purposes. Whatever purpose they have. We modelers are objective. We have to be.

So, there is a responsibility to do work in a technically correct way, not to question the inputs or assumptions behind that work. The analyst is like the bombardier over Vietnam in Joseph Weizenbaum's illustration: just running a machine according to the specifications of the equipment and instructions from outside, and not responsible for the consequences of doing so, as long as "orders"—whether mathematical or oral—have been correctly followed.

Former LACTC member Christine Reed would agree with the staff view that since the users of advice were ultimately responsible, they could lay the ground rules to be followed by the "prostitutes" they hired:

> I guess it's subject to manipulation sometimes, and the responsibility is the people who are asking for the technical advice, they're accountable in the end. And they're the ones that if they only want one number instead of a range of numbers and their whole system doesn't work, then they're the ones with egg on their face, they're the ones with nothing but one rail line and no more Prop. A. I mean, you know it will be the commission members and the supervisors who will have to live with that in the future and, you know, if SCAG didn't like the kind of work they were doing, they should not have bid on it. They were doing that under contract. I mean, they have no right to complain [that they were told to work with inappropriate assumptions]: they were a prostitute just like everybody else, in a sense.

If those seeking advice specify model assumptions, however, this conflicts with the aim of contracting with outside "experts" to conduct a supposedly "impartial" analysis. This is especially problematic when the client lacks the knowledge to understand the technique for which assumptions are being supplied. It is even more troublesome when the client not only cannot understand the procedure, but is more interested in the results than in the way they are obtained. As the SCAG staff member commented: "Those people [the LACTC staff] never came over and asked me any questions about this report." When asked why not, he replied:

> I don't think they understood it. I don't know. Maybe they understood it and they accepted it . . . They just took the patronage numbers and went with them. That was it. They were only interested in getting those numbers in the EIR report there . . .
>
> In any modeling endeavor, or any situation where you have an agency like the LACTC engaging either SCAG or a private consultant—anybody—generate patronage numbers, it is up to the LACTC people, including the Board, the Commission, the committee members, you know—Jacki Bacharach and the others besides the staff. They have to evaluate the model. They have to make judgments . . .

They gave us the definition—this is the line, these are the stations, and we ran the model and said this is the patronage . . .

They liked the numbers—the numbers were not too low that it invalidated their basic premise that they need to build a line in the L.A.–Long Beach corridor. I mean, there are political considerations.

If I came out with a number like 12,000, they would have said SCAG doesn't know what it's doing. They would have gone somewhere else. It's like if you want to get a diagnosis, you find a doctor whose going to give you the pill you want, right?

Dan Caufield, then project manager for the Long Beach light rail for LACTC, was asked about the reliability of the LARTS modeling system SCAG had used to forecast ridership. "I don't know," he replied. "The fact is that SCAG is the regional transportation agency. We want ridership, we go to SCAG. I'm not going to question the way SCAG does ridership. I don't need to know if it's reliable, or any other thing . . . We don't have time to argue. I've got the biggest light rail in the States to build. We've got to get something built. And provide the public with it, whether it's right or wrong . . . What is significant about the number, it's only the first digit." How could that first digit be verified, Caufield was asked. "It's confirmed by horse-sense," he replied.

Another LACTC staff member, Richard Stanger, indicated that responsibility for the results lay with SCAG, "So we used other peoples' information, which took the monkey off our back when we were questioned about it. When we were questioned about it, we could say well, if you can get them to change their numbers, then we might change our priorities." In short, we have a total denial of responsibility. SCAG can convince itself that its work is "impartial" simply because it followed procedures "correctly"; LACTC can meanwhile lay responsibility with SCAG. The model would certainly not have been understood by commission members, most of whom had not even seen the SCAG report, and LACTC staff showed little interest in questioning the model's inner workings: the main thing was to get a result.

Understandings of Modeling

Few of those interviewed had a purely naive belief in the powers of modeling. Princess Goldthwaite, transportation aide to then Los Angeles City Council Transportation Committee Chairwoman Pat Russell, however, painted the traditional picture of technical information informing decisions: "We have various agencies that we have turned to, professionals, who give us the data that we need to come to our conclusions," she said. Asked how she knew the SCAG forecast of fifty-four thousand daily riders for the Long Beach line was reliable, she replied:

> Well, they use professional techniques, the latest technologies that are out there, not just speaking off the tops of their heads. They have used the most comprehensive technology to project these kind of things that we have available. Any time you're projecting, you're not going to make the mark, I mean, we're not infallible beings. Hopefully, they're not too far off the mark.
> *[But you thought it was a fair and valid way they went about doing it?]*
> Yes.
> *[Suppose forecasts suggested few people would use the Long Beach trolley. Would you still support it?]*
> I don't think so, but I think we have gone beyond that point now.

Debbie George of County Supervisor Deane Dana's office believed in the possibility of framing the problem in terms amenable to modeling, and allowed for the possibility that forecasts "could be perfectly accurate." Jacki Bacharach, then chair of LACTC, also talked of the possibility of "accuracy." "I guess I believe in the science of modeling, but I think you have to use it in sort of a common-sense fashion," she said.

Long Beach Councilwoman Eunice Sato, an opponent of the light rail project accepted the possibility of "accurate" modeling, too "I think they should have done a better analysis . . . I'm not a technician in that regard, but I know there must be someone who knows how to make projections. Not me, not me, but I think there's somebody who can do it. Not somebody in the company who wants to put the system in." Others, though, supported modeling less as an absolute arbiter, than as a "safety net" against making bad decisions on other grounds. As LACTC Commissioner Marcia Mednick said: "I think that if the scenario was negative, the line would not get as far as it does." And, to Commissioner Christine Reed, the value of technical analysis was that:

> You've narrowed the realm of possibility with the technical documents. They'd be all over the map otherwise . . . The technical work narrows the realm of possibility and forces the politicians to think a little bit about what they're doing and makes them choose between reasonable alternatives, and maybe they don't choose the most A1 four-star choice, maybe they choose the two-star choice, but they don't choose the goose egg.

Technical work "establishes some parameters and some limits," said Jim Sims of LACTC staff:

> It puts some limits on it. In other words, we couldn't have gone with a rail system which was a total dud. If the model showed it was a total dud, it showed that it caused extensive environmental damage, couldn't have gone with it, established some limits.
>
> What you had in essence was you had a range of opportunities there which the modeling and technical process shows ok these are—the output from the technical process shows that these are an acceptable package of alternatives. The political process then decides which one of those to take, even though, in the scoring system, one of those may score higher than the one selected. But in the political scoring system, something else scored higher. But I think the technical process is essential in order to weed out the things that really don't work . . . We as technicians and bureaucrats, our job is to establish sort of some limits and some parameters.

Christine Reed put it this way, when asked if people weren't interested in finding out what the benefits were,

> Well, yuh, they were sort of. But I think that basically they had, unless there was overwhelming evidence to the contrary, a horrendous environmental decision, or horrendously negative statistics, but given some ooshee jello bowl realm in the middle of so-so statistics, then basically nobody gives a shit. So that the work of the statisticians and the analysts is important to the degree that it keeps you out of horrible trouble on the negative side. That's what it does. And me as a politician. It keeps us out of horrible trouble.
>
> So we do all this work and we get all this analysis, and they run all these numbers and they do all this environmental stuff and then you have all that stuff and you look at it all and you see a realm of possibility that's quite large and maybe out on the far edge, some realm you know realm of the negative that you don't want to get into. To me that's the value that all this has, and I mean I am saying this in a general sense, it has to do with other decision-making other than transportation, I mean I've been 10 years now as an elected official, and this doesn't just go to the county decision-making process, but works here in the city pretty much the same way.

Forecasting as a Political Tool

Gerald Leonard, who had worked for former Los Angeles County Supervisor and rail proponent, Baxter Ward, stressed the "self-serving" role of forecasts. "You let them prove your own point for you." "I thought SCAG was looking into a crystal ball," said LACTC Alternate Commissioner Ted

Pierce, "and it was kinda cloudy. I don't really perceive right now having 55,000 riders a day . . . And how they came up with these figures I don't know. That's why I say I think it's a glass ball. Fifty-five sounds good." "You could have changed assumptions to produce 100,000 or 25,000," he was told. "I know," he replied.

Alternate Commissioner Blake Sanborn knew that "probably the 54,000 was a medium ground of a number of different ideas. I think it's a pretty safe number; that's why it's being used. No one's going to go out and say 100,000 and no one's going to say 25,000 so they're going to say 50." In that case, he was asked, why do the computer work in the first place? "Or draw it all from a hat and take an average of it," he replied.

Carmen Estrada, member of the RTD Board, questioned the forecasts, since "it's clear that they're high in almost every other city where the projections have been made." Los Angeles Councilman John Ferraro meanwhile chipped in: "Of course, they weren't going to hire somebody that's going to say, wait a minute, you're going the wrong way." Long Beach Councilman Edgerton, like Eunice Sato a light rail opponent, compared the political pressures on modelers to the force put on Galileo to conform. "Political pressure can make you believe a lot of different things," he said. Sato, despite her beliefs in the potential of modeling if honestly done, was similarly distrustful "What did they feed into the computer to make it come out that way . . . ? You can put in what you want to . . . You and I know that if somebody wants to do something they hire a consultant to prove what they want to do." And, said Compton Councilman Maxcy Filer: "Whatever I want to, I can put into a computer. A computer will spit out anything if I program it right."

Others saw forecasting as a marketing device, rather than a tool for evaluation. "Often ridership projections are being done strictly as sales tools," said LACTC Rail Construction Committee member Manuel Perez: "The forecast is public relations . . . It's a little indelicate, but basically true: A patronage forecast is just like a marketing plan for a development . . . The selection of the L.A.–Long Beach line, it was chosen because it had already been deemed that this would be the first line."

Craig Lawson, transportation adviser to Mayor Bradley, saw things the same way:

These guys are building these buildings down here with the assumption that they're going to be leased. And that's the way that they get their financing. They tell their bank and their investors this thing is going to be leased and we're going to have the whole thing filled up. Well, how do you know that? Do you have signed contracts? Most of the time they don't. So, it's like any construction project. You have to have a certain assumption that you base it on, and the assumptions I think for transit in Los Angeles are we are going to continue to grow, we are going to continue to need this. And that the studies that I've seen show that this is going to be a worthwhile project.

What makes this view of forecasting as an advocacy tool disturbing is, that while private developers seeking financing can be expected to maintain that their buildings will be leased, public bodies using public funds are expected to promote the public interest, rather than to "market" one particular view of it. While any banker will be expected to check the reasonableness of developers' claims, furthermore, there is no similar check on the light rail claims, which are made by the very agency entrusted to protect the public good.

The common practice of subsequently amending forecasts to ensure targets appear to be met is perhaps the ultimate political use of forecasts. An initial study of rail ridership in Buffalo, for example, predicted 160,000 daily weekday riders, while a later one used for the project environmental assessment reduced the length of the proposed line as well as the forecast to 92,000.[2] Sharply lower forecasts of 45,500 were, however, adopted in 1981. Actual ridership as of 1995 was 26,115 daily weekday passengers.[3]

In Portland, initial forecasts of 42,500 daily weekday riders used for staff recommendations, "probably low due to a number of purposely conservative assumptions used in the simulation process" were replaced by a pre-opening estimate of 19,270.[4] Actual 1996 ridership of 27,000 was far short of the promise which went along with light rail's adoption. Sacramento tells a similar story, with initial forecasts "more likely to understate than overstate future transit demand levels" of 40,000 average weekday boardings reduced to 26,000 two years later, and actual attainment as of 1996 standing at 25,017.[5]

According to the Los Angeles MTA, "Original EIS forecast ridership for the Green Line was 100,000 boardings per day. However, shortly before the opening, the forecast was revised downwards to 10,000 per day."[6] While the initial forecast had been for a mature system—not the first year of operation—and was done without knowledge that reduced federal defense spending would negatively affect employment in the western part of the area served by the line, the very low new forecast enabled Metropolitan Transportation Chairman Larry Zarian to declare exactly one year after the East-West Green Line's August 12, 1995 opening that "The Green Line carries nearly 15,000 passengers each weekday, which is more than we projected for our first anniversary when the line opened last August. This is exciting news for all of us."[7]

In many ways the Los Angeles Blue Line has proved the exception to the rule, since actual average weekday ridership in the fiscal year 2000 of 61,714 beat the forecast of 54,702. Similar political posturing nonetheless took place shortly prior to the line's opening. LACTC had initially estimated that first year weekday patronage would be 35,000, based on growth curves observed on other systems.[8] A January 1990 report on staffing and operating/maintenance costs cited expected first-year Blue Line weekday ridership of 30,400. A RTD report later that year, however, assumed ridership at only 15,000 per day for the fiscal year 1991.[9] Just before opening, the forecasts were reduced again: "We are conservatively estimating 5,000–7,000 riders daily initially, building to 12,000 with full system operations."[10] It was based on this last-minute estimate that the *Los Angeles Times* concluded that actual rider levels of eighteen thousand achieved by January 1991 "wildly exceed expectations, climbing to three times the projections of six months ago when Blue Line set off on its inaugural run."[11]

Forecasting as Legitimation and Self-Delusion

While forecasting can be seen as a tool for salesmanship or for misrepresentation, there is a more subtle view of the role it performs: forecasting can serve a function of making us feel secure about what we are doing, of deluding us into thinking that our decisions—in fact based on other grounds—are legitimate and correct.

"The use of doing it," said one member of SCAG staff,

> is that you've got to have some justification that is acceptable to the public and the politicians beyond an intuitive feel. And engineering has established itself as a valid profession that people respect. Every time people drive over a bridge, people are putting intuitive faith that the engineer knew what happened when he designed the bridge. When public money pays for a project,

people have intuitive faith that the engineering methods are advanced enough that it's worthy of the investment. The model is used to validate what is intuitively acceptable.

The modeling does not come—as pictured in the ideal world of Stopher and Meyburg—before and as the spur to decision-making—but afterward, and serves to give the aura of science to decisions which have been made for other reasons. As one other analyst, who preferred not to be identified, commented "They [LACTC] liked these numbers. The numbers were not too low that it invalidated their basic premise that they need to build a line in the L.A.–L.B. corridor . . . The fact that L.A.–L.B. was going to be built was a foregone conclusion. I think this is fundamental to your whole study."

And, as we shall shortly see, many politicians only accepted those elements of the forecasting that concurred with their prior beliefs, rather than using the forecasts to form their beliefs. People feel uneasy, however, about letting "intuitive" beliefs, hold sway, and like to feel they are backed by the wisdom of science. As Manuel Perez said, "Basically [forecasting] is to establish a comfort level for decisions being made . . . PR is a comfort level. This is the best available information."

"How they came up with those figures, I don't know," said Ted Pierce. "That's why I say I think it's a glass ball." When asked whether there were benefits to doing the forecasts, he responded:

> There can be. I think they were a positive force for the Commission and the people who were really pushing the L.A.–Long Beach line in the very beginning to say, ok, looking back, here's SCAG, I mean it's a credible organization, and they've said that their 55,000 people daily will ride this thing. That's incredible. With that ridership we ought to do it. So what it does is it gives confidence to the people who are promoting it as well as to the general public who are listening to the pleads to build this thing.

Former LACTC Executive Director Rick Richmond also said there was a need to have something solid: "But I just think we need some numbers. You've got to do some work based on some numbers. So you produce the numbers." As Alternate Commissioner Blake Sanborn said, "The politicians seek to have a technical backup so that they on the face of it look as though they are analyzing information by the experts and making a determination that this is what should happen." "You have to have some comfort level," said Barna Szabo. "You can't go on gut feelings; you can't go on politics . . . You're going to have legal challenges. You're going to have editorials. And, you're going to have to respond to those with some of this data. You can't go by the seat of your pants."

Forecasting legitimizes and gives confidence to rail promoters. The numbers appear solid, and can provide a basis for self-delusion, grounds for believing in what you are doing.

Reactions to SCAG Negatives

The need to believe in what you are doing is basic, and one human reaction to maintain security in one's judgment is to filter out contrary information. One of the most interesting sets of responses collected during interviewing came after respondents were informed of the negative findings SCAG had made relating to the Long Beach light rail project; in particular, of the low volume of travelers expected to switch from automobile travel to light rail, and of the negligible improvements to highway congestion expected as a result of the new rail service.

In their answers, respondents tended to reflect their prior beliefs, rather than to accept other conclusions resulting from the information provided. And, if negative information was in some cases accepted by those who already believed light rail should go ahead, other reasons were given as to why the line would be valuable, despite such findings.

Two LACTC staff supported SCAG's view that few people would desert their cars for the train. As then Executive Director Rick Richmond said, "I don't think we've ever contended that [traffic would be immediately reduced on freeways]. That argument will cause you to never make any kind of transportation investment of any significant scale . . . Any public transit investment, be it for express bus or railway . . . primarily draws its usage from existing transit use."

LACTC's Jim Sims (who previously worked for SCAG) also accepted the negative SCAG results, "I think it's clear that the real impacts on air quality will be minimal, it will be positive, but it will be minimally positive, and I think the impacts on congestion will be minimal direct impacts. I think, probably, at least initially, the number of people attracted from automobiles into transit, who would not otherwise have used transit at all, would be a small number." But Paul Taylor, LACTC deputy executive director at the time, credited light rail with "having the capacity to do more travel without doing harm to the air quality." He was then shown the SCAG study, and immediately discredited it: "What is this? I don't even know what this thing is. I don't know what they're trying to do with this either. Never seen it. But then you and I know that it's simply what SCAG is up to most of the time, anyway."[12]

Taylor was given one SCAG estimate on light rail versus heavy rail costs, which indicated that the LACTC light rail projects did poorly compared to the heavy rail.[13] "That's so outrageous as to be laughable," he replied. "One can only hope that no one reads this stuff." On the question of how many light rail passengers would be former automobile travelers, rather than bus users, Taylor said: "I don't believe that the people are mostly going to come out of the buses." "Although that's what the documents for your EIR said," he was told. "Yuh," he replied.

Richard Stanger of LACTC also refuted SCAG's finding that only a minimal number of people would be enticed out of their cars: "I'm not sure that that can really be believed," he said. "If that was the case, then I do think that it may be a questionable investment."

Debby George of Supervisor Dana's office was one of those who had not previously seen the negative SCAG information, even though some of it was contained in a recently released SCAG glossy magazine, *Crossroads*.[14] Told that only 1,600 people per day were expected to transfer from automobile to light rail, she replied: "Only 1,600? Where did they get those figures? I think it's much more than that. It has to be much more than that . . . I would challenge if that is right, personally."

LACTC Rail Construction Committee Citizen Member, Allan Jonas, told that SCAG was forecasting that light rail would mostly take people out of buses rather than cars declared: "I don't believe it." Alternate Commissioner and Norwalk Councilman Bob White felt similarly: light rail will "reduce tremendously the number of cars that will be on our local freeways," he said. "I would think that about 10 to 20 percent would leave their cars home and take the light rail." Presented with the SCAG data indicating the small expected transfer from automobiles to light rail, he said:

I don't agree with them. I don't agree with them on that because those statistics like, er, it almost makes me think that SCAG is in the other corner, maybe fighting this, I don't know why they would . . . I'm just saying that I can't believe that it would affect only 2 percent of the cars off the freeway . . . My answer to that would be that because of advertising and publicity and working on this and having a good public relations firm or public relations within the LACTC, that you could certainly improve that 2 to about 10.

Long Beach Mayor Ernie Kell also reacted against the SCAG findings. When he was told that SCAG only expected a minor positive impact on traffic, he said. "I would take exception to that. Number 1, if you take a bus off the freeway, you're removing a vehicle, a good sized vehicle at that. So you are taking traffic off of that, and I think that once this line comes in and people find that they can park their car and ride up and enjoy the paper, and have less traffic to worry about, I think you're going to find more people riding it." Ted Pierce also gave SCAG limited credibility when SCAG produced findings which did not go with his beliefs, since SCAG Executive Director Mark Pisano "is totally against rail, and he just generates that to a lot of his staff people . . . He likes buses, I guess."

One way in which uncertainties about forecast-year performance were rationalized, was to argue that if there might be problems at the start, all would be well in the long term. On the SCAG projections for total rider-ship, Allan Jonas said: "But if they fudge on the side of optimism, I can for-give them, because in the long-run they'll be true." Christine Reed agreed that if only the small expected number of passengers transferred from auto to light rail, "it wouldn't be worth it," but added that "I think we'll get more than 1000 minus trips [out of automobiles], I really do . . . When the line to the airport goes in, then it will get used more and when there are more lines it will be used more." Burke Roche, transportation adviser to Supervi-sor Kenneth Hahn gave a similar argument: the line would do better once interconnecting lines were built, he said. "And I do think if you provide a decent system, that they're not going to be coming in the same bumper-to-bumper."

When told that SCAG was forecasting that only sixteen hundred current daily automobile users would use light rail instead, Barna Szabo, Commis-sion alternate and employee of Wrather Corporation (in development) did not refute the claim, but found other grounds for supporting light rail: "That to me is the wrong way of looking at what light rail can do for you . . . What light rail can do, it relates very well to certain land-use pat-terns, and can also enhance certain land-use developments."

In short, those findings of SCAG which were unsupportive to the light rail interest were easily dismissed by those who supported light rail.

The Ethical Problems

To some extent, there are clear-cut ethical problems in the way forecasting work is conducted. The assumptions LACTC provided SCAG were politically driven to be to the advantage of rail. And one SCAG staff member saw his organization as acting like a doctor who prescribed the pill the patient wanted, whether or not it was the best medicine for the job.

More problematical—because ethical issues thereby get pushed to the background—is the degree to which technicians are required to act according to the routine of their part of the pie, rather than consider the perspective of the pudding as a whole. The prime responsibility of the technician appears to be as an efficient cog in a large machine, performing the prescribed task without asking questions outside the routine imposed. Under this view it is possible for a member of the SCAG staff to both be aware of gross deficiencies in the work produced and to say that an "honest" job has been done. The modeler's place is seen as somewhere on a production line: the inputs coming from the stage before, to be processed and passed on to the next stage. "I'm trained as an engineer."

The question of asking if the right question is being asked—Churchman's first task of ethics—is not even relevant.[15] The machinelike way in which the task proceeds blinds its participants to such matters, allowing them to delude themselves into thinking they are doing the right thing just by following the given instructions correctly. In many ways, forecasters operate like the computers they employ: there's an algorithmic understanding of the work to be done. And the uniquely human ability to go beyond given procedures and look at things differently when the need arises is overlooked. Like that of the Vietnam bombardier Joseph Weizenbaum cites, the modeler's brief is pre-defined and restricted. And that restriction enables the modeler to escape responsibility.

The model itself becomes a political actor; it carries powerful political messages, although neither model developers nor users may be aware of them. The use of car ownership data to gauge demand for travel, cited in chapter 4, which biased results against poorer areas, presents one such example of the model-as-politician.

The model is in many ways part of a larger system of avoiding responsibility. Use of the model presupposes that if a certain "demand" can be found for the given technology—in this case light rail—relative to costs, then the project should go ahead. The situation is comparable to Leamer's death penalty case discussed in chapter 4: Just as the desirability of execution is assumed if execution results in less murder, the potential for the light rail project to be desirable, given the results of certain ridership/cost tests, is assumed.[16] Questions about where people in different neighborhoods actually want to travel; of how they might best be served; of who should be given preference in being served; of how present travel patterns might desirably be changed by new transport services; these questions are not so much ignored as given presupposed answers by the model.

The model assumes, for example, that people in neighborhood A will travel to workplace B, even if the work there is unsuitable. The assumptions of today's travel patterns, furthermore, are implicitly carried forward to tomorrow's, with discrimination against those with fewer travel opportunities today, and without consideration that they might in fact be better served by some system other than rail.

Total system impacts and opportunity costs are ignored: The focus is on the number of passengers that might be attracted to rail, not on the potential financial viability of the bus system which, as we have seen—once deprived of resources—was to lose far more passengers than the rail system can be expected to gain, resulting in a net loss of transit system-provided mobility.

The politicians receiving the results are supposedly responsible for them, even though they lack the technical understanding to evaluate them. But to all accounts, that does not seem to matter terribly. From a look at their understandings of modeling, a curious picture emerges, one in which modeling serves as a visible symbol of rationality at work, while at the same time it contributes little to actual decisions: given the way things work in practice, the lack of understanding seems of little consequence to the functioning of the daily routine.

Perhaps more disturbingly, forecasting was seen in some quarters as a marketing, not an evaluation device, a means of persuading "investors" that a project was sound (in fact, when SCAG had something negative to say about rail, it was as if they were "in the other corner," rather than fighting the good fight). This is the theme taken up throughout the studies of Hamer.[17] But is this a question of outright dishonesty? It may be seen as quite the opposite: there is a need for something solid to anchor pre-existing beliefs. A "comfort level" beyond intuition is required, and forecasts provide something which appears to be substantial. "You can't go by the seat of your pants." A strange double standard seems to exist: there does appear to be awareness of the fragility of forecasting; yet, because that very fragility is concealed in hard-looking numbers produced by chunks of electronics and steel, there is something reassuring about the hardness of those numbers, something to suggest that science, rather than intuition, is being invoked.

Forecasting—A Mythological Ritual; an Unethical Exercise

From the above it is clear that the belief in modeling as a scientific means of driving the "transportation planning process" described by Stopher and Meyburg and embodied in the procedures promulgated by UMTA and its successor FTA, and employed throughout the nation, is mythical. While Pickrell expresses concern that a project's benefits may vary from "the expected level that led to its selection from among the alternatives under study," the evidence from the Long Beach case shows that decisions did not follow from technical analysis, but that technical analysis was used to support an already existing predisposition to rail transit.[18] Interviews conducted for this project in San Diego, San Jose, Sacramento, and Portland, Oregon, also indicated that decision makers had made up their minds prior to the conduct of technical analysis.[19] Technical analysis, then, appears to serve a ritual function: it gives an aura of respectability to decisions which have been reached on other grounds, rather than focusing the decisions themselves.

Forecasting is an expensive exercise, with high costs for extensive computer processing. Yet, while all this money is spent to produce meaningless results, virtually no attention is paid to asking questions concerning the most pressing questions of human and social needs and of how government should address them. While the computer provides an easy answer, the real problems go ignored. And that is a real ethical problem.

The model is solid and provides easy if expensive answers. Its very makeup lies in the concrete easy-to-get-grips-on physical world. Trips are "generated" and "distributed" like electricity. Sight is lost of the fact that the trips are made by people, and that social decisions have to be made about who is to be served and how. Issues of poverty, education, and race should be part of the equation, but they are ignored in an approach that simply justifies a project which came into focus as what Altshuler refers to as a "preselected solution."[20]

If forecasting is simply a ritual, and the espoused "rational" processes for technology selection are not actually being used to make decisions, what does lead decision makers to choose rail? This is the question to be addressed for the remainder of the book. We will see that the choice of rail ultimately results from a mythical understanding of the benefits to be thereby derived; and that the myth of technical reason provides legitimation for the believed mythical powers of rail.

CHAPTER 6

The Political and Institutional Makings of the Los Angeles County Light Rail Program

I'm the author of Proposition A. Which is the first tax measure that was
ever voted for successfully by the people of Los Angeles County. Terribly
unpopular for a politician to advocate a tax. Verboten. But I did it.
And the people supported me. I was born in Los Angeles. I've been in
public office 39 years, and I've seen all these measures cemented in talks;
in talks and fancy charts, and studies and plans—would fill this room—
but there's been no action. Faith without works is dead, the Bible says.

—Supervisor Kenneth Hahn

RIGHT UP TO THE ENACTMENT of Los Angeles County Proposition
A of 1980, which was to provide the funding to build the Long Beach Blue
Line light rail, Los Angeles transit proponents associated the development
of rail facilities with the promotion of a downtown-based urban form;
highways were, in contrast, associated with a dispersed metropolis. In many
ways, the political struggle for road as against rail development has been
about what sort of Los Angeles was desired by the protagonists in question.

As early as 1909 local government began taking responsibility for road
construction. The Los Angeles County Board of Supervisors approved a
$3.5 million bond issue that year. The next year the state began allocating
funds for highway construction.[1] As people poured into Los Angeles and
auto ownership grew, so did traffic congestion, particularly in downtown.[2]
In response, the Traffic Commission of the City and County of Los Angeles (despite its name, a voluntary membership association, consisting
mostly of local businesses) retained three renowned planners, Frederick

Law Olmsted Jr., Harland Bartholomew, and Charles Henry Cheney, who completed *A Major Traffic Street Plan for Los Angeles* in 1924.[3]

The plan saw congestion as a function of "unscientific" street width and design, and "improper" use of existing street spaces. It sought to produce a "balanced scheme for handling a tremendous traffic flow" by establishing different classes of roadways for different traffic needs to avoid the "promiscuous mixture of different types of traffic," which the authors said caused congestion. Of particular note, the plan called both for roads focusing on the central business district and roads which linked other places but steered clear of the CBD itself. The plan authors saw the concentration of activities as a stimulus to road congestion, and called for limits on building heights to control its effects. A network of through roads, boulevards, and parkways was proposed in support of a dispersed and multicentered urban form.[4]

The road system was not to be a mere response to transportation need, but vital to shaping the city. As stated up front in the letter of transmittal, "We believe the *Major Traffic Street Plan* here presented provides a broad, practical, well-balanced scheme for handling traffic towards which *the city can advantageously grow, and to which it may gradually adjust itself*" (my emphasis). A pattern of opening and widening of streets had been established by the end of the 1920s which was to be critical to shaping the future.[5] The Regional Planning Commission began work on a county plan, producing reports emphasizing the relationship between highways and land use. It specifically recommended a minimum lot size "consistent with the development most desired in California cities . . . to avoid overcrowding in the ultimate stage of developments."[6]

By 1930, Los Angeles' urban sprawl had spread on a vast scale. Robert Fogelson documents the outward expansion of suburbs into the countryside in parallel with the decline of downtown Los Angeles. The process, "which reflected the newcomers' preferences, the subdividers' practices, and the businessmen's inclinations, was also self-perpetuating. Dispersal devastated the central business district, and decentralization spurred outlying subdivision." During the 1920s the Los Angeles central area's share of retail trade had dropped from 57 to 25 percent. As Peter Hall documents, by the mid-1930s, 88 percent of new retail stores were opening in the suburbs, in places like the San Fernando Valley, Westwood, and Hollywood as well as along Wilshire Boulevard.[7]

The changing form of the city created a tension as the transportation infrastructure became increasingly out of gear with urban needs. By the late 1930s, "as increasing volumes of cross-town traffic tried to move on a street system that still radiated from the city centre, there was a threat of total breakdown. And, portent of what was to come, congestion increased in outlying business districts also."[8]

Nineteen thirty-three had seen a commitment to highway construction and maintenance within cities, with a quarter-cent of the state gas tax set aside for that specific purpose. This increased to a half-cent in 1935. In 1934 planning got underway for Los Angeles' first freeway—the Arroyo Seco Parkway—which was built between 1938 and 1940.[9] It was paid for from a variety of sources, including the state, city funds, and gas tax revenues from Los Angeles and South Pasadena.[10] Public transportation remained financially a private venture.

The first published proposal for a comprehensive freeway system came from the Automobile Club of Southern California in 1937. Its representation of the club's auto-owning members and their preferred lifestyle was clear from the start:

> We wish to emphasize that the Los Angeles area has grown up with the automobile. Motor vehicle transportation has shaped its growth to the extent that the business and social life of the area is today vitally dependent upon the motor vehicle for the major part of its transportation. If street and highway congestion continues to increase, the day is not far distant when the automobile will in many parts of the area have lost its usefulness. At this time, the economic loss resulting from readjustment alone will have reached a staggering total . . . Future orderly growth is vitally dependent upon the establishment of a system of transportation serving all parts of the area.[11]

The report called for a system of freeways based on traffic flow and projected population patterns, and the relationship of growth patterns to transportation links is made clear by the routes which were to completely bypass the traditional central business district.[12]

By the 1941 publication of the *Master Plan*, the need for freeways was made explicit in a public plan: "In fact, the inescapable conclusion is that as the population of the Metropolitan Area passes from four to six million, one of the *two eventualities will have to be faced—a drastic reduction in the proportion of automobiles to population, or the relief of the highway system by supplemental freeways.*"[13]

Brodsly describes the Los Angeles County Regional Planning Commission publication, *Freeways for the Region,* as a propaganda device to impress government officials and the public of the urgent and necessary need for freeway construction. "All motorists in Los Angeles County," the report said, "and this means all of us, have felt the need for some superior form of motorway in this region to supplant the existing highways." Once more the Los Angeles ideal, decentralized and inescapably tied to the automobile, was associated with highway development: "Satellite communities, well planned within themselves and in relation to a freeway system [would] ... provide a better way of living and still preserve the social and economic advantages of the urban center." Freeways would be "facilities which are deliberately designed for the decentralized community."[14]

The pressure for freeway building grew yet higher by 1946, with local and state governments, automotive organizations, and local business interests coming together to support the Los Angeles Metropolitan Parkway Engineering Committee report, *Interregional, Regional, and Metropolitan Parkways,* which contrasted the urgent need for freeways with the lack of sufficient funding.[15]

The 1947 California Collier-Burns Highway Act provided the means for this vision to be accomplished, with funding for a 14,000-mile state highway system costing $2.4 billion.[16] Significantly, the system was to fund itself from an increase in gasoline tax from 3 cents to 4.5 cents a gallon. Thus, while the state freeways were a response to overwhelming public desires, they were also to be funded by users. Fogelson writes that Angelenos found their rail system "expendable, and, with few exceptions, they wholeheartedly accepted dependence on the motor car without fully comprehending the implications of this commitment."[17] While the highway program raced ahead, there was no parallel general public interest in financing what was seen as a declining private transit industry being increasingly abandoned for the car.

Road-building activity in California was paralleled by highway enthusiasm throughout the United States, as demand for automobile purchases and usage escalated following the close of World War II. This led to the launch of the National System of Interstate and Defense Highways in 1956, for which a national highway trust fund—much like the one established in California and other states—was set up to reimburse 90 percent of the planning and construction costs of qualifying routes.[18]

Alan Altshuler tells the story from a national perspective, placing both the great highway-building programs and government's neglect of mass transit until recently in a tradition of following the private market. The highway program, in particular, served both a market demand and demands from the voters. By the early 1950s, the auto industry, concerned that the acceleration in automobile sales since 1945 would slow down without highway system expansion, began orchestrating demands for strong action to combat congestion. Allies in the oil, steel, rubber, and trucking industries and their labor unions joined in. "It is important to keep in mind, however, that there was widespread public support and that the market forces to be served by highway construction were already sweeping the field in the marketplace. Thus those who lobbied successfully for increased highway construction were able to operate within a highly congenial framework of popular taste, market behavior, and apparent political predisposition." Altshuler stresses that the pro-roads campaign was uncontroversial with virtually no dissent in either extensive media reporting or congressional hearings. Highway interests may have been one of the most powerful lobby groups, but their goals were in harmony with the growing desire for automobility which was as yet unassociated with air pollution, oil shortages, or community disruption.[19]

Nineteen fifty-six saw the approval of the Interstate Highway system. The same year, California state gas tax increased by 1.5¢ per gallon (it rose a further one cent in 1963, making state gas tax 7¢, with an additional 4¢ levied by the federal government).[20] The California Freeway and Expressway system followed in 1959, its authorizing legislation providing a freeway grid for metropolitan Los Angeles which would put all urban areas within four miles of an on-ramp.

The building of California's freeways was, in short, a politically obvious choice. In fact, as Brodsly puts it, "what is most apparent in so many of the planning reports is the perceived lack of real alternatives."[21] While highway building was also taking place in other large American cities, the freeways went particularly well with the dispersed Los Angeles lifestyle, and had as their supporters all who wanted to be a part of that dream. What could be closer to the ideal of the "American Way," than a government acting to provide what its citizens clearly wanted? Awareness of the negative environmental impacts of highways had yet to emerge as a policy issue.

The Political and Institutional Development of Rapid Transit

While Southern Californian highway building was a response to public demand, rail transit system development proved to be far more controversial, lacking sufficient political or popular appeal to stimulate major action until Los Angeles County approved Proposition A in 1980. At stake were the quite different political associations of the two transportation concepts. Freeways went along with the dispersed, low-density dream which brought Angelenos to Southern California and had mass political appeal. Rail transit systems focused on downtown and for a long time enjoyed only localized support, seeming foreign to those who came to Los Angeles to avoid the concentration and congestion of the traditional cities of the East.

Los Angeles did get an early rail subway in 1925. At a cost of $9 million— paid for by the Pacific Electric as a private venture—and 4,325 feet long, it provided a double track express right-of-way for the Pacific Electric Red Cars, terminating in the basement of one of the largest office buildings ever built in downtown. As a transit journal of the time relates, its opening was seen as a great day for Los Angeles, and specifically for downtown:

> The christening of the new subway terminal was a memorable event in the story of Los Angeles' progress. It was witnessed by 1,500 persons who swarmed down glistening ramps into the heart of the earth itself to jostle each other for honors of making the first memorable passage . . .
>
> The official dedication of the subway and its terminal was preceded by the greatest luncheon in the history of the Los Angeles Chamber of Commerce and by an inspiring parade, led by a brass band, through the downtown streets.[22]

While in 1923, Pacific Electric management had proposed a whole series of subways, radiating from Pershing Square in downtown Los Angeles, the decision to terminate the Hollywood tunnel at Hill Street for financial and engineering reasons largely ended the plan.[23]

Pacific Electric was not, however, the only actor at play, and interests were to emerge to promote the case for rail even as Pacific Electric's financial condition worsened and moved it to discontinue rail services. As Vey-sey relates, downtown merchants and realtors had formed the Central Business District Association, the Los Angeles Traffic Association, the Downtown Businessmen's Association, and the Chamber of Commerce, and these organizations were to support the first plan for a comprehensive system of grade-separated rail rapid transit, produced by the Chicago consulting firm of Kelker, De Leuw in 1925.[24] The Kelker report was to carve the way for future attempts to bring rail transit to Los Angeles. It set a tone of breathlessness, citing the development of rail as necessary for the health and growth of Los Angeles; it presented a solution to social and economic problems in engineering and construction terms; and it acted to impose a vision of the traditional city marked by a concentrated center and suburban periphery on a city which had already shown it was different.

Kelker, De Leuw cast the response to the Los Angeles transportation problem in terms of growth per se, rather than looking for the implications of the dispersed form of growth already well in evidence:

> Los Angeles has become a large metropolitan center and it is of vital importance, at this time, that transportation facilities be planned upon a scale commensurate with the present and prospective development of the City and County. The phenomenal growth in population and industrial activity, together with the tremendous increase in street traffic, makes the construction of rapid transit lines not only necessary but imperative if an adequate, quick and convenient means of public transportation is to be provided and traffic conditions are to be improved.[25]

To meet the need, a system of rail rapid transit lines was proposed:

> providing high speed service with few stops—for long haul traffic between the central business district and the intermediate and outlying sections of the urban area . . . Rapid transit lines should be the backbone of the transportation system of a metropolitan city. They furnish facilities for high speed train service and make possible the transportation of large numbers of people over great distances in short periods of time.[26]

An emphasis was thus put on speed and capacity. That speed and capacity was to be focused on the downtown core.

Kelker, De Leuw called for improvements to interurban lines too, stressing speed of service once more:

Facilities for high speed operation of interurban trains in urban territory are essential and can be supplied in Los Angeles, without duplication of expenditure, by constructing additional tracks on the structures of the proposed urban rapid transit system. Such improvements within the urban areas, coupled with the elimination of grade crossings at the important highways in the territory between them, will make possible the maintenance of high speed service.[27]

In all, the plan called for the construction of twenty-six miles of rail line in subway, and eighty-five miles on elevated structures. Kelker did recognize that "The desire of the average citizen to own his own home has caused the single family dwelling to predominate and the absence of large apartment buildings is noticeable. Such a condition is very desirable, but it is one of the prime factors which makes the construction and operation of rapid transit lines on a self-sustaining basis, a difficult financial problem." The consultants did not, however, therefore conclude that a rail network would be unsuitable, but that it would need to be subsidized: "If the city's *unequalled position*, when compared with other large cities with respect to the number of families per dwelling, is to be maintained, it must continue to spread and this spreading can be accomplished only by providing rapid transportation at *a reasonable rate of fare.*" If the cost of rapid transit construction were shared by riders, by "the property benefited" and by "the public at large," the report continued, "then the extent of the rapid transit system may be proportionately increased."[28] Accordingly, funding through city and county bonds and assessments was called for.

To understand the consultants' rationalization of rail, it is important to note their eastern orientation:

> In comparing Los Angeles with other large cities we find the closest analogy in the city of Chicago. Chicago, like Los Angeles, is the center of a vast agricultural territory, and the principal factor in bringing about its rapid growth was the construction of a large number of railroads serving this territory and terminating in Chicago. Los Angeles is the terminus of three transcontinental railroads and in addition has the advantage of a splendid harbor.[29]

Kelker, De Leuw's work implicitly holds that like Chicago and other cities, Los Angeles did revolve around a core, even if its periphery was built at a lower density. It ignored the evidence which even then was strongly developing to show that this was untrue. This mistake, nonetheless, accorded with the political interests of the downtown organizations which had backed Kelker's efforts.

The report attracted critical review from the City Club of Los Angeles, which represented the interests of a dispersed Los Angeles and warned that the downtown focus of the rail plan would simply attract congestion. "The great city of the future will be a harmoniously developed community of local centers and garden cities, a district in which the need for transportation over long distances at a rapid rate will be reduced to a minimum."[30]

Shortly after publication of the City Club's review, a fight ensued over two proposals: one involving the construction of four miles of elevated line in downtown Los Angeles, which would have aided the Kelker, De Leuw plan, the other calling for a Union Station near the original Pueblo site and without elevated railway connections. The rail companies, Los Angeles and other regional chambers of commerce, many civic groups, and all major local papers except for the *Los Angeles Times* supported the elevateds.[31] The *Times* opposed having "hideous, cluttering, dusty, dangerous, street darkening trestles in our downtown."[32] Voters narrowly defeated the elevated proposal, and the Union Station plan went ahead instead. The Kelker, De Leuw proposal was shelved along with the elevateds. As Brodsly points out, the Kelker plan followed approval of the *Major Traffic Street Plan* with its ideological and financial commitment to automobility, and would have been unlikely to appeal to voters, especially given the likelihood of strong media opposition to a bonding issue for elevated lines.[33]

The politics of central-city support was outgunned by those who demanded a dispersed metropolis. Support for rail was not, nonetheless, to die. Despite the continuing decline of the Pacific Electric and the conversion of rail services to bus operation, a 1933 proposal called for grade-separated rapid transit on a system of four basic routes.[34] Its conception addressed the concerns of downtown business by advocating patterns of transportation focused on the core, rather than encouraging further decentralization.[35] *A Transit Program for the Los Angeles Metropolitan Area* followed in 1939, calling for long-term rail rapid transit development, including routes on freeways, and shorter-term operation of express buses on Freeways.[36]

In 1945, city officials brought in leading transportation consultants to address business and civic leaders. This led to the formation of the Los Angeles Chamber of Commerce's Metropolitan Traffic and Transit Committee which, in turn, formed the Rapid Transit Action Group (RTAG). The group invited eight hundred business, civic, and political leaders for the release of *Rail Rapid Transit—Now!* in February 1948.[37] The cover showed a train traveling at speed and urged that "It's needed," "Now or Never," and that "It Costs Less." In the foreword, the report announced that "Autos are too expensive for most people. Both autos and buses congest the streets." Turning to page 1, the reader is informed that:

> Our people must have rail rapid transit to take full advantage of the still limitless area where we make our homes. It is every man's desire to have a plot of ground free from the grind of factory and office. He wants to make his family secure. He wants time to play and he has pride in his own fireside. Rail rapid transit will develop many new communities and will enhance the growth of old. Our people need not huddle in the shadow of office buildings nor gather close to the factories. Rail rapid transit will make it possible for us to live where we like and work where we please.[38]

Speed is stressed in the report, as is acceleration, comfort, and quietness. Several benefits of rail are listed, including the reduction of congestion. "Those who do not have to use automobiles will be attracted to rail rapid transit service," and as a result, "This will make driving easier and will reduce competition for available parking in congested areas." System ridership is forecast at 220.5 million per annum, bringing in revenues sufficient to cover costs. To provide financing, bonds were called for, to be issued by a new "Metropolitan Rapid Transit District patterned somewhat after the Metropolitan Water District." In addition, the district would have the power to "recommend a tax levy only to pay any portion of principal or interest which is not paid from revenue," but not to subsidize operating costs.[39]

Not surprisingly, downtown interests favored the RTAG proposal, while business and property-related groups who had already invested in substantial office/commercial centers outside downtown were opposed.[40] Those favoring concentration wanted rail; those after a dispersed metropolis did not, and this was to be the persistent pattern of support and opposition until the lead-up to Proposition A of 1980.

Legislation was proposed to create the Rapid Transit District RTAG wanted.[41] The League of California Cities adopted a resolution to oppose it. Long Beach Councilman Ramsey declared that "local shoppers would travel to Los Angeles to buy a spool of thread if this high speed rail line should be operated," and the Long Beach City Council unanimously opposed the legislation. The *Santa Monica Evening Outlook* issued the editorial opinion that "It is designed to save the downtown shopping district of Los Angeles at the expense of other districts and at terrific cost to all taxpayers. No real economic need for it exists beyond the need of downtown Los Angeles merchants to reverse a 25 year old trend."[42] In April 1949, the Los Angeles City Council voted eight to six against creating a rapid transit district, reflecting the many council votes from outside the downtown area. RTAG's plans were dead.

In 1950 the California Assembly Interim Committee on Public Utilities and Corporations released a report based on public hearings and a survey. The survey found that while 40 percent of respondents favored one of three fixed-rail options offered—a monorail system, and two types of conventional rail system—47 percent called for buses on freeways, and this latter proposal received the committee's endorsement. In mid-1950, however, a former Air Force pilot, George Roberts, proposed a monorail to link Van Nuys (in the San Fernando Valley) and Long Beach. Governor Warren appointed a state commission to look into this, and in July 1951 the Los Angeles Metropolitan Transit Authority (LAMTA) was created, and empowered to construct and operate such a monorail line. Nothing further came of the proposal during the next eighteen months. But other and more fantastic schemes continued to be advanced, including the Hastings Electric Railplane System, involving cars "sliding between an upper and a lower rail at speeds of one hundred fifty miles per hour."[43]

In 1960, the Los Angeles Central City Committee and the Los Angeles City Planning Department issued *Los Angeles Centropolis 1980: Economic Survey*, in which great concern was expressed over the decline of the downtown:

> Should the present trend continue without any improvement, twenty years from now the central section will account for an extremely small share of the total assessed values of either county or city properties. Clearly, in view of the tremendous public and private investment in structures and public utilities in the area, it is a matter of direct concern to the entire city and metropolitan area that the value of these equities be preserved.[44]

One of the advantages of the central city was said to be that it supported rapid transit, rather than merely that rapid transit supported it: "A strong rapid transit system requires an active and well-populated Central City Area." The report also called for "high density multiple dwelling units" close to or within the central city and for corridors along the transit routes to be "rezoned for higher density uses."[45]

The Development of Federal Policy

Federal policy was to develop in ways supportive of capital-intensive rail rapid transit, rather than the support of bus systems struggling to maintain failing urban services. Private bus companies which continued in operation did so without subsidy, while it was assumed that publicly owned systems would also continue to cover all of their operating costs with fares. With the increasing incidence of red ink, however, there was little reason to feel that an investment in transit—unlike in highways—would be repaid by users. Patterns of declining transit ridership contrasted with a market-led demand for highways. And while there was a tradition of publicly provided roads, transit had traditionally been a private industry, with the role of government limited to regulation. Transit not only lacked public support. There were no reasons why it should find advocates in either corporate America or military interests. In addition, urban transit generally took place within the confines of state lines and did not logically seem to be a federal responsibility.[46] While highways were of interest to all states, furthermore, transit was important to only a relatively small number of urban areas, while Congress was dominated by rural and small-town conservatives.[47]

Federal involvement in transit was eventually spurred by concerns over failing commuter railroads at the end of the 1950s. The mayor of Philadelphia called a meeting of twelve mayors and seventeen railroad executives in 1959. They stated that "it is apparent that high-speed mass transportation, particularly in the form of rail, both surface and subway, must play a vital, important part in furnishing transportation to the great urban areas." Fearing that failing immediate action, "nearly all urban areas are in danger of actually losing their commuter railroad facilities," they called for the formulation of a national policy to provide subsidy support, with an emphasis on the provision of new commuter equipment and improved facilities.[48]

Significantly, bus operators as well as public urban transit properties were left out of this discussion as the precedent was formed for emphasizing the rebuilding of rail. Bus operators, in fact, for the most part kept themselves at arms' length. As James Dunn has observed: "The bus industry was represented by the American Transit Association (ATA), which was not inclined to seek help from the federal government because the majority of its members were private businesses suspicious of the strings likely to be attached to federal subsidies."[49]

A February 1960 approach to the Senate Commerce Committee Subcommittee on Transportation by New York, Philadelphia, Cleveland, and St. Louis city chief executives, the state of Pennsylvania, and the Pennsylvania and New Haven Railroads did not secure funding, despite an emphasis on seeking only long-term, low-interest loans for maintenance and expansion of urban mass transportation facilities with a study of the desirability of grants-in-aid to follow.[50]

As Jones explains, a tactical decision was made to seek hearings from a committee more responsive to urban concerns. The proponents of federal aid therefore framed draft legislation to fall within the jurisdiction of the Housing Subcommittee of the Senate Banking, Currency, and Urban Affairs Subcommittee. Thus, federal aid for transit would follow the Model Cities/urban renewal federal-city partnership model, rather than the federal-state-local route followed by highway funding. "Thus the tactical requirements of obtaining a receptive hearing dictated that transit aid be sought within the framework of federal aid for urban renewal rather than federal aid for transportation."[51]

The emphasis on capital intensive projects was founded on the principle that increased efficiency would result. In 1960, Senator Harrison Williams of the Senate Banking, Currency and Urban Affairs Committee called for legislation that would limit federal assistance to low-interest loans rather than long-term subsidies, creating "the imagery that a relatively small federal commitment would provide the transfusion necessary to restore the transit industry to a competitive footing."[52]

The Senate passed the Williams Bill on June 27, 1960 by a voice vote. Senator Strom Thurmond, a conservative southern Democrat, made the only opposition speech—complaining that the federal government should stay out of state and local affairs; that the urban areas which would benefit were richer than rural ones which would nonetheless contribute to the costs; and that federal expenses should not be allowed to expand excessively. His comments are important in that they set a trend for future transit program development: opposition would be on the grounds of unfair advantage to large urban areas or excessive expenditure in general, but rarely on the grounds that the particular programs to be advanced were ill-designed and could be replaced by better ones.

Support for federal aid developed from central-city political and business leaders and the Eastern commuter railroads as well as from planners, metropolitan transportation experts, academics working in urban affairs, and the metropolitan press.[53] The call-to-arms by central-city interests and involvement of big-city mayors was to be of central importance in shaping policy, for unlike the regime for highway programs, funding for transit revitalization was to mainly bypass the state level of control. With an emphasis on allocation at the city level, larger metropolitan areas with their advanced lobbying systems and congressional connections were put in a disproportionately powerful position to acquire federal money and fund large-scale endeavors. With a state funding funnel, the money might have been more evenly allocated, with a greater share to the smaller population centers in each state, and a lesser emphasis on expensive capital-intensive projects concentrated in a very few places.

How was the problem seen? Supporters saw traffic growing while automobile ownership increased and railroads fell into disrepair. Population decline in America's central cities was also of great concern to those active in lobbying for transit money. As documented by Jones, the decline of central-city populations was only a focus of attention as an undesirable trend rather than as a cause of transit's problems to be accounted for in designing a solution. The emphasis was one of providing capital to sustain or expand existing patterns of service rather than to inquire if declining demand for transit was an indicator that a different public policy response was in order.

The theory of transit's decline implicit in the legislation emphasized the loss of patronage due to the sorry condition of the industry's physical plant. The inability of private carriers to attract capital investment for the modernization and extension of service was seen as a consequence of the imbalance of federal policy and its bias toward freeway construction. It was also interpreted as a consequence of the failure of localities to manage urban development and coordinate highway, transit, and land-use planning. This, in turn, was seen as a by-product of the fragmentation of governmental responsibilities in metropolitan areas.

Virtually no attention was given to the economics of transit operations. Only passing notice was given to the impact of state regulatory policies on transit's effective earning power. Little attention was given to the operating costs that would be associated with expanding peak-hour transit services as Senator Williams was to propose. Nor was much attention given to fundamental changes in lifestyles that were altering the spatial arrangements of metropolitan areas, the geography of trip making, and temporal rhythms of transit's residual traffic.[54]

A jurisdictional battle between Commerce and Housing was to prove highly significant for the future of transit programs. The Housing and Home Finance Agency, which won jurisdiction, was used to dealing directly with cities and largely bypassing the states, and the transit program was handled likewise, providing the opportunity for major cities to become directly involved in lobbying for federal funding.

Legislation in 1961 provided for $50 million in loans, underlining the emphasis on providing a temporary form of rehabilitation support for the return to profitable commuter operations, and $25 million for "demonstration," projects. This latter concept was to prove vital to making feasible propositions of the large rail projects that were to come. The idea of a demonstration, as the name suggests, was to test the success of a new concept. The idea was to provide a boost for projects which would ultimately support their own operations financially.

A study initiated by the President and commissioned by the Secretary of Commerce and the Housing and Home Finance Administrator, was conducted by the New York based Institute of Public Administration (IPA), a body long associated with the view that strong central-city cores relying on mass transportation, metropolitan government and metropolitan planning were desirable. The Institute's report compounded the impression that there were advantages to emphasizing the provision of short-term capital money for urban rail.[55] The findings they presented to Congress pointed to the technological advantages of grade-separated rail rapid transit which allegedly gave it a performance advantage over transit in general.

The IPA report recommended the provision of federal grants "predicated on the development of comprehensive regional transportation plans."[56] It thereby set in motion an emphasis on planning for systems to accommodate future growth, rather than to stabilize established systems. President Kennedy's "Message on Transportation" affirmed the view that transit was about more than transportation: "To conserve and enhance values in existing urban areas is essential. But at least as important are steps to promote economic efficiency and livability in areas of future development."[57]

The Administration now tried to court support from highway supporters, emphasizing that transit programs would cause no harm to anyone, while providing widely distributed benefits.[58] In what was to become a familiar theme, advocates stressed that transit could provide added capacity which was equivalent to highway capacity. Atlanta officials, for example, cited a 1959 report which found that without rapid transit Atlanta would need 120 expressway lanes radiating from central Atlanta, together with a twenty-eight-lane downtown connector. Conservative congressional opposition "hostile to 'spending,' 'socialism,' and federal intrusion into local affairs," delayed action until 1964, when the Urban Mass Transportation Act authorized $375 million in capital grants.[59] The nature of this opposition is noteworthy: the opposition was of an ideological nature rather than critical of specific ways the money would be spent or interested in offering a reformulated or better alternative.

BART

With Los Angeles' rail plans on hold, it was only natural for its competitor to the north to take the plunge. San Francisco's BART was to be the product of the growth issues that had been rooting for rail in Los Angeles as well as one of the first popular protests at the perceived environmental damage of the freeway. Furthermore, as Peter Hall says, it provided an example for the future to BART's many admirers, and particularly first-time visitors, who see the system "as the archetype of everything a modern urban rapid transit system should be, and a model for the great majority of America's large cities that still lack one."[60]

As in Los Angeles, rail was promoted as a means to revitalizing downtown given concerns about increasing suburbanization and the connected decline in importance of the central business district. With congestion hitting the gateways of San Francisco's core, "civic leaders were convinced that the problem with which they were contending was a transportation problem." The leadership felt that increasing the capacity of transportation links to downtown would help rescue San Francisco. "Given the limitation by topographic constraints to six points of entry, the answer seemed obvious: rail transit."[61] Rail transit, it was felt, "would do for San Francisco what the subway did for Manhattan, where highrise construction and transit became necessarily linked."[62]

The private nonprofit Bay Area Council was created in 1945 to replace the San Francisco Bay Regional Council which had been created and initially funded by the State to guide postwar development.[63] With initial funding from Bank of America, American Trust Company, Standard Oil of California, Pacific Gas and Electric, U.S. Steel, and Bechtel, membership of the organization was dominated by firms headquartered in San Francisco

Figure 6-1. BART train at Lafayette Station. (Photo supplied by BART and reproduced by permission)

rather than the Bay Area as a whole.[64] The Bay Area Council was to be the prime mover in convincing the California legislature to establish the San Francisco Bay Area Rapid Transit Commission in 1951 to study transit problems and come up with a coordinated rapid transit plan.

The commission initially retained the engineering firm of DeLeuw Cather, underlining an emphasis on the provision of increased capacity and construction as the way to achieving it. The commission's 1953 preliminary report called for separated right-of-way mass rapid transit. Parsons, Brinckerhoff, Hall and Macdonald (PBHM), whose founder had been heavily involved in the development and construction of New York's subway system, was hired that year and produced a report on "Regional Rail Transit" in 1956.[65] The report gave the Bay Area the following choices: "whether to accept the stagnation and decline of interurban transit and to prepare for drastic decentralization and repatterning of its urban centers to meet the avalanche of automobiles that will result—or whether to reinvigorate interurban transit so as to sustain the daily flow of workers, shoppers, and visitors on which the vitality of these urban centers depends."[66]

The cause of declining transit patronage was erroneously understood in terms of poor quality of service rather than regional change in response to the wide range of new opportunities permitted by the flexibility of automobile transportation. "Since the existing systems had been found wanting, it followed that a proposed transit system must be new."[67] The consultants pointed to the cost of additional travel time and freight movement and to the property depreciation resultant from congestion. "We do not doubt that the Bay Area citizens can afford rapid transit; we question seriously whether they can afford NOT to have it" (emphasis in original).[68]

PBHM, now Parsons, Brinkerhoff, Quade & Douglas, was also a major partner (together with Tudor Engineering and Bechtel Corporation) responsible for the second major planning document, the Composite Report of 1962, where the benefits of BART were more heavily connected with maintaining desirable growth patterns than with transportation benefits. The first benefits listed were "1. preservation and enhancement of urban centers and sub-centers; 2. increase property values; 3. help to prevent disorganized urban sprawl; 4. improved employment conditions because transit will attract industry."[69]

BART received early support in 1959, when the California Legislature gave approval for Bay Bridge tolls to be used to build BART's transbay tube. A group known as the "Blyth-Zellerbach Committee" then played a major role in the lead up to the 1962 ballot measure which approved bonding for BART. It included "the heads of such San Francisco-based corporate giants as the Bank of America and the Pacific Gas and Electric Company. Without their campaign, it is unlikely that Bay Area residents would have approved BART."[70] The committee in turn set up a campaign organization called Citizens for Rapid Transit, headed by top San Francisco banking interests. Citizens put out advertisements describing BART as "the low-cost cure for the traffic mess," while "the main concern of its supporters and planners was the creation of profitable patterns of urban economic development."[71] San Francisco downtown interests were to be joined by construction interests in funding BART-campaign activities, and the firms that pitched in money were to subsequently benefit from BART contracts.[72]

Despite the evidence of a business lead role in promoting BART, it is unreasonable to suggest that there was some sort of conspiracy against the public. As Hall points out, buses were losing traffic to cars and a

> "salable" system had to be rail-based. The notion of low-capital-intensive system was unknown. The 1956 consultants' report was received with almost unalloyed enthusiasm by the public, the press, the professional press and the California legislature. During the whole period from 1951 to 1957, no one apparently suggested any serious alternative to the BART concept.[73]

Perhaps most significantly, however, a citizens movement to protest freeway construction took place in parallel with business pressure for measures to maintain the competitiveness of San Francisco. The movement, which was to subsequently mushroom to the other major urban areas of the United States, was vigorous. Protests of the social and environmental consequences of freeway building began in 1957 and culminated in a 1959 unanimous vote of the County Board of Supervisors to stop construction of seven of the ten freeways planned for the city.[74]

With the freeways stopped, BART offered an environmentally acceptable alternative. As one consultant claimed, thirty-two new freeway lanes would be needed by 1975 in the absence of BART.[75] In contrast, "each two-track rapid transit line would have a seated carrying capacity equivalent to 30 freeway lanes."[76] This clearly seemed a far more attractive prospect. According to K. M. Fong, business leaders saw their campaign as in the public interest and their contribution was in line with a general public belief in the benefits of rail rapid transit.[77]

BART's 1976 ridership was only 51 percent of what had been forecast in 1972 for 1975, while only 35 percent of passengers were people who previously drove—it had been assumed that 61 percent would come from cars. Not only were capital cost projections far exceeded, but the notion that the system would make an operating profit was to prove wrong. Hall ultimately puts the rise of BART down to a general belief in a flawed conception:

> Some alternative to the private car was needed and rail rapid transit was the one seriously considered. Sums were done to show that drivers would divert from their cars and make the system viable. But no one apparently considered whether this was plausible. Similarly, the cost estimates were accepted without even elementary skepticism. The fact was that everyone wanted to believe the predictions, because they seemed to offer a way out of serious present problems. Because of this desire, there was a mass suspension of belief, and almost ideological commitment to a new system.[78]

Rail Develops Momentum in Los Angeles

As Hamer notes, every move of the Los Angeles MTA in the early 1960s "took place in the shadow of the victorious automobile and the one-upmanship of that enemy in the north, San Francisco." With approval of BART's high-tech, high-speed transbay rail system in 1962, the pressure was on for a response from the Southland. "The euphoria generated by BART finally brought about the 1964 creation of RTD and the end of MTA."[79]

RTD had a state mandate to build a rail rapid transit system, using sales tax financing if voters would approve it. As Sy Adler points out, however, the City of Los Angeles was given only two out of the eleven seats on RTD's Board of Directors.[80] Four were given to other cities in the county, and the remaining five went to the County Board of Supervisors. SCAG, created in 1965 for "discussion, study, and development of recommendations on problems of mutual interest of orderly development of the Southern California region," was given the task of coordinating regional transportation.[81] It represented all Southern California counties except San Diego, and Los Angeles did not contribute a dominant part of its membership.

In the same year, Daniel, Mann, Johnson and Mendenhall (DMJM) completed a report for RTD which, as Hamer puts it, "proved to be an extraordinary effort that attempted to reconcile existing and projected locational behavior in Los Angeles with the necessity for rapid transit."[82] The study provided an alternative to the inefficiencies of sprawl in which "high densities and high levels of economic concentration are developed in an organized manner. This would permit substantial economies of time as well as services such as utilities, police and fire protection . . . and could easily be accomplished by overlaying the current 'spread city' pattern with very high capacity travel arteries in the form of [rail] rapid transit."[83] DMJM estimated that in the absence of rail transit, between eighteen and forty-three inbound freeway lanes would be needed into the Los Angeles core in addition to the twenty-four inbound freeway lanes already planned between 1964 and 1980.[84] The suggestion was that rail and freeway capacity were substitutable, irrespective of origins and destinations—experience has shown this to be quite wrong.

Also in 1965, the Los Angeles Chamber of Commerce set up a Citizens Advisory Council on Public Transportation, with fifty out of sixty-four members representing business and including representatives of some of the largest businesses in Los Angeles.[85] Two years later the council published *Improving Public Transportation in Los Angeles*, a report which called for a new rapid transit system, citing benefits of increased property values as well as greater mobility and the reduction of traffic congestion. "In some cases, such as Downtown Los Angeles and mid-Wilshire," the report declared, "rapid transit may make the difference between sound economic growth and ultimate stagnation." The report saw the central area as being "expected to continue to have the greatest population density and employment concentration of any area by 1980," while "high rise construction activity is expected to continue a strong tendency toward centralization."[86] A principal benefit of the proposed transit system was believed to be the encouragement of centralized development by raising land values as well as increasing accessibility.[87]

With continued attention on the successful funding of BART to the north, RTD sponsored a ballot proposition—Proposition A of 1968—for a half-cent per dollar sales tax to fund an eighty-nine-mile rail rapid transit system, to be both larger and more expensive than San Francisco's BART. The *Los Angeles Times* reported that "Meeting at the Los Angeles Chamber of Commerce's headquarters, the new citizens committee drew a nucleus of 75 top business and civic leaders. It set a goal of $750,000 to mount a major public information program."[88] In the event, $458,612 was raised, with 85.6 percent of these campaign contributions in favor of the proposition coming from business, and 46.6 percent of contributions from five central zip code zones.[89] The City of Santa Monica—located well away from the core and seeking to preserve an identity separate from Los Angeles—predictably came out in opposition. In the event, the proposition won only 44.7 percent of the vote and, needing 60 percent, failed.

A year later, a principal from Kaiser Engineers (one of the consulting firms involved in the 1968 proposition) backed establishment of the Committee for Central City Planning, an organization funded by downtown corporations and committed to downtown renewal. In 1972, *Central City Los Angeles 1972/1990* was published. "As always, the failure to build the rail rapid transit system is assumed to lead to stagnation and decay."[90]

In the interim, RTD had turned its attention away from rapid rail. Indeed, on November 12, 1971, state Senator Mills accused RTD of a "negative attitude" to rail and called for the start of construction of at least one major rail transit line.[91] As the *Los Angeles Times* put it, RTD had "emphasized buses and forgotten rail." But then, the *Times* reported, "political pressure" had resulted in the hiring of consultants—including most of those involved in the Central City Los Angeles report—for a new study.[92]

The consultant team recommended a 250-mile rapid transit system, with an initial rail system of 116 miles, to be accompanied by twenty-four miles of busways, for a cost of $6.6 billion over a twelve-year period (See map 6-1). A ¾¢ sales tax was recommended for funding (this became 1¢ on the ballot). The consultants' report indicated that in 1990 1.05 million people would use rapid transit, and 706,000 of them would otherwise have driven, "As a result, congestion on freeways and arterial streets will be reduced . . . Much of the attractiveness of transit will be to people traveling to centers of activity. Today, for example, about 38 percent of work trips made to the L.A./CBD are by transit. With rapid transit, this number will jump to 65 percent."[93] The highest usage corridor was said to be Wilshire Boulevard, with 42,000 one-way riders per day estimated at the maximum loading point. SCAG followed up by doing their own forecasts, in which the Wilshire maximum load was only 30,900 daily passengers. Further study by Voorhees & Associates (who had produced the initial forecasts) reduced this still further to 17,900.[94]

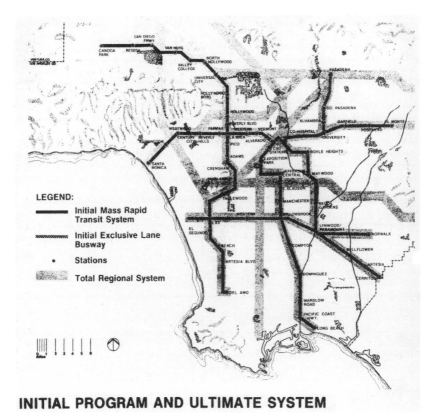

LEGEND:

▬▬▬▬ Initial Mass Rapid
Transit System

〰〰〰〰 Initial Exclusive Lane
Busway

• Stations

▓▓▓ Total Regional System

INITIAL PROGRAM AND ULTIMATE SYSTEM

Map 6-1. The consultant team's recommended system. (Source: Southern California Rapid
Transit District)

As Hamer observes, these results shocked other consultants on the team.
"Given both the commitment of RTD and its downtown lobby to a rail network, the Voorhees effort had to be put into a "proper" perspective," one
founded on a "common sense" which supported political advocacy of rail
and which was not documented by any analysis.[95] "However, all the results . . . are contrary to common sense and seem to ignore the realistics
[sic] of everyday life in the Los Angeles area. At a minimum, it should be
expected that the . . . [Phase III] system . . . potential would be on the
order of 2,500,000 patrons rather than the 1,500,000 to 1,600,000 . . . calculated on a daily basis."[96] Completion of the Phase III study produced network maps which did not specify particular routes; "instead, broad bands

of color overwhelm county maps and provide the illusion of rail transit terminating in every homeowner's garage."[97]

SCAG, the regional agency, approved neither specifically of rail transit in particular nor of the program as a whole, although it did approve the creation of specific transit corridors. The proposed system, SCAG said, "would use much of the region's transit improvement funding to heighten the mobility of a small portion of the population—and that improvement would come in the long run." "The high capacity fixed guideway system encourages long work trips, and thus the conversion of open space into low-density urban communities. Employment would continue to locate in major centers, creating even greater disparities between the residential and employment centers. *This is counter to regional goals.*"[98]

Citizens for Better Transportation, chaired by Los Angeles Mayor Tom Bradley was formed to campaign for the sales tax ballot measure, Proposition A of 1974. Thornton Bradshaw, president of Atlantic Richfield Company (ARCO), owner of the massive Atlantic Richfield Plaza in downtown Los Angeles, was appointed Chairman of the Executive Committee. Bradshaw was recommended to Bradley by the Los Angeles Chamber of Commerce, which also supplied recommendations for membership of the finance committee. $562,827 was collected for the campaign this time, with 93.6 percent contributed by business. Most of this support came from the central area of the city.[99]

Once again, the proposition pitted central-area political interests against those from the many smaller centers scattered across the Los Angeles metropolitan area. A group of mayors and chambers of commerce from Pasadena, Glendale, Beverly Hills, and various smaller communities bought a full page advertisement in the *Los Angeles Times* in which Proposition A was described as "regressive and inflationary [and . . .] for the benefit of major corporations along Wilshire Boulevard and downtown Los Angeles who will profit most by so-called rapid transit."[100] The next day only 46.3 percent of Los Angeles County voters supported the proposition, and it therefore failed.

A Change in Political Climate

The 1960s and early 1970s had been a time of fundamental changes in national attitudes to transit financing. The freeway revolts in San Francisco and the related authorization of BART were precursors for national patterns to follow. A grassroots movement spawned concern at the environmental and social problems related to the highway and the automobile. According to Jones, this was part of a movement involved in the fight against reckless growth, corporate exploitation, and neighborhood powerlessness.[101] Factors which attracted attention included:

> automotive pollution, the disruption of neighborhoods by freeway construction, the loss of urban open space, and the loss of neighborhood amenity. Media attention to environmental issues generated broad support for mass transit development and automotive pollution control, with the two policies frequently paired in media accounts. Most important, *support for transit was established as a litmus test of the environmental credentials of elected officials,* and congressional support for additional federal aid grew in bandwagon fashion [emphasis in original].[102]

The mass transportation sessions at the December 1969 National League of Cities Congress in San Diego were popular, with public officials from cities of all sizes in attendance. As Smerk says, the public acquisition of traditionally private transit had finally alerted many of these officials to the transit issue and the federal transit program. Meanwhile, a broad constituency was inundating Congress with letters, phone calls, and personal visits, providing the nationwide support needed to carry forward a major program. Smerk notes that this effort began to pull in even conservative legislators without big-city constituencies. The duality of appearing to serve objectives of enhancing both transportation and welfare was particularly attractive. Welfare programs were unpopular given the apparent failure of much of the Great Society effort to quickly solve complex problems, but a program that provided access to opportunities and helped provide an environment for positive change was altogether more attractive.[103]

In 1970, Congress overwhelmingly passed the Urban Mass Transportation Assistance Act with $10 billion in federal aid over a twelve-year period, and did so with the blessing of the Nixon Administration. Altshuler attributes the endorsement of transit expenditures by a Republican president to the policy's broad appeal, whether to those concerned over economic vitality, the environment, stopping highways, energy conservation, assisting the elderly, handicapped, and poor, or simply with getting others off the road so they could drive faster. "This is not to say that transit was an effective way of serving all these objectives, simply that it was widely believed to be so."[104]

With anti-highway sentiment spreading from environmentalists to a generally popular cause, it was also time for highway advocates to adapt. An anti-transit stance could threaten to count highway interests further out of a decision-making process which was developing toward considering a position of considering highway and transit options together. "Reflecting the orientation toward inclusiveness of the American political system, they have judged advocacy of increased transit spending to be an effective way of protecting their own vital interests."[105] This was particularly the case, given that the strength of the anti-highway faction was enough to block the highway aid bill in 1972.[106]

The main congressional actors behind the 1973 Federal Aid Highway Act were Senator Jennings Randolph and Congressman Jim Wright, both strong highway advocates with little interest in transit. By bringing transit into the highway camp and sending disputes about whether money should be spent on highways or transit to the local level, they allowed harmony to return to congressional highway policy.[107] The following 1974 Act set the federal matching share for transit projects at 80 percent—making it similar to the contribution to the now-maligned highways—and allowed the substitution of interstate highway construction funds for transit construction, should this be the desire at the local level. As Jones says, the focus on capital meant a commitment to rail, as only rail projects could absorb so much funding restricted to capital investment.[108]

In parallel with developing national thinking, California state Senator Jim Mills, a major advocate of light rail and the father of the San Diego light rail program, had authored Proposition 18 in 1970, which was to make available highway funds to be used for building rapid transit. The highway lobby had given generously to the campaign against it, and the proposition failed.[109] Mills tried again with Proposition 5 of 1974, which allowed local areas to vote part of the state highway trust fund for transit purposes. It was more restrictive than Proposition 18, excluding operating and maintenance costs from its funding, and it included the greater appeal of a tie-in with the new federal legislation to also permit transfer of highway funds for transit use. It specifically provided money to match the federal contributions to transit development made possible by the 1973 Federal Aid Highway Act.

The proposition passed, and gave birth to the idea of a Los Angeles "starter line," to be built without the need for a popular vote as the first segment of a regional transit system to be completed later.

In March 1975, RTD formed a Rapid Transit Advisory Committee (RTAC) to choose the route for the starter line. Argument now revolved not so much over whether or not to have rail—with the CBD for and the more dispersed communities against it—but over whether the CBD should be the primary beneficiary or whether the more dispersed centers should be counted in, too. There were two major competing system possibilities, one involving extensive tunneling via Wilshire Boulevard, with terminals in North Hollywood and South Central Los Angeles, the other making extensive use of existing rights-of-way and stretching the far longer distance from Canoga Park in the San Fernando Valley to Long Beach, via the Los Angeles CBD. The shorter line was preferred by Los Angeles Planning Director Cal Hamilton, the Los Angeles Chamber of Commerce and SCAG. The longer route was backed by Supervisor Baxter Ward (who represented the San Fernando Valley) and by the many cities the line would serve. RTAC opted for specifying the San Fernando Valley–Long Beach corridor—but not a specific route.

Argument over route selection continued. At California Senate hearings the Los Angeles Chamber of Commerce supported the Wilshire subway, which would best serve its members.[110] The study the chamber had sponsored was said to have shown that the Wilshire route was "technically" if not "politically" better, attracting more riders by going through a denser corridor. RTD President Byron Cook, Supervisor Baxter Ward, Long Beach Mayor Clark, and Rolling Hills Councilwoman Dorothy La Conte called for the longer route. Los Angeles Mayor Bradley gave his support to the Wilshire subway in August 1975. Long Beach Mayor Clark criticized Bradley for endorsing "a system that will be totally within Los Angeles."[111]

The RTD Board now voted for a rail line from downtown Los Angeles to Long Beach as the first priority, avoiding for the moment the question of which route should be taken to the north, and placating politicians of the county's second largest city, Long Beach, whose congressional representative—Glenn Anderson—was a key player in the fight to get federal support for rail transit in Los Angeles. UMTA Administrator Robert Patricelli, however, advised RTD to study the whole corridor. The study—performed by

Caltrans—found for the downtown Los Angeles–Long Beach route, with the northern routes to both follow later.

More local-level disagreements ensued, a SCAG member finally recommending that RTAC endorse a combination known as "U" to UMTA. "U" was a four-part program which included the Wilshire subway; a people mover for downtown; bus improvements in general; and freeway bus services in particular. This program was adopted by the RTD Board in September. There was little rail involved, but the bus-on-freeway program would provide service to the dispersed areas of Los Angeles County and beyond.

As Adler relates, there was still no local consensus on what to do. When UMTA Administrator Patricelli visited Los Angeles in December 1975, he found Supervisor Hayes and a Citizens Advisory Panel appointed by Mayor Bradley criticizing the people mover; Supervisor Kenneth Hahn—representing low-income areas such as Watts—calling for spending money to improve bus services and keep fares low as a top priority over rapid transit; and the executive director of the Central City Development Corporation endorsing both the subway and the people mover.[112]

County Supervisor Baxter Ward was now becoming increasingly visible on the transit playing field. Ward had unsuccessfully run for mayor of Los Angeles in 1969, but he was elected a county supervisor in 1972. According to one report, Ward "seeks no campaign contributions, receives no campaign contributions, and points out that he is therefore absolutely beholden to no one."[113] "Familiar to Southland residents as a well liked television reporter for 17 years," Ward entered the fray with a clean image.[114]

In 1974 Ward had convinced the county to purchase and refurbish a train for $2 million to be used for commuter service. It had been referred to in media criticism as "Baxter's choo-choo." Amidst the confusion over how to proceed with transit proposals in Los Angeles, he now prepared his own plan to present to the voters. Known as "The Sunset Coast Line," it presented a futuristic vision of a rail transit utopia unified by two-hundred and thirty miles of heavy rail main line operating at 85 mph, and fifty-one miles of light rail (see map 6-2). Monorail service would also be provided as feeders. The estimated cost was between three and four billion dollars in 1976 dollars.[115]

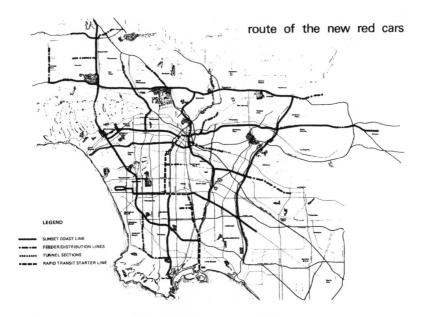

route of the new red cars

LEGEND

———— SUNSET COAST LINE
·—·—· FEEDER/DISTRIBUTION LINES
········· TUNNEL SECTIONS
▬▬▬ RAPID TRANSIT STARTER LINE

Map 6-2. Proposed route system for Baxter Ward's Sunset Coast Line.

Our headline is TRANSIT—281 miles of it.

And the bottomline is COST—ten cents a day.

Ten cents a day will be the average cost for each resident in Los Angeles County in the sales tax issue that we propose.

Those ten cents will build the finest transit system in the world. It will be the newest, fastest, and most comfortable.

And to both the headline* and the bottomline*, there is this footnote:

*It will improve the air considerably.

Ward's ballot measure—asking for a half-cent sales tax to fund his plan—failed.

SCAG included the four-part program in its Regional Transportation Plan in April 1977, with Los Angeles City Council endorsement following in September. The state, County of Los Angeles, City of Long Beach, and RTD also agreed to the program, without full agreement on the rail element.[116]

Institutional Change

The continuing failure to reach agreement on what to do at the local level, and the resulting reluctance of federal administrators to provide transit funding for Los Angeles, had meanwhile attracted the attention of state legislators. As Shaw et al. put it, Los Angeles' many transportation agencies, mixed with regional, state, and federal agencies, made for an "intergovernmental maze."[117]

California Assembly Bill 1246 of 1976 created the LACTC as an answer to the transportation stalemate, and the Commission began work on January 1, 1977. The commission included the five county supervisors, as did the RTD Board; it also included three representatives from the City of Los Angeles (RTD had two); one from Long Beach, and two from the smaller cities (RTD had four from the smaller cities). It was given powers for short range planning, resource generation and allocation, policy and program development, project selection, and new system development.[118] Bringing decision makers to act in concert rather than conflict, the agency's creation was often referred to as "Los Angeles—get your act together."[119]

"Unlike SCAG," a staff budget memorandum announced: "We are not regional planners; unlike Caltrans, RTD and other implementing agencies, we are not builders and transportation system operators. To the Commission is entrusted responsibility for policy setting, programming of federal and state funds, and priority-setting, among competing projects."[120]

The legislation called for transportation policy to "avoid undesirable duplication of transportation services, achieve the operation of a coordinated and integrated transportation system which will reduce auto usage and dependency, reduce the consumption of scarce and expensive energy fuels, and reduce the levels of auto related air pollution."[121] In working towards this, LACTC was to coordinate the operation of all public transportation services with the county so as to "achieve efficient operation thereof and shall resolve all jurisdictional disparities between public transit operators."[122] "Fiscal authority" for rapid transit or other guideway systems was given to the LACTC.[123] As the Commission was starting up, BART was opening its first sections in San Francisco. Its spirit of "it can be done" reached Los Angeles.[124]

Los Angeles County Supervisor Kenneth Hahn quickly emerged as a key actor on the commission, and was ultimately to serve as the public entrepreneur who pulled all the threads together to forge consensus on rail—even though rail was not one of his top priorities. In a January 1978 statement of goals, Hahn reflected a feeling that the public "want their leaders to act—not to study and restudy things endlessly . . . It is time that we act decisively together to get on with the needed transportation improvements."[125] He called for agreement on the construction of "one segment of rapid transit" as well as a number of bus programs and closure of gaps in the freeway system.

Caltrans had leadership responsibility for freeway transit, and in September 1978, came up with a proposal for "significantly increased bus service on the County's freeway system and the construction of busway and station facilities to encourage bus and carpool riding." The study found that:

> Buses appear to be more suitable than any other form of mass transportation to serve the entire County with a reasonable capital expenditure and within a reasonable time. They are also competitive in terms of operating cost. Exclusive guideways with buses, carpools, and vanpools operating along them will be able to accommodate peak hour travel forecasted for 1990 that would otherwise require construction of at least two equivalent freeway lanes in each direction.[126]

At the LACTC meeting on January 10, 1979, Supervisor Hahn praised his fellow commissioners. "I'm amazed to find out that when I've asked each one of you, 'I want you to put for the first time not any political interest or subdivisions at heart, but the goal of transportation for 7 million people,' each of you has done it."[127]

Handing over the Chair to Commissioner Ed Russ, he warned that a further oil crisis could put

> our entire economy and transportation needs . . . in jeopardy. The automobile and the trucks will be put aside. There will be rationing and people will say, "why didn't you build a system that people can use to go to work, go to school, go home?" That's why it's so important to have these alternate plans for rail and buses on the Century and Harbor freeways . . .

> And this Commission, if you keep the same dedication, the same spirit, will solve every major problem and somebody someday will say, "We're thankful for the Commission. The Los Angeles County Transportation Commission had the vision and had the will to provide good transportation for our people." And I thank you for this beautiful experience this year.

Hahn's references to energy problems reflected recent experience of oil shortages, while issues of congestion and pollution—and Los Angeles' failure to deal with them while federal money went to other cities to build rail systems—became increasingly visible. The commission's four-part program continued to reflect a focus on bus-on-freeway development, with only the Wilshire line included for rail development. Baxter Ward, of course, wanted to change that and, ultimately, with his influence on Proposition A of 1980, he did.

Ward had remained actively involved in the rail development program despite his ballot proposition defeat in 1976. In 1978, he came up with the "Sunset Ltd." a scaled-down version—56.9 miles long—of the "Sunset Coast Line," and put an "advisory" on the November ballot, asking which route voters advised LACTC to develop, given a choice of three.[128] The results were:

> RAIL transit from LOS ANGELES INTERNATIONAL AIRPORT to UNION STATION via elevated and at-grade structures along the Century and Harbor Freeways, and subway from the Convention Center to Union Station 52%
> BUS and carpool guideway from LOS ANGELES INTERNATIONAL AIRPORT to the LOS ANGELES CONVENTION CENTER, via elevated and at-grade structures along the Century and Harbor Freeways 26%
> SUBWAY from UNION STATION to the WILSHIRE DISTRICT 21%

Ward used this outcome to claim greater public support for a longer, lower-cost rail route, compared to the Wilshire subway or bus rapid transit. He wished to try another ballot proposition in November 1979, this time for an assessment district (levies being placed on homes and businesses) to fund a proposal of the same scope as the original "Sunset." "The assessment district would guarantee [local matching funds for] the federal funding for Wilshire," he noted, in addition to providing the rest of the extensive system planned.[129]

At the July 25, 1979 Commission meeting, Ward reminded his fellow commissioners that their organization was "titled the LACTC. It is not titled the Wilshire Subway Transportation Commission. It is the Los Angeles County and I am a Los Angeles County Supervisor. And I have regretted additionally that in all of the considerations for transit we have never viewed this county-wide." But LACTC Executive Director Jerome Premo, pointed out that there were other (albeit not rail) projects which benefited other parts of the county, while Mayor Bradley urged the "doing of the possible," reminding Ward that he had supported Ward's "Sunset Coast Line" ballot proposition, which had failed.[130]

On August 6, 1979, the LACTC Finance Review Committee recommended against placing Baxter Ward's proposal on the November ballot for a variety of reasons, including the perceived regressive nature of Ward's financing method and the fact that "a Countywide consensus on the desirability of the proposed rapid transit system is not evident."[131] Two days later the Commission approved the committee's negative recommendation nine to one.

On August 8 the Mayor of Long Beach, Thomas Clark, writing in response to a letter from Supervisor Yvonne Burke, also expressed the view of the City Council—reached unanimously the previous day—that Ward's measure not be endorsed. Among his reasons were that "Long Beach has higher priority transportation programs that would require far less funding than our 'share' of the Sunset Coast Line. These programs include the downtown transit mall, improved and expanded local bus service, the completion of Route 47, increased bus service to Los Angeles International Airport, and improvements to the 'Iron Triangle' intersection." The Board of Supervisors subsequently voted three to two against putting Ward's proposition on the ballot.

The Wilshire proposal—in contrast to Ward's—could proceed with state and federal funds without the risk of another ballot failure. The commission accordingly adopted the Wilshire proposal as part of its four-part plan, and lobbied for it in Sacramento and Washington. Premo said that while rail was preferable to bus on Wilshire, in all other corridors bus-on-freeway transit was superior to Baxter Ward's proposed rail services, and that such busways could be converted to rail later, should demand justify it.[132]

Nineteen eighty was to be a dramatic year. An LACTC staff report once more emphasized the role of buses:

It is staff's view that the transit system alternative which best suits the needs of Los Angeles County is one which allows for significant short term improvements in our bus system, our supportive ridesharing efforts and the efficiency with which our entire transportation system (including our streets and highways) moves people, *and* long term investment in rapid transit in our high volume transit corridors.[133]

An eighty-two-mile rail transit system was cited as *"one option* [emphasis in original] among many for a rapid transit system objective." The report found that "The half cent increase in the sales tax is the most desirable financing alternative of those presented, based on the amount of funding it would likely provide and the flexibility we would have in the use of those funds (i.e. short term vs. long term improvements, bus vs. rail, operating vs. capital)." On March 26, a motion was carried "that the Commission request the Legislature to increase by a half cent our sales tax with a 50 percent vote of the electorate in November."[134]

Ward continued to push for his image of a railborne Los Angeles:

> Ward has a strategy for the Sunset Coast Line. Every Tuesday when the supervisors meet, the chairman asks his colleagues to put his Sunset Coast Line financing proposal on the county ballot in June. And every Tuesday they do no such thing. With the exception of a single supervisor, Kenneth Hahn, Ward is convinced his colleagues don't want to see the Sunset Coast Line back on the ballot because they know it will win the voters' approval the next time around.[135]

That "single supervisor," Kenneth Hahn, now emerged as the primary pusher of a ballot measure that was to become Proposition A of 1980. It was to be quite different in emphasis from the one LACTC staff had proposed.

The Commission ran public hearings in July 1980, and paid for full-page advertisements in both the *Los Angeles Times* and *Herald-Examiner* of July 20. Readers were told that 50 percent of funds raised from the increase in sales tax would be for mass transit guideways (no specific mode mentioned), 40 percent for immediate bus service improvements, and 10 percent for other transit improvements. Mail-in coupons were included for readers to express their opinions. Ninety-three percent of the eight hundred respondents said that Los Angeles County needed to improve its public transportation; given a choice, 20 percent called for a rail-based system, 16 percent for a bus-based system, 60 percent for a mixed rail/bus system, and 4 percent gave no answer. The rail/bus mix clearly seemed to be the best recipe for a proposition's success.

On August 6, the commission passed a motion calling on county counsel to draw up language for the sales tax measure. Russ Rubley put forward a motion for 50 percent of sales tax funding to be allocated to guideways and major transit centers, 25 percent to regional bus improvements including regional TSM (Transportation Systems Management, a number of low-cost measures), ridesharing, and fare relief; and 25 percent returned to local government on a population basis for local public transportation purposes including paratransit.[136] There had been strong support at the public hearings for returning a share of sales tax proceeds to local governments.[137] The motion was carried seven to three, Ms. Killeen, Baxter Ward's voting alternate in his absence, voting against a measure which would not guarantee rail in particular.

Following the motion staff recommended "that the exclusive guideway element of the program include both rail and bus modes as well as other technologies should they prove feasible and desirable, but that specific modes for specific corridors not be identified."[138] Barna Szabo, representing Supervisor Burke, made a motion that this recommendation be approved; it was seconded by Los Angeles City Councilmember Pat Russell, and carried.

Consensus Is Forged

A tense meeting on August 20, 1980 elicited the final decision to go for the ballot, and radically changed the language to be used. Supervisor Hahn quickly took the lead role, reminding the commission of

> great men who have built a great municipal water and power department . . . We all know that there have been studies made for mass transit since 1946 and even before that time. Probably, $25 million have been used in public funds for studies. We have had studies in depth for years, and what we have to do is have action, and this Commission has the obligation under the law to submit to the voters a plan to raise revenue.[139]

Hahn now proposed a plan modeled after one in Atlanta, where voters had been given a package to vote on which included a stipulation that the transit fare be held at fifteen cents for seven years, in addition to building a rail system with sales tax revenues:

> I propose that today the Commission go ahead and give to every rider a guarantee in writing that for five years from this day forward the rate on the RTD bus will be 50¢ anywhere in the County with a 10¢ transfer, for senior citizens $4.00 per month, and for students $4.00 per month. The half cent sales tax will raise approximately $225 and this subsidy will take about $75 million. The other two-thirds will be used for operations, buses, and rail rapid transit. I predict that there will be a favorable vote in November and

the federal government will provide matching funds to Los Angeles County. With this proposal, I am willing to make a motion on this subject.[140]

Supervisor Hahn's motion (which included a provision for 25 percent local return of funds) was seconded by Supervisor Ward, and discussion followed.

Ronald Schneider, Principal Deputy County Counsel, had a question about the word "rail," included before "rapid transit" in Hahn's motion (on the copy of the motion it is actually penned in to the typescript). "That would change the definition of rapid transit in the proposed Ordinance," Schneider said. "Because right now, rapid transit is not limited to rail."[141]

Ed Russ, representing the League of California Cities, which had voted seventeen to fifteen against a ballot measure said he was now in a "tough spot," though he recognized that the 25 percent for the locals would be "good for the League."[142] After some further questioning, but before a decision on the exact provisions or wording of the proposal had been made, a motion was put by Russ: "Shall the .5 percent sales tax proposal ballot be placed on the ballot in the November election?"[143]

It was a close call. The League of Cities had instructed their representatives Ed Russ and John Zimmerman to vote against. As Christine Reed (then the alternate to Ed Russ) explained in an interview "they regarded [this] as a nebulous rail project that would not benefit them: I mean, this was like eighty cities talking to each other. And most of them were not going to derive any direct benefit out of this. And that's why they voted to oppose it." The two votes from the City of Los Angeles were against. Pat Russell explained in an interview that "I had great concern about our ability to pass it. And I thought if we put it on the ballot and it failed it would be really disastrous for the future of public transportation." Supervisor Hahn's chief of staff, Burke Roche, explained it differently: the "major motivation," he said was that "they're all scared to death of an increase in taxes." Ward, on the other hand, claimed that the Los Angeles votes were against because "a rail line would be built some place where it would drain off funds from Wilshire." In the event, John Zimmerman voted for the motion. According to a nonattributable source, "under pressure from Baxter Ward and Kenny Hahn [he] went against his instructions from the League of Cities."

When the roll was taken, five were for, six against. Ray Remy—one of two City of Los Angeles votes against allowing the measure to be placed on the ballot—dramatically reversed his position. The *Los Angeles Times* reported that Remy, who was representing Mayor Bradley, said that "although Bradley had deep reservations about the half-cent sales tax proposal, 'the mayor did not want our vote to be the one that would have kept it off the ballot.' "[144] Accordingly, as the meeting transcript reported, "Mr. Remy then stated that he wished to change his vote and voted in favor of placing an issue on the ballot in November."[145] The motion thereby passed.

Supervisor Ward now called for a minimum of 50 percent of sales tax revenues to be used for rail transit.[146]

There was some opposition to Hahn's proposal, and questioning of the amount of money to go to bus subsidies, but Barna Szabo said that:

> Given the lack of time to educate the public, perhaps more thoroughly on the more sophisticated proposal, I think we ought to go with this one. This one offers certain benefits to the bus riders, it offers a program for those who want to have rapid transit, it offers a protection for the cities who want to retain some revenues for their own purposes for transit improvement. I think that is the genius of the proposal that it provides for the expansion of the possible constituency of those who will go out and vote for the measure and, hopefully, give us the two-thirds vote.[147]

In a later interview, Szabo said of Hahn's packaging of bus with rail: "I wouldn't say it's very pure and honest . . . What Kenny Hahn's thinking was if [we] provide a perceived public benefit, which is the lower bus fare, the public then will be willing to accept a public obligation for something else, which hopefully will work out." Hahn recognized the importance of the subsidy in a *Los Angeles Times* interview: "If we do not give the rider a subsidy, you can kiss it (the sales tax hike) goodbye. I wouldn't support it myself."[148]

In addition to providing the attraction of bus fare subsidies, Supervisor Hahn added a component to try to draw the League of Cities representatives to the LACTC to vote to put his proposition on the ballot "I might as well tell everybody that this proposal would give and take to try and bring a community of 81 cities together with all of their various interests and that is why we put the 25 percent for the cities that don't use RTD."[149]

Commissioner Wendell Cox suggested an amendment to the motion language to refer to "regional transit," eliciting the response from Supervisor Ward: "Let me second that motion please. *Under the assumption that it implies and states 'rail'*" (my emphasis). Ward continued to stress the importance of rail: "It is my belief that anything less than 50 percent offered the people for rail transit will cause the issue to fail, not even get a 50 percent vote."[150]

In an interview Ward said of the proposition:

I thought it was badly flawed because it chose to split up the money. In 1976, if I could have gotten a one-cent sales tax, I wasn't going to give any money for buses . . . In 1980, Kenny—Kenny Hahn—was fearful that there was going to be a reduction or loss of most of the subsidies for the buses in the county, and the bulk of the riders of the buses in the county are people from his supervisorial district and he likes to protect them in as many ways as possible. And one protection was to see if he could derive extra funds for the bus system. He had to compromise. He drew up a plan that let the buses drain off funds from a half cent sales tax for a limited period of time.

Then he had a political problem. How the hell do you get the rest of the members of the County Transportation Commission, of the RTD, to vote to put this proposition on the ballot unless you give them something? And in effect he bribed them politically by saying: and a major portion of the proceeds will go to the cities. And they'll use it for their local transportation purposes. They don't have any transportation purposes. They have taken that money and wasted it. Not one foot of track has been built.

At the August 20 meeting Heinz Heckeroth, ex officio representative from Caltrans, questioned the exclusion of nonrail guideways from the ballot language:

My understanding, when we first discussed the ballot measure, was that we were looking for a measure that (1) was not modal specific, and (2) that we wanted to implement it on the basis of addressing the needs in each corridor at the time that the funding would become available for implementation on a corridor-by-corridor basis. In my best judgment, rail is an option; light rail is an option; high-occupancy vehicle, exclusive lanes are options; all serving the regional requirement for public transit in terms of line haul supplemented by our definition of regional transit which in effect is an enhancement of line haul systems on arterials as well as freeways. I would like to see us go back to that original concept. It seems that we have lost it.[151]

Commissioner Russ supported Ward: "I think we have to guarantee the voting public, as Supervisor Ward says, that there is going to be so much for fixed rail; they are tired of seeing it going down the tubes of RTD."[152]

Supervisor Hahn continued to emphasize the importance of the fare reduction: "Unless there is a reduction to fifty cents, they will say, 'Why should I vote for it' . . . You cannot take out the subsidy that will get the big vote." He nonetheless agreed to reduce the five years of fare reduction to three. Bob Geoghegan, alternate to Supervisor Edelman, objected that the siphoning off of funds to keep fares low would be at the cost of improving bus services: "Practically, this amendment would limit any further expansion of the RTD even though we are going to have increased ridership because we will have the low fares. So for that reason, I am opposed to it." Supervisor Hahn countered that "We don't have a penny now unless the people vote for it."[153]

Other commissioners tried to get their concerns addressed, but the consensus drifted toward accepting the restriction on rail, along with the fare reduction and the local return. "I am going to bow out because it is rail," said Baxter Ward. "I have believed in that since I have come to Los Angeles. I think the people are entitled to it." A roll call was taken for Supervisor Hahn's motion, which passed seven to four.[154] A discussion of whether a map was to be included with the ballot, and which lines would be included now ensued. Various commissioners asked for particular routes to be included. "Yes, just generally go along the San Diego Freeway route," urged Commissioner Russ. "Go along the San Diego Freeway from the Airport down to the Harbor route or if you like to Long Beach."

An objection was raised by Ray Remy, alternate to Mayor Bradley, that some of the proposed lines might not be built for twenty-five years: "What we are going to wind up is coloring the County gray and black with lines . . . You might as well extend the line to every community and say that is the ultimate outcome." Corridors were nonetheless placed on the map, broad-brush fashion, irrespective of when they might be built. The more places which appeared to be getting service, the better the prospects for the proposition passing. As one member of LACTC staff put it "basically you just take a small map of L.A. and a big felt tip" (see map 6-3).

A motion was made by Supervisor Hahn to accept the ordinance with all the changes, and it passed seven to four. The ballot language was then approved unanimously. The resolution was then read, and passed ten to one (resolution, including ballot, attached as appendix C). In the final version, 25 percent of revenues were to at all times go to local jurisdictions;

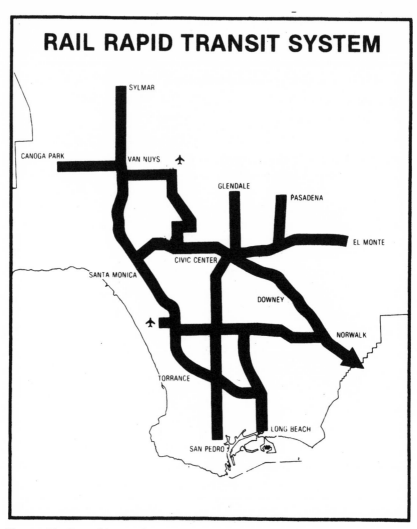

RAIL RAPID TRANSIT SYSTEM

Map 6-3. The Proposition A rail system, as indicated to voters in official election pamphlet.

"such sums as are necessary" for specified bus fare subsidies were mandated for three years only, with the remainder of revenues during that time going for construction of the rail system. After the three years, 35 percent was earmarked for rail system construction alone, with the remainder—after deduction of the local return—to be allocated for "public transit purposes" in general, including the rail system. The resolution was approved three to two by a vote of the Board of Supervisors on September 2, with Supervisors Hahn, Ward, and Edelman in support, Schabarum and Burke against.

Walter King, Hahn's Alternate Commissioner on the LACTC and a close and longtime friend described the struggle to advance the proposition in an interview:

> Proposition A: we couldn't get one politician in the City of Los Angeles, not one Chamber of Commerce, to even endorse it. I didn't see how it could possibly pass, but Kenny has a knack of knowing, and he knew he was right, and only Kenny and the people wanted it, so they voted.
>
> Now, the reason they voted was that Kenny was very much bus-oriented and . . . he insisted that we have a subsidy for the bus, so I think he made it very attractive, we would have the 25 percent, and the 40 percent would be used for three years just for subsidy of the buses, and that all the time we would have 35 percent going to [rail] transit.

The business community played a low-key role, and the campaign for Proposition A had an extraordinarily low budget of only $21,000. According to the October 27 *Los Angeles Times*:

> Of 100 major Los Angeles companies asked to financially back the proposed half-cent transit sales tax on the Nov. 4 county ballot, only Arco and Mattei, Inc., have made contributions.
>
> Supervisor Kenneth Hahn . . . said he is "ashamed" of the business community for failing to support the measure.

Former LACTC chair, Jacki Bacharach, explains why the business community distanced itself: "We had tried transit measures for so long that nobody believed it would actually happen." Not only did business wish to avoid backing a losing horse, but the Los Angeles Chamber of Commerce had become "more and more ineffective," according to Bacharach.

The League of Women Voters, AFL-CIO, and the Los Angeles Taxpayers Association (an organization of nearly 250 of the county's major business and civic leaders) provided endorsements, as did the *Los Angeles Times* and the *Los Angeles Herald Examiner*.[155] An election pamphlet was put out in support of Proposition A by Citizens For Effective Public Transit, a lobby group co-chaired by Supervisor Hahn, LATAX (the business-supported Los Angeles Taxpayers Association) Executive Vice-President Burke Roche (who was later to be Hahn's chief of staff), Bill Robertson, head of the County Federation of Labor, Rita Barschak, President of the League of Women Voters of Los Angeles County, and RTD Board President Thomas Neusom. It had an American flag on the cover and the following words alone:

Reduce Bus Fares Now!

Vote ☒ YES On Proposition A

Figure 6-2. Voting advertisement for Proposition A.

Inside, readers were told that:

Prop. A will cost you only about a dime a day, or the price of a couple of tanks of gasoline a year. But here's what you get back in return . . .

50¢ countywide fares on RTD buses, a flat 10¢ transfer charge, and a monthly transit pass that costs only $20. That's a 15¢ reduction in the fare alone!

For senior citizens, students and the handicapped—a 20¢ countywide fare and a monthly pass for only $4.00.

Lower express fares for commuters!

Money for local transit improvements countywide, including more buses and better security!

The start of a *real* countywide rail rapid transit system . . . not some boondoggle, fancy experiment or pipe dream!

The clout we need to get our income tax dollars back from Washington to help build an effective public transit system right here at home!

Later, the flyer said that:

Prop. A will help unclog the freeways and help in the fight against smog by building a real rapid transit system that by law will have to serve at least
The San Fernando Valley
The San Gabriel Valley
West Los Angeles
Long Beach
South Central Los Angeles
The South Bay-Harbor Area
The Santa Ana and Century freeway corridors.

Although Mayor Tom Bradley had told the *Valley News* that "he would vote for the proposition, he indicated he would not campaign for it."[156] He was quoted in the flyer, nonetheless: "Prop. A is an innovative approach to our public transportation problems. I support it," he said.

Arguments for Proposition A contained in the California election pamphlet, distributed to all voters (see appendix D) and signed by Hahn, Roche, Robertson, and the League of Women Voters, contained similar arguments. Conservative Supervisor Pete Schabarum, RTD Board member Mike Lewis (later RTD chairman and chief of staff to Schabarum), and Edward Martin, described as President of the Southern California Transportation Action Committee, supplied arguments against the proposition, warning that Proposition A:

will raise your taxes every time you make a purchase. In return, if you live in the suburbs, it offers you nothing . . . This new tax money for transit won't necessarily put more bus service on the street and it won't put an end to transit strikes . . . By the Transportation Commission's own admission, it will take 75 years to construct the system envisioned on their map. That's a long time to wait for an alternative to our next energy crisis.

Interestingly, while the time needed to construct the rail system was criticized, the system itself was not attacked.

At the November 4 election, 54 percent voted for Proposition A. "It was a lengthy fight. Tuesday November 4 was a historical day for better transportation," Hahn declared afterward.[157] The largest percentage positive vote was in Supervisor Hahn's district—with 67.9 percent of voters in favor.

The *Los Angeles Times* carried an analysis of the vote on November 16, 1980, stating that "voters living along the proposed rail rapid transit corridors in Los Angeles County tended to favor Proposition A, the transit tax measure in the November 4 election, while those who would not immediately benefit opposed it." The largest positive vote by a city was in Compton, 80 percent of whose votes were in favor. "The City of Los Angeles supported the measure by 61 percent but its central area, where unemployment and use of Southern California Rapid Transit District buses is high, gave it 68.5 percent."

Out of the five supervisorial districts, Baxter Ward's put in only the fourth highest vote for the Proposition: 50.2 percent. And Baxter Ward, himself, failed to be re-elected. He lost to former state Assemblyman Mike Antonovich, who had used the purchase of the $2 million train by Ward as a campaign issue against him.[158] As Ward's swansong, he persuaded Supervisors Hahn, Edelman, and Burke to lease this train to Caltrans for three years so that the state could use it to operate service between downtown Los Angeles and Oxnard (in Ventura County). Opponents, Supervisors Antonovich and Schabarum, were absent from the meeting. On December 23, however, with neither Ward nor Burke remaining on the Board of Supervisors, the *Los Angeles Herald-Examiner* reported:

> Antonovich introduced the motion . . . to derail the contract. Joining Antonovich in putting the train out to pasture were Supervisors Pete Schabarum and Deane Dana.
> Supervisor Ed Edelman was absent, and the lone dissenting vote came from Supervisor Kenneth Hahn.[159]

Los Angeles "Comes of Age"

Because of the question that state Proposition 13 might apply, requiring a two-thirds rather than 50 percent vote for Proposition A to become law, the matter went to the California Supreme Court, which took until April 30, 1982 to reach the five-to-one decision that the proposition had legitimately been approved by the electorate. Reacting to the court's ruling, Supervisor Ed Edelman declared that "Los Angeles has come of age. We are entering a new era of support for rapid transit. We are closer to getting something done on rapid transit than at any time in our history."[160]

At this point any discussion of whether rail or some other alternative should go ahead vanished—since rail was mandated—and argument shifted to the question of the order in which the rail lines should be constructed. By this point the Long Beach line had already been chosen as the most appropriate route with which to begin. It had already been the subject of study, much of the right-of-way needed was already in place, and it appeared to be doable at low capital cost.

The Decision to Go to Long Beach

The Long Beach possibility really came into focus with a public hearing before the California Assembly Committee on Transportation entitled "Light Rail Transit in Southern California—Return of the Red Cars," and held in Long Beach. The hearing took place before the Supreme Court had ruled on the legitimacy of the Proposition A vote, and at a time when state financial support for light rail in Los Angeles was under consideration.[161]

The committee chairman, Bruce Young, whose district included part of the Long Beach rail route, made an opening statement, saying that the demise of the Pacific Electric had been the result of "an absolute bankruptcy of public policy . . . Like lemmings we all followed the popular belief that the era of local rail passenger service was gone. Unfortunately, gasoline no longer costs twenty-five cents a gallon and modern technology has provided solutions to many of the operating limitations which plagued the Pacific Electric." Young pointed out that the San Diego light rail had just opened at a low capital cost, and "in my mind San Diego has pointed the way to the potentially enormous possibilities available in many other communities throughout California . . . The Starter Line of the San Diego system is almost identical to the last surviving line of the P.E.—the L.A. to Long Beach route."[162]

Assemblyman Elder—representing Long Beach—spoke next:

> I'd like to thank you for holding the hearing here in Long Beach because as you pointed out on April 9, 1961, the last trip the Pacific Electric Red Cars, as they were known at that time, occurred between Los Angeles and Long Beach. It's appropriate, therefore that we begin with what was then, and frankly is now the most viable corridor for transportation. I would also ask that in light of the fact that I fly, Bruce, routinely out of Los Angeles International Airport, that you might think about a spur which goes from Long Beach to LAX because getting up at 5 in the morning to go and catch an air-

plane is an inconvenience that not only I share but a number of my constituents have complained about.[163]

Assemblyman Elder now offered to co-author AB 1460, which "funds a portion of [the Long Beach route]," with Assemblyman Young. He complimented Long Beach city officials for their interest in doing

> something about the right-of-way on the P.E. line between Ocean Boulevard and Willow Street on Long Beach Boulevard. I think that that is a very positive indication that there is much support for this kind of alternative. I think we have, as has been pointed out, the opportunity to right a wrong decision made over twenty years ago that brought about the demise of the world's most extensive rapid transit system. The eventual rebuilding of the old Red Car line would be a giant step in not only alleviating our area's mass transportation problems, but in helping to accelerate the already fast moving redevelopment of downtown Long Beach.[164]

Long Beach Councilman Marc Wilder reaffirmed one of Supervisor Hahn's basic points, the need to act:

> The 60s and the 70s were times when we evaluated the perfect systems and what society needed in relation to those perfect systems. The 80s presents us with a very different picture. It presents us with the necessity to get on with the possible, to take those plans and use them as a road map for what really needs to be done and bring them about quickly because if we don't, the impact is going to happen in spite of us. If we don't get on, three and a half million more people will come to Long Beach, Los Angeles, Orange County, the entire region as a result of immigration and natural birth.[165]

The ensuing discussion—which focused on the characteristics of various rail technologies and the feasibility of implementing light rail-proceeded on the assumption that passenger rail would be beneficial in Southern California. Light rail was seen to be low in cost while attractive to passengers, a politically potent recipe for success. Assemblyman Frizelle, added the sole critical element. But, while he was concerned that there be local—not just state—financial support for light rail, he supported the principal of building light rail, per se, contingent on that local support.

An industrial historian evoked Snell's conspiracy theory, talking of the "untimely death" of the Red Car system, claiming that "the automobile was accepted by the public to a large degree because the alternatives were even less desirable."[166] While buses were "dirty, smelly, and definitely not rapid," he said, the rail system was deliberately made unattractive "by eliminating trains from the schedule, slowing their running times, and performing little maintenance on increasingly dilapidated cars or tracks." No mention was made of the changing form of the city, or the reasons the automobile had become the preferred mode of the dispersed metropolis, compared to fixed public transportation. No members of the academic community of transportation specialists (almost all of them vehemently opposed to rail) were present to testify.

The chair of the LACTC, Long Beach Councilman Russ Rubley spoke, pointing out that although "I am of course, very much intrigued with the idea of a rail line between Long Beach and downtown Los Angeles . . . that is just one proposal and as Chairman of the County Transportation Commission I want to take a look at the competing proposals before saying this is where we should begin."[167]

Assemblyman Young responded: "and certainly as a City Councilman from Long Beach I would hope that you and I could put our shoulders together and try to push the Long Beach line only because it seems to be the first one available."[168]

Long Beach Assemblyman Elder echoed that the Long Beach line:

can be done the quickest and if it is done the quickest, there will be literally, in my opinion, an explosion of interest that overlaps not only Los Angeles County, but Orange County. And the people when they see something as viable as this could be in operation, the support, the political support, the financial support will follow. And I think that's one of the reasons we have to go because it will be a physical demonstration for people in this county to see what can be done when all levels of government cooperate.[169]

Board of Supervisors and LACTC member Deane Dana—another politician representing Long Beach—also supported light rail. A representative from the office of Congressman Glenn Anderson of Long Beach, was acknowledged, but did not speak. Assemblyman Young returned with comments at many points, stressing the low cost of light rail, and its ability to be built quickly, and criticizing the high cost of the Wilshire Metro Rail. He concluded the meeting by declaring that "today has marked a new beginning" and calling for "our collective decision to move now on some form of light rail transit."[170]

The 1980 election brought the Reagan Administration and a fundamental change in federal politics. Reagan brought a new conservatism to federal government, with opposition to virtually all domestic programs and particularly to those that provided direct aid to cities and other local jurisdictions. While the president wanted to cut back on transit expenditures in general, rail transit programs quickly became a popular Reagan target. The Administration found that "The construction of new rail transit systems or extensions has not proven to be as cost-effective as less capital-intensive projects," and called for a cutback in capital grants as well as the federal operating subsidies which had by then become common.[171] The new urban mass transportation administrator, Arthur Teele, imposed an indefinite moratorium on new rail starts.

The relatively lower cost of the Long Beach line, compared to Wilshire, had the important strategic advantage that it could be built without the need to apply for federal funding or go through lengthy federal environmental review requirements and an appropriations process in a climate of the Reagan federal administration's opposition to rail. As former LACTC chair Jacki Bacharach said in an interview, "The goal of the Blue Line was to build rail and get it in place and make it a reality." Bacharach proposed using the full local investment in Long Beach light rail (to be built without any federal assistance) as a match for the Wilshire Line's federal funding, thereby maximizing local dollar leverage power, while avoiding the federal funding and supervisory process and associated delays to the Long Beach project.

The Long Beach line had other undeniable political advantages. Not only did the Chair of the Assembly Transportation Committee show his strong support, but it passed through two county supervisorial districts—those of both Supervisors Hahn and Dana. It served the district of Congressman Glenn Anderson, campaigner for federal rapid transit funds for Los Angeles. Anderson's district was not to be served by any of the lines for which federal funding was sought, and it was necessary to give him something in return. Anderson "loved pork barrel projects and he was a trader in that sort of thing," according to Jacki Bacharach.

Assemblyman Young's Long Beach hearings were referred to in an LACTC staff report shortly afterward: "That hearing focused on a potential Los Angeles–Long Beach line. While that line may have the best potential for early implementation, that choice should not be made in Sacramento."[172] By then, however, Caltrans was already conducting light rail engineering feasibility studies for the Long Beach line (and also one to follow Santa Monica Boulevard).

On September 23, 1981, LACTC unanimously passed a motion to "perform a detailed evaluation of the Long Beach to Los Angeles light rail corridor . . . to determine whether it could be the initial line for construction of a low-cost rapid transit system complementary to the Wilshire Starter Line," and to "evaluate other available rapid transit rights-of-way and transit technologies countywide."[173] While other corridors were certainly to be considered, it was clear that the Long Beach line was to be the focus of attention.

As if this was not enough, Assemblyman Young appeared at the October 14 commission meeting to present his proposal, under which "The Los Angeles to Long Beach line, covering a distance of 21 miles and costing approximately $168 million, would be the first line constructed."[174] After Chairman Russ Rubley had presented Assemblyman Young a "Red Cars" T-shirt from the commission, he made a motion to authorize the L.A.–Long Beach light rail line. The motion was seconded by Supervisor Hahn, replacing a resolution Hahn had himself prepared calling for the same thing. Both motions were, however, referred to the Ad Hoc Rapid Transit Committee, rather than receiving a direct commission vote. "Mr. Remy asked the Commission to reflect on Assemblyman Young's proposal."[175]

At the October 28 meeting, Chairman Rubley's motion was brought back to the Commission, with wording amended by the ad hoc committee.[176] It now moved to "authorize evaluation of the L.A.–Long Beach light rail line; such evaluation shall be carried out in a timely and expeditious fashion which will enable the LACTC to program the project' if it so chooses, in the next State Transportation Improvement Program (STIP) cycle."

The motion was adopted without objection.

Analytic work was performed by consultants Parsons-Brinckerhoff, but as one then LACTC staff member speaking not for attribution said:

> You can do those reports and find any conclusion you want to find and we knew before we started, at least I knew in my own mind, that the conclusion was going to be light rail Los Angeles to Long Beach. No two ways about it . . . Here's a line. And maybe it's not the best line, but it's the one that the

technical people are saying can be done relatively cheaply because you've got all this wonderful right-of-way. We don't know whether anybody'll ever ride it, but that's another story. It goes through a minority part of town, which is good. It goes downtown, which is good. It goes to Long Beach, which is good. Politically it worked . . .

I've always believed that if you could have had a secret vote as opposed to a public vote, the project never would have been approved. I don't think there were more than three votes on that Commission for that thing.

Long Beach Mayor Kell's explanation for the choice of the Long Beach corridor to go first was quite consistent with this: "It probably was the first corridor," he said "because initially they thought it would be the cheapest one, one they could get in place easiest initially, probably with the least amount of political problems, cover the longest distance, and the cheapest. So they could show accomplishment, also." A senior member of LACTC staff conceded that he thought the "Long Beach line was chosen not because it is the most cost-effective line. I think there are many other lines that are more cost-effective. Perhaps it was the most politically necessary line."

Commissioner Christine Reed also took up this point:

The reason that the Long Beach line got picked is because it ran through the district of the Chair of the Assembly Transportation Committee, and Supervisor Kenneth Hahn—who is the father of Proposition A—wanted it there . . . and the City of Long Beach wanted it there. And the Commission—or the transportation community of Los Angeles—does not feel brave enough to stand up to the Chair of the Transportation Committee of the Assembly. Or was not at that time. I personally don't think it still is today brave enough. So that what it was; was a small example of pork-barrel politics . . .

And so the decision was a basically straightforward—in my opinion—pragmatic political decision, which was never discussed in public, but you know that is the clear reason that this line was picked, because of the political forces, the elected officials that were lobbying for it, and also the city reps.

Were people not willing to question this? I asked longtime RTD Board member Marvin Holen "They don't have the moral strength to stand up and say no," he replied. "Everybody's got their agenda. And you begin to look at the things on the RTD agenda, which become endangered because you may get into a squabble with the L.A.–Long Beach line."

On March 17, 1982, Executive Director Rick Richmond transmitted a staff report to the Commission, saying in his cover letter that "Staff believes that our intensive four-month evaluation has shown that the Los Angeles–

Long Beach line can serve as a start toward a more comprehensive system, that other potential lines also show promise for relatively low cost rail transit it would not be prudent for us to pursue alternative transit technologies [to light rail] at this time."[177]

With the legality of Proposition A yet to be confirmed, the staff report concerned itself with funding availability. The Long Beach line, assuming a light rail technology, had been estimated by Parsons-Brinckerhoff to require total capital costs in the range $254–280 million, although two other corridors were seen to be less expensive. A series of criteria were ranked for each project under consideration:

> Each has its own strengths and weaknesses, making it difficult to single out only one as being "best" in some absolute sense.
>
> We would argue, however, that it is neither necessary nor desirable to choose the "best" at this time, although we feel that Corridor No. 1 [Los Angeles–Long Beach] probably ranks highest *by virtue of the strength of expressed support* [my emphasis] and its potential for serving as the "spine" of a larger system which could serve all of Southern Los Angeles County.[178]

On March 17, the Ad Hoc Rapid Transit Committee reviewed the staff recommendation, concurring with staff, and issuing joint recommendations that "The Commission should proceed with engineering, environmental assessment, and right-of-way negotiation on the Los Angeles–Long Beach light rail line and declare its intent to construct this line presuming cost, patronage, construction schedule and environmental impact presently known is confirmed on completion of the above activities."[179] The committee also called for engineering, environmental assessment, and right-of-way negotiations for three other lines.

On March 24, the full commission met to reach a decision. Supervisors Dana and Hahn—into whose districts the Long Beach corridor falls—were particularly vociferous. "I feel very strongly about the light rail," said Supervisor Dana:

> I think it is absolutely mandatory to begin some transportation development in metropolitan Los Angeles County. The Long Beach to Los Angeles proposal is a good starting point because it connects two major population centers . . .
>
> The Long Beach to L.A. rail line will provide a key transportation link to recreation and employment centers along the way. This does not mean we do not want an underground rail system. The Wilshire Subway is highly expen-

sive, offers no connections to other areas and both are very important given the consideration the need to make maximum use of limited dollars. Light rail, in contrast, can be more readily extended to other areas of the county, thereby serving far larger population segments. We need to look no further than our clogged freeway system any working day to see the need for transportation alternatives. A commitment to a light rail system puts us on the road to providing fast, safe, and modern transportation in a system that can be tied with existing bus transit programs and thereby relieve the heavy load that exists on our freeways today.[180]

Supervisor Hahn spoke next, sounding a catalytic rallying cry:

I think today can, I hope, be a historic day . . .

We can do something or we can have another study. We as public officials cop out having a consultant tell us what we already know what to do. We spend millions of dollars on plans to say let's have mass transit again. We have the great harbor of Long Beach/San Pedro, it's a part of our defense, our economy. We have Union Station in the center part of Los Angeles. In 22 miles distance, we could have a mass, fast, even a bullet-type train for transportation in light rail. It can go in 22 minutes with proper grade separation. We could really break through the barrier of transportation in Los Angeles if we had the will to do it. But, there has been so much jealousy and bickering as to what route should go first. Then nothing moves . . .

Figure 6-3. Artful design in the Red Line. (Photo courtesy of author)

The Long Beach/Downtown through the Long Beach Avenue, the right-of-way is there, enough for four tracks, two in each way, you could have an express, fast bullet-type train and a local one to serve all the local communities. We've got the money sitting in the bank. If we don't spend it, it will go back to the State. The people, by 54 percent, voted for better transportation. It was the only measure voted for a tax measure that has passed. The freeways are jammed every night. I don't know how any of you feel when you go home on the freeway. It is bumper to bumper still . . . I would like to see this program go ahead with dispatch.

I know this is a compromise . . . I've been in politics long enough to know that sometimes you have to give a little to get a lot. If we can bring in these other engineering studies as a symbolic sign that we are concerned with these areas next time, in the future, then this is a good program. This means in effect that I hope the policies, Mr. Chairman, because you will be the chairman in history when you say so, ordered for the first time the correct authority to get ahead and get going on it. At the same time, *you protect your area, every other route receives a piece of the action as they say. You can go back to their constituents and say we have a route planned for your area some day. But we must give first priority to the route that people can actually start riding on trains* [my emphasis]. I hope that we can unite on this and I'm glad that Deane Dana came down to speak also on this. He represents both Long Beach and San Pedro Harbors. You know what the Navy is doing down there. You know the potential tourist trade that will be there. You know the downtown interests. This is a real shot in our economy, and I think a signal to the United States that Los Angeles will not talk anymore, but doing something for mass transit. Thank you very much, Mr. Chairman.[181]

Alternate Commissioner Pierce, a member of the ad hoc committee said that "as a resident of the San Fernando Valley, [he was] happy to see that the commission is looking at the possibilities of light rail in the San Fernando Valley as well as in the other areas." Chairman Edelman responded by saying that:

We do have, I believe, a recommendation of the Ad Hoc Rapid Transit Committee that will answer the needs of not just the Long Beach/Los Angeles area, but these other areas shown to be at least preliminarily feasible . . .

Certainly, today is a historic day, as previously indicated by Supervisor Hahn and my colleague Deane Dana, and we are indeed fortunate that the Commission has been able to hammer out a consensus. That has not happened in the past. We've had people in high office fighting over one route or another route, sometimes sidestepping this Commission, going around this Commission, but because of this Commission, we have brought different interests together and we have hammered out a consensus position that we can all agree to and that we can get public support for.[182]

Commissioner Rubley said he saw the day as "Red Car Day" and also as the commission's "graduation day, in graduating in the fact that we are now stop talking and we are going to act." Supervisor Hahn now returned:

> Just to conclude, I've just been handed a petition that I would like to file. The list is about 500 names supporting this action for the Long Beach/Downtown rail line priority. The signatures are from cities including Whittier, Downey, Glendale, Manhattan Beach, Monrovia, La Habra, Arcadia, Montebello, Reseda, Glendora, Woodland Hills. I think there is a community-wide interest in this project.[183]

Chairman Edelman followed by calling for a vote: the motion to proceed with the Long Beach line, while doing preliminary work on others, passed unanimously. The decision to bring light rail back to Long Beach had been made.

Bus Fares Down; Rail a Priority

As preparations for the Proposition A bus fare reduction to take effect on July 1, 1982 were underway, LACTC staff expressed concern over publicity being given to the fare decrease: "They assume ridership will go up 10–15 percent without such promotion. If the promotion results in a 25–30% increase, the cost of providing the increased bus service will subtract substantially from the funds available to the Los Angeles/Long Beach rail line and other transit projects during the next few years."[184]

It was clear where priorities lay—with rail—and made clearer from the choice of music references in the *Los Angeles Times* report of June 28: "To the background strains of 'Chattanooga Choo-Choo,' political friends and former foes of Proposition A stood together last week in Arco Plaza, taking and sharing credit for what banners and straw hats called 'the new era in public transit.'" But, as KTLA-TV (Los Angeles) reported, "to commemorate this giant step in local transportation, Hahn, Edelman and Bradley hopped onboard a bus to try out the new fare. Hahn had to borrow his fifty cents from an aide."

Documentation shows that, while his own proposition formula had directed funds toward bus fare reductions in preference to bus service expansion, Supervisor Hahn did make constant efforts to have bus service added to serve his constituents. In a letter to then RTD General Manager John Dyer, dated July 19, 1982, for example, he stated: "I noticed on both Saturday and Sunday that the bus line on Vermont Avenue from 120th to Los Angeles is heavily loaded and this reminded me to ask you if you have added any additional buses on any heavy line where the passengers have to stand." On October 13, he wrote again: "I am concerned that you have not added enough buses for the public. There seems to be a reluctance on the part of RTD's staff to encourage the people to use the buses."

Dyer responded on November 11, citing "restrictions associated with Proposition A funding." These "restrictions" were cited in a letter from UTU General Chairman Earl Clark, to Supervisor Hahn, dated March 14, 1984:

> The current level of RTD service is appalling. Schedules are impossible to meet . . . thousands of riders are being passed up daily . . . and buses are overcrowded to the point where the possibility of a major catastrophic accident is very real.
>
> It is our understanding that the Commission's refusal to permit the RTD to schedule the service needed to accommodate the increasing ridership is based upon a Memorandum of Understanding which limits the number of service hours the District is permitted to operate. This was apparently done to permit the Commission to squirrel away as much as possible of the Proposition A funds for the Metro Rail Project.

Environmental Impact Assessment work for the Long Beach line was carried out; then, on March 27, 1985, the LACTC "basking in the glare of television lights and amid self-congratulations" unanimously approved the Long Beach project for final design and construction.[185] On October 31, 1985 ground was broken on the project. On July 14, 1990, to great fanfare (see epilogue), the line was declared open.

The Wilshire Line Follows

With the Reagan Administration firmly opposed to transit funding, congressional action played an increasingly important role in pushing funding and projects forward. A 1982 reauthorizing act set federal transit funding $1 billion higher than the Administration had opposed for each of four years.[186] The pattern of previous legislation limited the effectiveness of Reagan opposition. Ironically, a preference by the Nixon Administration for having specific program decisions made at the local level resulted in the federal government being presented with specific projects to fund, rather than with the role of developing alternative strategies which might have been more effective on a national as well as a local scale.

Because of the principle of local decision-making on preferred projects, local interests could still call for rail even when technical documents showed advantages to bus over rail operations. In St. Louis, indeed, environmental work found a TSM alternative based on improving existing bus services, providing traffic preemption for buses and other modest improvements "more cost-effective in meeting transportation objectives than the other build alternatives." Both the TSM and a busway alternative were found to generate higher ridership than light rail, and at lower capital costs, while operating costs were similar. This did not stop the Preferred Alternative Report calling for light rail on a number of vaguely expressed grounds such as "more potential for economic development," "Improved image increasing St. Louis' ability to compete for convention and tourists," "greater improvements to transit reliability," "travel time savings," "increased accessibility to jobs," "increased accessibility for the transit dependent," and "reduced bus congestion in St. Louis."[187] It was not for the federal government to tell St. Louis their assessment was wrong but only to decide whether or not the city should receive funding for the rail system it wanted.

The Reagan Administration tried to present road blocks to local rail preferences, with the Transit Administration frequently critiquing and demanding changes to technical documentation. Local preferences nonetheless prevailed in determining preferred projects and, if the Administration was not going to cooperate with project funding, local officials could go directly to Congress. The pattern of providing for extensive local, rather than state, involvement, and the direct tie between local officials and their federal elected representatives provided the strongest link in making rail projects a reality.

While UMTA's program had been largely unaffected by congressional earmarking in the pre-Reagan years, the situation changed with the imposition of the rail-building moratorium. The appropriations committees' conference report for the fiscal year 1982 directed UMTA to fund alternatives analyses for six specific cities and to fund actual construction in five others (two in Miami alone). The 1983 report earmarked funding for alternatives analyses in seventeen cities and for engineering or construction in eleven—with a distinct bias to committee members' home states. UMTA and its authorizing committees opposed this earmarking, but could not stop it.[188] Notable rail openings during the Reagan years included heavy rail and a peoplemover in Miami, and light rail in Buffalo, Pittsburgh, Portland, Sacramento, and San Jose. San Diego, the first line of which had gone ahead without federal aid, opened a federally funded extension, while Washington METRO opened extensions and planned more of them. St. Louis had money earmarked for work on its light rail, and Los Angeles began building its subway, despite particularly vehement Reagan opposition.

Local opinions and lobbying from Los Angeles interests were critical to having Reagan's opposition defeated. Cliff Maddison, the Washington lobbyist who coordinated Los Angeles' lobbying effort, said in an interview that Democrats, Republicans, the Sierra Club, the Los Angeles Chamber of Commerce, the San Fernando Valley Industry and Commerce Association, and the *Los Angeles Times* all came together to support the project. Unlike the Long Beach Blue line, which leaves downtown to head through low income neighborhoods, the Red Line was planned to serve prime areas of central Los Angeles real estate as well as the San Fernando Valley, providing a substantial draw to business and real estate interests. Many construction companies which stood to gain were members of the Los Angeles Chamber, adding to business support. Unions were also highly involved, with testimony from Bill Robertson of the AFL-CIO stressing the advantages of job creation in addition to the environmental benefits of proceeding with rail.

According to former Chair of the RTD Mike Lewis (who was also deputy to conservative County Supervisor Pete Schabarum), "the whole Reagan position on anything was that we have to cut spending," and that lobbying directly in Congress was the only way to get around this. Regional loyalties were played upon and Republican support garnered too, from Representative David Drier of the East San Gabriel Valley, for example. "I think ultimately the argument that prevailed with the Republicans was that Congress had appropriated the money and if it didn't get spent in California it would be spent somewhere else," Lewis said.

There was a period when Chairman Lewis and General Manager John Dyer would take the Sunday night Red Eye flight to Washington every week for a day of intensive lobbying each Monday. The round of visits would include the California congressional delegation and members of key authorizing and appropriating committees. Glenn Anderson's position on the Appropriations Committee was leveraged as much as possible, Lewis said. The presence of Lewis was especially important given his conservative credentials. The most important local pressure was brought to bear, however, by Los Angeles Mayor Tom Bradley, and Maddison says Bradley's efforts were critical to getting rail financed.

Visits to Los Angeles were arranged for an assortment of members of Congress, with an emphasis on the chairs and ranking members of critical congressional committees. "We'd take helicopter tours of where it's going and, as you can see on the Hollywood Freeway, it's bumper-to-bumper: it's jammed," said Maddison. Lewis says that the Los Angeles Chamber of Commerce, the Atlantic Richfield Company and other principal downtown employers backed these private trips and paid honoraria to members of Congress in addition to covering their travel expenses. Engineering and construction firms that stood to benefit from project contracts such as DMJM and Parsons Brinkerhoff chipped in as well, Lewis said.

Substantial efforts were made to provide a display of local unity. "Ralph Stanley had asked us to demonstrate that there was support for this thing besides the downtown business interests, so we put together a dinner with about forty different organizations—the State of California, Assistant Secretary, people from the homeowners' groups in the Valley, the Valley Industry and Commerce Association: we knitted together a pretty impressive coalition."

The 1983 Surface Transportation Assistance Act brought it $700 million for the first phase environmental study and planning. Given administration criticism of the Red Line proposal, RTD proposed an initial 4.4 mile "Minimum Operating Segment" from Union Station in downtown Los Angeles as far along Wilshire Boulevard as Alvarado Street. The idea was to get that portion funded and secure funds for further extensions later on. A federal full-funding agreement for the initial segment was signed in 1985.

President Reagan did not falter in his efforts to reduce transit financing, and vetoed passage of the Federal Mass Transportation Act of 1987 with its five-year reauthorization and generous transit funding. This act was to be critical in getting funding appropriated for the next stage of the Los Angeles subway, which President Reagan specifically identified as a reason for vetoing the act. Intense lobbying effort from Los Angeles and other interested rail-building cities had the president's veto overridden by the Senate. Reagan attempts to sway Republican Senators supporting the override were in vain.[189] Ralph Stanley, President Reagan's UMTA Administrator from 1984 to 1987 who tirelessly opposed new rail starts but was forced to proceed with them under congressional mandate, got his reward from Senator William Proxmire: The Golden Fleece Award for useless federal expenditure.

RTD's 1983 plan consisted of an 18.6-mile subway route that went out along Wilshire Boulevard before turning north to Hollywood, bending back eastward, and then heading northwest to the San Fernando Valley.[190] The route, known as the "wounded knee" alignment, bringing in Hollywood and San Fernando Valley interests as well as those in downtown and along Wilshire, was designed on the basis of maximizing local political appeal rather than transportation amenity.[191]

Controversy was to arise over methane and other potentially hazardous gases underground the Red Line's route. A 1985 fire in a store located sparked a particular scare. The actions of one Congressman, Henry Waxman, threatened to derail the Red Line's extension, allegedly over the gas issue, but Waxman was clearly worried about the views of wealthy constituents who would rather keep the project away from their residences. In 1986, Waxman succeeded in having a rider added to an appropriations bill to ban Metrorail construction until local officials could prove the system safe from underground explosions.[192]

A new locally preferred routing was adopted by RTD in 1988. The Wilshire part of the line now terminated at Western Avenue, short of the methane gas zone. The Hollywood/ San Fernando route was rerouted to leave the Wilshire Line at Vermont Avenue and head north there to avoid the methane zone, creating two branches. The involvement of Congressman Julian Dixon, who had a political interest in having the line detour through his low-income district to the south of Wilshire (at the same time taking it away from the area Waxman wanted avoided), was critical to the removal of barriers to federal funding.[193] Dixon agreed to include the subway line with such a detour in language for the Intermodal Surface Transportation Efficiency Act (ISTEA) of 1991. The LACTC 1992 Thirty-Year Plan referred to the new routing as having scope for revitalizing the Pico/San Vicente area, which is in the Congressman's district.[194]

Republican Representative Bobbi Fiedler joined Representative Waxman in providing Congressional opposition to funding Metrorail but, unlike him, was concerned with ideological issues of wasteful spending. She not only opposed a project which stood to serve her San Fernando Valley constituents, but campaigned and was reelected on the basis of it, doubtless helped by the generally bad reputation into which the project had fallen.

The Waxman/Fiedler efforts temporarily blocked Wilshire Line authorization, for which approval could not be found on the House Appropriations Committee until, Maddison says, "Tom Bradley came back and met with Jim Wright, the majority leader/speaker, a Democrat from Texas. [Then] we won." The new route was approved and received federal funding in August 1993, with thought of further extensions directly along Wilshire Boulevard dropped, at least for the time being. That same year, the Los Angeles County Metropolitan Transportation Authority was formed from a merger of LACTC and RTD.

The MTA's first CEO, Franklin White, appointed by Mayor Bradley in 1993 as he was leaving office, only lasted two years. Faced with the daunting task of creating a unified agency from two former agencies, which had frequently been in disagreement, White failed to stand up to the board, which was not only bickering amongst themselves but interfering with his ability to manage, according to two MTA sources. While funding uncertainties loomed increasingly large, nothing was done to face a gap which would at best require curtailment of the rail program. The twenty-year plan, *Transportation for the 21st Century*, warned about a need for retrenchment in the face of funding realities, but included a San Fernando East-West rail line to the 405 freeway and Red Line extensions west on Wilshire to the 405 freeway/UCLA and east to East Los Angeles, in addition to the northern extension of the Blue Line to Pasadena already a part of the baseline rail system. An assortment of other rail lines were also included for consideration, with moves to consider low-cost options for constructing them.[195]

In September 1995, faced with a civil rights action which alleged that the MTA had improperly run down its bus system through its focus on rail transit, and thereby hurt the interests of low-income and minority people, the MTA board adopted as its "highest priority improvement of the quality of bus service in Los Angeles."[196]

Further problems with Metrorail, including subsidence on Hollywood

Boulevard—with buildings beginning to sink because of subway construction and creation of a sinkhole—plagued White, who fired the project contractor. The show *60 Minutes* ran a slot ridiculing the subway, the federal government suspended all Metrorail funds; and White was fired. Joe Drew was appointed the MTA's next CEO in 1996, but did not last out the year, leaving amidst a scandal over a one-third of a million dollar consulting report, the recommendations of which were rewritten the day before a critical board meeting.

Nineteen ninety-six was also the year of signing of a consent decree which resolved the civil rights lawsuit. The decree froze fares (already the subject of a court restraining order), which MTA had been seeking to increase, and required MTA to purchase buses to reduce overcrowding and improve services. Bus service improvements were to be specifically targeted at the transit-dependent communities of Los Angeles County.[197] The demands for bus service improvements together with the loss of a planned fare increase and a recession which was hitting sales tax revenues made the MTA's financial situation tighter.

On January 6, 1997, Federal Transit Administrator Gordon Linton wrote to MTA Chair Larry Zarian to express his concern over whether the MTA could fund its twenty-year plan. The MTA's ability to meet its obligations under the Red Line funding agreement with the federal government appeared particularly questionable "given the insufficient sales tax revenues, the overruns in operation budgets and the recent court-ordered consent decree for additional bus operations." Also expressing concern over "the inability to obtain competent construction management staff," Linton said he was "alarmed by the lack of consensus of purpose among members of the Board and the lack of political will that is inherently necessary to address these kinds of challenges if the Red Line is to move forward as originally envisaged over a decade ago."[198] The MTA adopted a recovery plan for the Red Line, transmitted by Linton to Zarian on January 15.[199] Also in January 1997, Linda Bohlinger, formerly deputy CEO was named interim CEO of the MTA. Bohlinger had come up through the former LACTC, and was an avid supporter of rail transit.

Administrator Linton wrote back to Zarian on April 9, 1997, complaining of "serious deficiencies and questionable assumptions in the proposed recovery plan, both technical and financial—and those deficiencies have yet to be rectified."[200] Linton expressed his concern at:

... the absence of cash reserved to accommodate increases in capital or operating costs, further shortfalls in sales tax or fare revenues, or reduced appropriations by the Congress ... We find, furthermore, that the proposed recovery plan assumes cost savings that are overly optimistic ...

To be candid, MTA has yet to demonstrate that its capabilities of long range planning have improved since previous efforts, which have fallen short of pragmatic financial constraint ...

We are incredulous that, despite the engineering and financial difficulties on the construction already underway, the Board is contemplating even more requests to the Congress for various costly extensions to your rail system. Moreover, MTA has apparently agreed with the City of Los Angeles that the City's future payments for MOS-3 will depend upon MTA's pursuit of a San Fernando Valley extension to the Red Line.[201]

Linton demanded a thorough analysis of the issues raised, of how the recovery plan would be revised to meet its deficiencies, and an indication of resources for this purpose.

Come August 1, 1997, Leslie Rogers, FTA Regional Administrator, wrote to Bohlinger stating that since Linton's April 9 letter, "numerous meetings have been held and countless documents have been exchanged among and between the FTA and its contractors and the MTA. However, as of this time the MTA has not produced a recovery plan that is financially and technically responsible from FTA's perspective."[202]

The same month, Julian Burke was hired as permanent MTA CEO, and put in motion a radical change of direction for the MTA. Coming from the private sector and with a background in corporate turnaround, he quickly took a businesslike approach to straightening out the MTA. In particular, he established credibility with the board, making it understood that they were not to interfere with his authority to manage. He appointed Allan Lispky, an expert in financial restructuring, as Deputy CEO, and set to work devising a realistic and financially balanced plan for the MTA.

Burke's first priority was to suspend all rail projects other than the North Hollywood extension of the Red Line, which was under construction. At a special board meeting on January 14, 1998, the board at first rejected a motion authorizing the CEO to suspend two Red Line extensions and the Pasadena Blue Line project (a light rail line traveling north from downtown Los Angeles, not actually planned to physically connect with the Long Beach Blue Line). It subsequently approved a motion allowing the CEO to suspend the projects for at least six months, while supporting efforts to preserve the federal and state funding programmed for those projects.

On May 15, 1998, MTA submitted its latest restructuring plan to the FTA, underlining the suspension and demobilization of all rail projects other than North Hollywood, while showing that funding sources for the remaining construction to North Hollywood was firm, and noting that as heavy civil construction was nearing completion, risks of large overruns were low.[203] On July 2, 1998, Gordon Linton accepted MTA's rail recovery plan.[204]

In further action, Proposition A of 1998 passed by a two to one public ballot, preventing the use of transportation sales tax revenue for any further underground rail construction. It had been sponsored by Los Angeles Councilman Zev Yaroslavsky "the minute Zev saw that the subway was not going to get to the Valley," according to Tom Silver, former deputy to County Supervisor Michael Antonovich. The proposition meant that while construction on the North Hollywood extension, of value to Yaroslavsky's Valley constituents, would continue, the other Red Line extensions could not. The local coalition could only last as long as there appeared to be something for everyone. Voters, subjected to the virtually daily lurid accounts of subway project mismanagement, found it easy to stop further subway construction.

Also in 1998, a member of the state assembly (and now a member of Congress), Adam Schiff, who represented Pasadena, introduced a bill to get round the MTA's suspension of the Pasadena Blue Line, now renamed the Gold Line, by taking it away from the MTA and giving its management instead to a joint powers authority, the Pasadena Metro Blue Line Construction authority, to which programmed funds, including the MTA's own sales tax revenues, were reallocated.

In November 1998, MTA staff presented the results of a study which responded to a requirement of the May Restructuring Plan to evaluate viable transit development options for all parts of the county, with an emphasis on corridors with suspended rail projects.[205] The presentation called for the adoption of Rapid Bus programs to improve service to the transit dependent in all parts of the county, with priority to the East Side, Mid-Cities, and San Fernando Valley corridors with suspended rail projects. Rapid Bus, as described in chapter 3, was essentially a concept to introduce higher-speed limited stop bus services with priority traffic signal treatment and improved frequency, vehicles, and street architecture. Not seen as a permanent replacement to fixed guideway transit, it was capable of enhancement over time with the provision of bus lanes on existing streets or the construction of new exclusive bus rights-of-way, although the board was assured that proceeding with the bus program did not necessarily preclude the construction of rail later.

The big advantage of a Rapid Bus program was the speed with which significant service improvements could be obtained, and the low cost of achieving such results. The board approved adoption of a countywide bus system enhancement program, with the preparation of a Rapid Bus demonstration project of three lines, including at least one line serving each of the East Side, Mid-City, and San Fernando Valley corridors. In March 1999, the MTA gave the go-ahead to the Metro Rapid Bus Program, following completion of a feasibility study.

The initial two Rapid Bus routes, inaugurated at a capital cost of $8.3 million, connected the East Side to Santa Monica via downtown Los Angeles and Wilshire Boulevard, and provided service to the San Fernando Valley from the North Hollywood terminal of the Red Line. Both services began on June 24, 2000, the date that the Red Line started operations to North Hollywood, and quickly achieved striking ridership results. The MTA's 2001 Long Range Transportation Plan for Los Angeles County called for the "aggressive implantation" of a total of twenty-two Rapid Bus lines covering four hundred route miles, and consideration of larger-scale capital improvements, either bus or rail, in a handful of corridors.[206]

Further underlining its commitment to bus service expansion, in July 2001 the MTA Board approved the San Fernando Valley Bus Rapid Transit Project, a fourteen-mile busway connecting with the Red Line North Hollywood station using a right of way it had already acquired with the previous intention of supplying rail service. The Pasadena Gold Line began service on July 26, 2003, while the MTA hopes to begin construction of the Eastside Gold Line in fiscal year 2004, pending approval of federal funding which is being requested on the basis of prior federal commitments to fund a Red Line extension. A six-mile light rail extension to East Los Angeles (with only 1.7 miles in tunnel) is being pursued, given the prohibition on using local sales tax funding for tunneling on what was formerly planned as subway route (sales tax funding can be used for the remaining 4.3 miles). There is local interest, also, along with support from Congressman Schiff, for extending the Gold Line a further 22 miles from Pasadena to Claremont. Light rail and busway alternatives for an east-west alignment south of the Wilshire corridor are being debated.

On January 22, 2001, the *Los Angeles Times*, forgetting its own previous strong editorial rail advocacy, looked back to the "imaginative fiction" of the "long-range transportation dreams of the MTA and its predecessor agencies. Where else, for example, could you find a $183-billion plan for 296 miles of subways and light rail lines in 1992, in the middle of a post–Cold War recession?" By comparison, the *Times* said, the MTA's latest plan "has its feet firmly on the earth." The *Times* approved of the MTA's modest-cost Rapid Bus plans, and agreed that "the best, least expensive and most flexible way of testing new routes . . . is with buses first, with studies on potential fixed-rail routes to come later." It was soberingly clear that Los Angeles' original dreams of grandiose rail building were dead in both political and practical terms.

But What Is the Draw of Rail?

If we have observed the political actions that put rail in place, we are left with the question of why rail was seen as so desirable in the first place and why, in particular was the Blue Line such an attractive proposition given the lack of the powerful special interest support that was to attach itself to the Red Line?

What understandings of rail put it into proposal after proposal since 1925? Why did Baxter Ward want trains, not buses, and wish to limit the scope of future generations on how capital funds might best be spent? What attractions did rail have that allowed the commission to rally around this mode of transportation as its principal program when, as we have seen from the analysis in earlier chapters that for the dispersed Southern California community, it makes little or no sense?

To probe the lure of rail, we need to examine the mythology which tells us why it seems attractive. We will then see how myth and political power can powerfully combine to obtain support for programs of little public benefit.

CHAPTER 7 **A Theory of Myth**

> We are often not aware of the ideological blinders that we wear because
> of our presence in a particular culture at a particular time and the
> particular training and experiences we have had.
>
> —Kramer

WE LIKE TO BELIEVE we do things for reasons. We study problems, come up with ways to tackle them, and choose the solution which seems best. Economists calculate which project has the best payoff—in economic terms. Politicians work out which one offers the most—in political terms. The economic and the political interact, and policy is supposedly made. As Graham Allison puts it, the central theme of such a "rational actor" view of policy is the applicability of aims and calculations to explain how decisions get made.[1] We have seen, however, that from an economic standpoint, the decisions made in Los Angeles don't add up. The Proposition A light rail system offers few benefits and at massive cost. The money could have been much better spent in other ways.

Powerful political processes have been at play and the actions of a few key actors made the Proposition A system reality, whether or not it was economically justifiable. Why politicians found rail to be in the best interests of either their constituents or themselves has yet to be explained.

The following four chapters aim to surface the generally tacit mechanisms by which decision makers inform themselves and make decisions. This involves developing and applying a theory to locate such mechanisms, and then testing whether these mechanisms do in fact guide decision-making. This chapter develops a theory of understanding which sees comprehension—and subsequent action—not in terms of the "facts" of economics textbooks and civics classes but of a potent mythology which explains the way things are and how they can be made better. Understanding is seen here as a function of one's position and experience in a particular culture.

Understandings are realized through cognitive processes which cut intelligible patterns out of what would otherwise be an unbearably complex and unstructured fog of perceptual phenomena. Symbolic processes—driven by symbols, images, and metaphors—play a central role in steering the cognitive knife, providing a basis for concept formation, for the understandings which those concepts entail, and for the formation of powerful myths which appear to represent reality.

The processes operating under these theories are not necessarily "illogical," except relative to the expectations of analytical reason, but conform to a different type of logic—a logic which both structures each symbol, image, or metaphor and binds them together to create myth. We must understand the nature and mechanism of this logic if we are to make sense of decisions made by humans, not analytical logic machines.

Cognition under Complexity

Experience

Everyday life would be impossible if every action or sensation were to be intensely examined or closely considered. "Like the centipede, confused by self-consciousness, everyone would be in-capacitated by complexity," writes Edward de Bono.[2] "The function of thought is to eliminate itself and allow action to follow directly on recognition of a situation."

Philip Morrison demonstrates both the value and the pitfalls of our everyday elimination of thought. "Routine physical chores, reading and writing, the complex web of relationships with family, friends, even pupils in the classroom, all those depend upon many skills of hand and eye and of habits of mind and speech, skills that are predictive, effective, and widely shared."[3] Common sense plays a vital role in permitting this to happen. Common sense is derived from experience and tells us which actions are most likely to succeed. Most of the time it serves us well. Such common sense tells us that the "air is invisible but always ambient; when one needs it freshened, the window is opened. Yet a glass ready on the shelf is, and always was, regarded as a matter of course to be empty, never as filled with air."[4]

Common sense actions

seldom need explicit calculation, nor is there any desire to pose sharp logical tests of the comfortable and usually adequate presuppositions for action . . .

What is involved are rough conclusions about a wealth of distinct details, rarely step-by-step paths to long-pursued ends . . . Neither generalization nor objectivity nor precision are very important to the common-sense frame of mind. . . . The light is almost always there in the room without delay once we snap the switch or kindle the candle. Whence and how light moves is not even asked and would indeed be hard to answer through our commonsense perceptions.[5]

While common sense interpretations and prescriptions guide us well much of the time, we can be deluded when extrapolating them to new domains, particularly when applying them to problems whose structure we have not probed, and which we do not understand except in terms of raw observation and "common sense" deduction. It was common sense, for example, to Melanchthon, with whom this study opened, that the Copernican argument was wrong, since "the eyes are witnesses that the heavens revolve in the space of twenty-four hours." While such decisiveness under conditions of limited knowledge is puzzling from an analytical perspective, it can be more readily understood through cognitive theories, which focus on the workings of the mind. As Christopher Alexander suggests, "The mind's first function is to reduce the ambiguity and overlap in a confusing situation . . . It is endowed with a basic intolerance for ambiguity."[6]

In a color-blindness test, a person with normal vision will pick out the structure of number or letter symbols hidden in a mass of multicolored dots; the surrounding dots are ignored as noise. To make sense of reality, the mind imposes structure and, as John Steinbruner suggests, much of such work is done "apparently prior to and certainly independently of conscious direction."[7] Such structures may seem to produce obvious solutions to complex problems, while the more intangible but crucial issues are never summoned before the mind's eye.

Simplicity

In Friedrich Nietzsche's words, we continually need a "narrower, abbreviated, simplified world."[8] The imposition of organizing and simplifying structure is basic to understanding. The interrelationships and overlaps of complex problems presents a confusing spectacle unless it is broken down in some way. Steinbruner gives the example of memorizing a series of digits, where a structure or rhythm is established and used as an aid to memorizing the whole.[9] In this case, the imposition of structure performs a useful heuristic function because it allows us to replay the original without incompleteness or bias.

With more complex problems, and when there is simply too much information for the mind to absorb all at once, the structuring can be less helpful. Parts, de Bono explains, are "extracted from the whole situation and then fitted together by means of fixed relationships to re-create the whole. . . . The choice of parts into which the whole is dissolved is dictated by familiarity, convenience and the availability of simple relationships with which to recombine them."[10] No conscious choice of the most important parts or how they are to be ordered gets made. A tacit perceptual process organizes information in ways which are easy to understand and which seem to make sense.

According to the "simplicity" or "Minimum" principle associated particularly with the Gestalt School, "we organize our percepts so as to minimize their complexity."[11] "Just as a soap bubble achieves the simplest possible configuration (e.g., the most symmetrical) without the need for goals or purposes, so does perception work automatically toward the good figure."[12] Figure 7-1, from Julian Hochberg, shows a series of "depth cues."[13] Each can

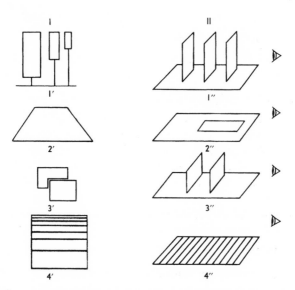

Figure 7-1. Depth cues. (PERCEPTION 2/e by Hochberg, J. Reprinted by permission of Pearson Education, Inc., Upper Saddle River, NJ)

be interpreted as rather complex forms in two dimensions or as simple forms in three dimensions. We invariably perceive the simpler alternative.

In fourteen out of the seventeen patterns tested in an experiment in which subjects connected dots of a pattern so as to reflect its perceived organization, "the most popular organization was also one with the shortest path length."[14] And, as Roger Shepard concludes, "Subjects uniformly use quite definite, though unconscious, rules to select (from the infinite variety of three-dimensional worlds consistent with a given retinal projection) the one that is most regular in some way—the one in which, for example, as many as possible of the lines are straight and parallel, the corners are right-angled, and the overall structures are symmetrical."[15]

As James Pomerantz and Michael Kubovy suggest, perception will fill in "missing" information so as to simplify representation.[16] It may also ignore or distort stimulus information toward the same end. Selecting the most parsimonious guess about the environment "may often be a useful heuristic. But sometimes the less parsimonious guess is correct." The concept that the heavens "rotate" around the Earth is a simple and intuitively attractive one; it is verifiable, furthermore, by the direct experience of the eyes. The correct explanation is far less mentally accessible and requires a higher degree of abstraction in order to be appreciated.

The desire for simplicity has the added effect of focusing attention on the concrete, rather than the abstract, as outlined by Martin Wachs and Joseph Schofer in their discussion of transportation planning:

> The evaluation of transportation plans has seldom, if ever, been based upon a rational inquiry into appropriate goals, meaningful objectives, and logical criteria which result from the chain of dependence relating these to the societal values. Objectives and criteria, rather, have been narrow in scope and have reflected the intuition and biases of planners who usually did not have explicit statements of goals and objectives toward which to work. This is not surprising, since it is undoubtedly simpler to deal solely with concepts for which there are physical referents than to try to relate abstract concepts such as security or belonging to the design of transportation systems.[17]

Common Sense in Transportation

The principles elucidated above are also central to the theories of symbolic forms to be developed below, and they are best illustrated later in the light of the full theory. Here, nonetheless, are some of the common sense arguments frequently given in favor of building rail transit, rather than taking other actions. Firstly, it is common sense that trains are preferable to buses because they can travel faster. It is common sense—and observed from daily experience—that the freeways are jammed up and that some alternative is needed. While buses must use roads, trains have their own right-of-way and they do not get caught in traffic.

Los Angeles is lagging behind the rest of the world: all successful cities have rail transit. Bringing rail to Los Angeles will make Los Angeles successful too. The monorail at Disneyland unfailingly works well and it's a lot of fun; rail systems in other cities provide comfortable, frequent, and fast service, and something like that could put Los Angeles on track, as well. People absolutely hate having to drive on the congested freeways, and will jump at the opportunity to ride on something clean, modern, and fast. It's common knowledge that buses are dirty, smelly, and slow, and that they attract crime.

Trains have only one driver to haul up to four hundred people. Obviously trains are a cheaper operation than buses which require ten times as many drivers to move the same number of people. Cities with rail systems, furthermore, have more orderly patterns of development. Rail transit can help contain L.A.'s urban form.

All these observations come from daily life, and they seem to make a lot of sense. They come together coherently, furthermore, making a powerful case for the construction of rail transit. "All I have to go on is my gut-level ordinary you know citizen, policy-maker housewife reaction," said then Mayor of Santa Monica and LACTC commissioner, Christine Reed, referring to an idea by SCAG Executive Director Mark Pisano to electrify and automate freeways. "It is too Star Wars," she said. Most people who make decisions are ordinary people, not technical specialists. They generally have many issues to deal with, and are liable to be heavily influenced by the experiences of everyday life. To the traditional academic mind arguments from such experiences may sound preposterous. Perhaps because of that, the reasons for the formation of such views have not been studied; but, if those views are to be changed, such an understanding must be reached. The traditional academic response is to simply tell decision makers with such views that they are wrong; as a result the traditional academic has often been ignored. We need to understand both the basis of the decision maker's thinking as well as that of our own in order to give good advice.

The psychological theories outlined so far have made an important contribution to our understanding of how we understand. Yet, they stop short of examining the conceptual building blocks with which we simplify the complex world we face and come to be "at home" with it. To this end, the next sections will outline a theory of understanding rooted in the idea of a "symbolic system." At the root of the symbolic system are symbols which lead us to conceive meanings. Symbols represent meanings in terms of images which, it will be argued, are themselves symbolic. Images act as symbols to create simplified world-views. As such they also behave metaphorically. Metaphor, it will be suggested, forms the basis of human conceptualization of knowledge, and metaphor, it will be shown, leads to the myths which provide the shared understandings which allow us to function as a society.

Symbols

What are symbols? How do they work? And by what criteria may they be identified in action? These questions must be addressed before using the concept of the symbol to interpret understandings at work in the formation of transportation policy in Southern California.

The Symbol Gives Rise to Thought

The symbol is the special preserve of the human. As Paul Ricoeur says, "the symbol gives rise to thought."[18] It does not, however, necessarily make humans more thoughtful. Indeed, Ludwig Von Bertalanffy claims that:

> Man is a symbol-creating, symbol-using and symbol-dominated animal throughout—but the use he makes of this remarkable achievement is only a minute part rational. Symbols—mink coats, flags, anthems, television advertisements, political catchwords, deterrence and what not—determine individual and social behavior only too often to the detriment of the individual, of society, and of humanity.[19]

As Murray Edelman observes "It is characteristic of large numbers of people in our society that they see and think in terms of stereotypes, personalization and oversimplifications, that they cannot recognize or tolerate ambiguous and complex situations, and that they accordingly respond chiefly to symbols that oversimplify and distort."[20]

Signs and Symbols

Although some commentators (Von Bertalanffy, for example) view symbols as a special class of sign, a helpful way to understand the symbol is to contrast it with the sign, as does Susanne Langer.[21] A sign, she says, is merely indicative. It announces the presence of something. Wet streets, for example, are a sign that it has rained. A smell of smoke signifies the presence of fire. "The interpretation of signs," writes Langer, "is the basis of animal intelligence," but humans do "the same thing all day long. We answer bells, watch the clock, obey warning signals, follow arrows, take off the kettle when it whistles, come at a baby's cry, close the windows when we hear thunder."[22]

A sign is associated with its object to form a pair. To each sign there is a corresponding object of greater interest than the sign itself. The fact that there is fire, for example, is more interesting than that we smell smoke.

A sign acts as an announcer to a subject. But symbols lead the subject to *conceive* their objects:

> Symbols are not proxy for their subjects, but are *vehicles for the conception of objects.* To conceive a thing or a situation is not the same thing as to "react toward it" overtly, or to be aware of its presence. In talking about things we have conceptions of them, not the things themselves; and it is the conceptions, not the things, that symbols directly "mean.". . . The sign is something to act upon, or a means to command action; the symbol is an instrument of thought.[23]

Our understandings of flags help show the difference. A raised white flag is a sign of surrender. A black flag is the ensign of a pirate ship. A dropped flag signifies the start of a race. A "flag stop" is a station where trains only stop upon being signaled. A flag sign in a railway timetable is an indication that one must signal to stop the train. These are all uses of "flag" as a sign in Langer's terms.

Yet there is also the symbolic in most of these flags. It is no accident that a flag of surrender is white: white symbolizes purity, innocence, nonaggression. Looking at a white flag, one receives more than the simple information that "the enemy is giving up." One gets a *conception* of that surrender. Similarly, the black flag does more than announce "pirates": it conceptualizes—it *symbolizes*—evil. What about the flag sign in the timetable? It may be no more than a conventional sign, but it can also have the power to be symbolic. It can trigger the conception of a sleepy wayside settlement, where most of the trains don't stop.

Only the flag which starts the race is nonsymbolic: It produces an almost reflex reaction. The flag drops and people start running. It has a transparent quality: one sees straight through the event of the flag dropping to the information it points to.

National flags are perhaps the most powerful symbols of all: they conceive a nation's identity, and are seen as something sacred and to be defended.

Symbols Act as Gateways

Although a symbol—unlike a sign—does not point to a specific object in a one-to-one relationship, symbols "may be part of a larger pattern or they may be connected to it. *They act as gateways to the larger pattern*" (my emphasis).[24]

Going through one gateway rather than another introduces one set of choices, rather than another. Although going through a gateway does not compel one to follow a particular path on the other side—Frederick Hacker notes that "symbolization has a compelling rather than a compulsory quality"—the likelihood is that once a particular threshold is crossed, the traveler will continue in the direction in which it leads.[25] The gate to the rose garden may lead to paths to several different varieties of rose. Once inside that garden, it is easy to become absorbed enough to spend the whole afternoon wandering around its offerings, perhaps trying several paths within its boundaries, oblivious to the yet more wonderful gardens which might lie through other gateways.

The Larger Patterns Are Culturally Established

As Ernst Cassirer suggested, symbols are creative functions of the individual mind and culture concerned. For Cassirer, as Von Bertalanffy writes, symbolic forms encompassed all activities characteristic of the human mind and culture: "They are not simply 'given' and committed for every human mind or mind in general, but develop in close interaction with the several fields of cultural activity."[26]

Symbols are interpreted individually; their meaning depends on the individual's membership in and experience of a particular culture: "The tensive symbol cannot be entirely stipulative, inasmuch as its essential tension draws life from a multiplicity of associations, subtly and for the most part subconsciously interrelated, with which the symbol, or something like it and suggested by it, has been joined in the past, so that there is a stored up potential of semantic energy which the symbol, when adroitly used, can tap."[27]

Despite the latitude for individual interpretation, symbols have the characteristic of demarcating the boundaries in which the "semantic energy" they hold can be released, thereby containing the scope of the conceptions they engender. Those boundaries are established historically and transmitted through tradition. Symbols thereby "permit contemplation of and action on objects that do not necessarily belong to the immediately accessible present. Thus symbolization is essentially connected with and constitutes the capacity to remember and to recall."[28]

Symbols Act as Blueprints

Symbols may come to be associated with particular conceptions through direct experience, but the blueprint for those conceptions may be determined historically by a series of events with which the present generation has no direct connection. In this regard, symbols can be said to act genetically. Just as an individual may carry traits of an ancestor he or she never knew, so symbols may carry forward a conceptual apparatus which arose in contexts quite unlike those in which they now operate, but which retains the power to direct meaning today. "The characteristically recurring feature," says Hacker, "is always the distance of the symbol from the symbolized, reflecting the autonomy of the symbol, namely its temporary or permanent emancipation and divorcement from the original context." The Homestead Act, for example, provided 160 acres of land free to anyone who settled and improved land, and 160 acres became and long-remained a symbol of self-sufficiency, "even though it is obvious that the amount varies according to the quality of the land and the type of crops."[29]

Roland Bartel's account of the symbol of the "Jordan River" in Negro spirituals also provides a compelling illustration of this principle. For the Israelites, the crossing of the Jordan was the culmination of a long and arduous pilgrimage from bondage in Egypt to freedom in Canaan, an experience that would suggest obvious parallels to the lives of the slaves. In the process of adaptation in the spirituals, the river and the crossing itself became the basic symbols, and the other details of the story vanished . . . Stripped of all localizing restrictions, the symbol acquired the flexibility and independence needed to fulfill a variety of functions.

Most frequently the Jordan River symbolized the dividing line between this world and the next, between servitude and freedom. It was the final step on the long way to salvation, and crossing it represented the final victory . . .

Sometimes the Jordan River became a symbol of difficulty in the struggle for a better life, as in "Stan' Still, Jordan," in "O, Wasn't Dat a Wide River?" and in "Oh the Winter, Oh the Winter." In other songs the river symbolized not the difficulty of reaching the promised land but rather the promised land itself, as in "Roll, Jordan, Roll."[30]

As Hacker notes, "In symbols, the idiosyncratic and personal is always intermingled with the general and universal."[31] The "Jordan River" symbol has an emotional impact on the individuals attracted to it. Through the sum of relationships of individual to symbol, however, the "Jordan River" becomes a general symbol of emancipation.

Symbols Convey Values

According to Von Bertalanffy, symbols can convey values and leave their imprint on emotions and motivations. "They enter into intimate connection with higher emotions or even animal drives."[32] The "Jordan River," for example, is tied to the value of freedom, provides a focus for the invocation of that value, emotionally revs up the congregation singing the spiritual, and motivates them to seek out that value. There are many examples of symbols arousing animal drives. Pinups of "sex symbols" and the Nazi swastika are cases in point.

Symbols often derive power from their simplicity of form, that is, from their manifestation of abstract concepts in easily identifiable, concrete forms. As Bartel points out, the phrases in spirituals that symbolize the difficulty of attaining the objects of one's dreams "possess a simple eloquence."[33] "Jordan River is chilly an' cold," and "Jacob's ladder is deep an' long," are examples.

Symbols Act in Specific Contexts

A discussion of the physical attributes of the Jordan River by two engineers would not likely arouse a yearning for freedom in them. Placing the Jordan River in the religious setting of a spiritual, however, unlocks its emancipatory semantic content.

Metonymy

"Metonymy is one of the basic characteristics of cognition," according to George Lakoff. "It is extremely common for people to take one well-understood or easy-to-perceive aspect of something and use it to stand either for the thing as a whole or for some other aspect of it."[34] George Lakoff and Mark Johnson point out a crucial difference between metonymy and metaphor. While metaphor is "principally a way of conceiving of one thing in terms of another," metonymy "allows us to use one entity to *stand for* another."[35] In a metonymic relationship between two elements A and B, B stands for A.[36] Lakoff gives as example a waitress saying "The ham sandwich just spilled beer all over himself." The ham sandwich stands for the person eating the sandwich.[37]

Pars Pro Toto

Martin Foss claims that "symbolism is founded on the relation of part to whole."[38] Abstraction, "the treatment of a part as if it were the whole, omitting the infinite differences which make up the real whole," is a basic function of symbolism, he says.[39] It is also an example of a metonymic function. Lakoff and Johnson demonstrate the focusing referential (or metonymic) function of the symbol in a *pars pro toto* relationship:

> There are many parts that can stand for the whole. Which part we pick out determines which aspect of the whole we are focusing on. When we say that we need some good heads on the project, we are using "good heads" to refer to "intelligent people." The point is not just to use a part (head) to stand for a whole (person) but rather to pick out a particular characteristic of the person, namely intelligence, which is associated with the head.[40]

"The Face for the Person" is another example:

> If you ask me to show you a picture of my son, and I show you a picture of his face, you will be satisfied. You will consider yourself to have seen a picture of him. But if I show you a picture of his body without his face, you will

consider it strange and will not be satisfied. You might even ask, "But what does he look like?" Thus the metonymy THE FACE FOR THE PERSON is not merely a matter of language. In our culture we look at a person's face—rather than his posture or his movements—to get our basic information about what the person is like. We function in terms of a metonymy when we perceive the person in terms of his face and act on those perceptions.[41]

Lakoff cites Richard Rhodes, a linguist who does fieldwork on Ojibwa, a Native American language of central Canada:

> As part of his fieldwork, he asked speakers of Ojibwa who had come to a party how they had got there. He got answers like the following (translated into English):
> —I started to come.
> —I stepped into a canoe.
> —I got into a car . . .
> What Rhodes found was that in Ojibwa it is conventional to use the embarcation point of an ICM [idealized cognitive model] of this sort to evoke the whole ICM. That is, in answering questions, part of an ICM is used to stand for the whole. In Ojibwa, that part is the embarcation point.
> Ojibwa does not look particularly strange when one considers English from the same point of view. What are possible normal answers to a question such as "How did you get to the party?"
> —I drove (Center stands for the whole ICM.)
> —I have a car. (Precondition stands for the whole ICM.)
> —I borrowed my brother's car. (This entails the precondition, which in turn stands for the ICM.)
> English even has special cases that look something like Ojibwa.
> —I hopped on a bus. (Embarcation stands for the whole ICM.)
> —I just stuck out my thumb. (Embarcation stands for the whole ICM).[42]

By Symbols Are We Made

"The human being uses and creates symbols," writes Hacker. "The individual is also made and created by symbolism."[43] Symbols direct our conceptions of people, things, or events. They illuminate and thereby make more inviting certain paths, leaving other avenues dark and out of view: "The reduction which is the characteristic for all symbolization and therefore for all rituals, leaves the greater part of the world, not closed into the symbol, outside of its emphasis. And so this greater part, left out, becomes unimportant and meaningless." Or, as Carlisle rather nicely put it, in a symbol is to be found "silence and speech acting together."[44] Thus, in Judith de

Neufville's example, the 160-acre symbol of self-sufficiency led to the restriction of many federal water projects aiding the arid West to farms not over 160 acres, despite the economies of scale demanded by modern technology. The symbol relayed the significance of 160 acres from a previous century, and kept silent on the more practical matters of the modern age.

Colonel North as Example

Is it an accident that Lieutenant Colonel Oliver North, accused of diverting funds from the illegal sale of arms to Iran to benefit the Contras in Nicaragua, gave testimony at Congressional Hearings in full-dress Marine uniform, even though he did not normally wear the uniform to work? "There is little doubt that North cuts an impressive figure in his uniform," wrote *Boston Globe* reporter Ethan Bronner on July 9, 1987. Bronner also quoted *Navy Times* editor Thomas Philpott: "The uniform carries a message," Philpott said. "It says 'I am someone who serves my country so show me the respect I deserve.'" As Edelman has observed, "emotional commitment to a symbol is associated with contentment and quiescence regarding problems which would otherwise arouse concern," and North's uniform was clearly worn to silence criticism.[45]

North and his uniform demonstrate the attributes of symbolism identified above. The uniform is a *symbol*, not just a *sign*. The uniform doesn't simply tell us that North is in the military. It is a vehicle for our conception of what North is. It serves as a gateway to a complex of associations of what it is to serve your country, to be courageous, selfless, risk your life, and brings into sharp focus these highly positive values. These associations have developed in human culture from far back in history, and have been transmitted from generation to generation by tradition as well as by direct experience. Throughout history the uniform has symbolized patriotism, selflessness, heroism. The uniform acts to recall to memory examples of military heroism.

The effect of the symbol will vary from person to person, since everyone has a unique set of lifetime experiences, but the uniform symbolically defines the boundaries in which its semantic energy is to be released. Whether or not the heroic associations of the uniform are relevant to the context of North's actions, they are brought forward and used to give a "reading" of his performance. By wearing the uniform (replete with medals from Vietnam, themselves imbued with symbolic significance), it makes it hard to see North in a negative light.

The uniform is a simple but dramatic symbol, and speaks eloquently for itself. It reflects only a part of North's identity, but it symbolically defines North as a whole: A man with a breastful of medals is a hero, and that is that.[46] To use the terms of Michael Polanyi and Harry Prosch, we "surrender" to the symbol, we let it "carry us away."[47]

It is important to note the "compelling" rather than "compulsory" quality of the symbolic uniform. The symbol successfully directs the understanding of North by the mainstream of the population, but those who do not adhere to the values the uniform represents will not be manipulated. Critics of the military, pacifists, those who have experienced the destructive effects of war may even be strengthened in their opposition by the presence of the uniform. To be effective the targets of the symbol must be receptive to the values it conveys.

It is interesting to note how Senate Committee Chairman Daniel Inouye attempted to deflect the impact of North's symbolic display: For the first time, the Senator wore on his left lapel a Distinguished Service Cross received during World War II. Symbolically it said: "You're not special, Colonel North, and you're not above answering for your actions to me."

MARLETTE—ATLANTA CONSTITUTION

'Colonel North couldn't make it to the hearings today, but he sent along his uniform for questioning'

Figure 7-2. His uniform symbolically defines North as a whole. (Copyright Tribune Media Services, Inc. All Rights Reserved. Reprinted with permission)

Interpreting Symbols

It is important to establish what shared meanings a symbol has come to evoke to a particular culture and how it came to have those meanings. In the case of symbols far removed from their initial context—"Jordan River," for example—it implies establishing a bridge across contexts. In the case of the Jordan River this is provided by a text of great significance to its audience, the Bible.

Understanding the roots and meaning of a symbol permits examination of how it might function in the specific modern-day context we wish to study. Knowing that the Jordan River is associated with the liberation of the Israelites explains why it might be such a compelling symbol for slaves yearning to be free. Understanding that the military uniform has been traditionally worn by people willing to sacrifice their lives for their country shows why it might be such a potent symbol of valor at the emotionally charged congressional hearings of Colonel North.

To interpret the symbols we suspect of being active in the formulation of transportation policy in Southern California, then, we need to establish their roots and the meanings they have thereby acquired. We then need to test if those meanings structure understanding in the texts under study. Do the symbols' associations translate coherently into the present context? Do they provide compelling summaries of complex situations? Are alternative explanations to the one suggested by the symbol absent when the symbol is present and active?

As will be established shortly, the most important role of the symbol is in the creation of metaphor, to give us a particular way of "seeing" a situation. Although not all symbols are metaphors, it is intrinsic to the function of symbols that they generate images and that those images act metaphorically. It is to the concept of the image, and to its function in metaphor, then, that we now turn.

Images

Images are the ammunition of symbols. They also act as symbols in themselves. Few images are as potent as those of life and death, and Colonel Oliver North came to his congressional hearings well-armed with these. North's uniform is a symbol, but the symbol only has meaning in terms of what it leads us to conceive, which is an image. North's medal-covered uniform evokes a heroic image, leading to a conception of North as a hero. The symbol produces the image, and the image sells.

The Nature and Functioning of Imagery

Images are a form of subjective knowledge capable of governing behavior.[48] They perform a "mediating" function to provide an escape from the confounding complexity of the real world.[49] We have an "intense hatred, for instance, of multidimensional value orderings," Kenneth Boulding says, and the image serves to render the many shades of complexity in black-and-white.[50]

While complexity is foreign to us, images come naturally. Images, writes Langer, are

> our readiest instruments for abstracting concepts from the tumbling stream of actual impressions. They make our primitive abstractions for us, they are our spontaneous embodiments of general ideas.
>
> Image-making is, then, the mode of our untutored thinking, and stories are its earliest product. We think of things happening, remembered or imaginary or prospective; we see with the mind's eye the shoes we should like to buy, and the transaction of buying them; we visualize the drowning that almost happened by the river-bank.[51]

What is it to "see" with the "mind's eye"? To imagine is "always to make something absent appear in the present, to give a magical quasi-presence to an object that is not there."[52] The image Oliver North left causes him to reappear before us after he has faded from the television tube. We identify with that image and it influences our actions.

The image is effective largely because those under its influence are unaware that it is an illusion: the magic of the "quasi-presence" is that it is taken for the real thing. "The imaginative consciousness of Peter is not a consciousness of the image of Peter," writes Jean-Paul Sartre.[53] "Peter is directly reached, my attention is not directed at an image but at [what appears to be] an object."

The way in which we "reach" Peter will depend on our personal associations with him and with phenomena related to him, and on the context of our life's experience in which those associations are placed. Someone who has had different dealings with Peter will have a different image of him. A third person who has never met Peter, but who has only seen a portrait of him in which he is smiling, will form an image of a smiling Peter, and that "Peter" might be quite a different person from the Peter revealed to someone who has known him in person. But, as Sartre says, "What we recover in an image is not this or that aspect of a person but the person himself as a synthesis of all objects."[54] That synthesis serves to generalize: to the person who has only seen the portrait of the smiling Peter, Peter always smiles. To the person who has only seen the heroic-looking televised Oliver North, Oliver North is always a hero.

Images Are Symbolic

The ability of images to present the partial as if it were the whole gives them a powerfully symbolic function. "An image is only an aspect of the actual thing it represents," writes Langer, "It may be not even a completely or carefully abstracted object. *Its importance lies in the fact that it symbolizes the whole—the thing, person, occasion, or what-not—from which it is an abstract*" (my emphasis).[55]

Though the heroic image which North's uniform engenders reflects but an aspect, and possibly a distorted aspect, of the colonel, it gives a picture of the man as a whole. The image "serves as a proxy for a set of unstated assumptions".[56] It obscures alternatives which "do not usually have the courtesy to parade themselves in rank order on the drill ground of the imagination."[57]

George Miller provides a helpful discussion of the mechanics of imagery. He quotes first a passage from Thoreau's *Walden*. This, he says, creates an image of Walden Pond: "You construct an image as part of the process of understanding the passage," and:

> The image helps you to remember what you have read. It helps you to remember in the following sense. If, after putting down the book, you were asked to repeat what you have read, you would probably not be able to repeat it verbatim. Nevertheless, you could reactivate your memory image and describe it, thus generating a different prose passage but one that (if your memory is good) would be roughly equivalent to the original passage that inspired the memory image.[58]

Lakoff and Johnson press the point that "[w]e typically conceptualize the nonphysical in terms of the physical—that is, we conceptualize the less clearly delineated in terms of the more clearly delineated."[59] In *Walden* the image we create of the pond is easier to commit to memory than the stream of words that make up the text. Where there is a lack of compelling imagery associated with a text, we have to resort to more abstract forms of memory, such as memory of the text itself. But, as Miller suggests, when vivid imagery does exist, it displaces abstract memory:

> One point that is sometimes overlooked when memory images are discussed: The vagueness of the image is critical to its utility. If memory images had to be completely detailed, like photographs, they could not preserve the incomplete information given by written descriptions. Thoreau did not describe every detail of the hills above Walden Pond; a reader who wants to remember accurately what Thoreau did describe had better not clutter the memory image with details that were not provided. It may be impossible to construct an image that does not contain some extraneous information that is merely suggested, not entailed by the text; a reader lacking any knowledge of the New England countryside, for example, would not construct the same image as a reader who knew it well. My point is simply that such additions from general knowledge are potential sources of error, even where they appear highly probable in the given context.[60]

Miller does not, perhaps, quite adequately press home the significance of this error. While he does stress the subjective nature of imagery and its dependence on the context in which it is formed, it is the *pars pro toto* symbolic nature of imagery that is of greatest importance here. We do not "see" only a part of the pond; we imagine the hillside, and how the pond is set in this landscape, and where the ice "not yet dissolved" remains on the surface. And although our experience will color our images with error, of that error we will be unaware; to us, however we see it, Walden Pond will seem very real and complete.

Images Reflect Emotions and Needs

An important function of imagery is to delve into "the vast storehouse of forgotten memories and experiences" that constitute the human mind, providing a channel for the subconscious to influence action, if not revealing it to us directly.[61] The act of imagination "is an incarnation destined to produce the object of one's thought, the thing one desires, in a manner that one can take possession of it," writes Sartre.[62] "In that act there is always something of the imperious and the infantile, a refusal to take distance or difficulties into account." Those craving water will see mirages; those in love will conjure up idealistic evocations of their loved ones.

Langer points out that while such behavior may not be logical "after the manner of discursive reason," the imaginative mode of ideation, as she calls it, "has a logic of its own, a definite pattern of identifications and concentrations which bring a very deluge of ideas, all charged with intense and often widely diverse feelings together in one symbol." The image thereby serves to select and to focus certain feelings. "In organizing itself into an imaginative form, desire becomes precise and concentrated," Sartre says. "Feeling then behaves in the face of the unreal as in that of the real. It seeks to blend with it, to adapt itself to its contours, to feed on it. Only this unreal, so well specified and so well defined, pertains to the void; or, if one prefers, it is the simple reflection of the feeling. This feeling therefore feeds on its own reflection."[63]

Images Anchor Inferential Logic

The image formed from the sum total of our experience serves to anchor "inferential logic."[64] It forms a compelling inferential mechanism because it is rooted in feelings, because it appears to reveal the whole picture, and because its built-in system of logic circumvents the need to employ "discursive logic."

Images Perpetuate Themselves

Images not only provide direct inferential mechanisms, but also feed the very subconscious on which they draw. The image, born of the subconscious, does not allow just any phenomena to be admitted back into the subconscious, but selects forms with which it is consonant. The image of Oliver North focuses feelings of patriotism and respect for people who serve their country, and in doing so reinforces the role those feelings have in our subconscious. The image behaves as an "inward teacher imposing its own form and 'will' on the less formed matter around it."[65] Images thereby fortify themselves and become resistant to change.

Imagery Is Intuitive, Not Reflective

The image, the very "incarnation of non-reflective thought," comes freely and helps endow "natural thinking" with its deceptive fluency.[66] Imagery selectively emphasizes certain aspects, omits others, and intuitively fills in for gaps in knowledge. By pretending that its partial impression is a rendering of the whole, it seduces us into accepting its intuitively attractive depiction as truth. By delivering its object as a whole, it also prevents any awareness of "image" on the part of its victims. By concealing itself from our consciousness, the image evades detection. Yet, working underground, it shapes our very understanding. Although, as Sartre says, it is only through reflection that the image may be foiled, the image, by so convincingly disguising itself in the clothes of truth, leads its prey to uncritical acceptance of its content and away from the path to reflection.

Metaphor

From Symbol to Metaphor

Aristotle defined metaphor in the *Poetics* as "giving the thing a name that belongs to something else." In terms of its Greek roots, a meaning is "carried over" from one thing to which it already belongs to another somehow like it.[67] Metaphor is an instrument of understanding, specifically "*understanding one kind of thing in terms of another.*"[68]

According to Langer, "In a genuine metaphor, an image of the literal meaning is our symbol for the figurative meaning, the thing that has no name of its own."[69] She illustrates the mechanics of metaphor with the example "The King's anger flares up":

> We know from the context that "flaring up" cannot refer to the sudden appearance of a physical flame; it must connote the idea of "flaring up" as a *symbol* for what the king's anger is doing. We conceive the literal meaning of the term that is usually used in connection with a fire, but this concept serves us here as proxy for another which is nameless. The expression "to flare up" has acquired a wider meaning than its original use, to describe the behavior of a flame; it can be used metaphorically to describe whatever its *meaning* can symbolize.[70]

Donald Schön characterized this symbolic meaning as a "program."[71] The meaning of a concept employed as a metaphor, A, is taken as a program for the exploration of its subject, B. In doing this, "expectations from A are transposed to B as projective models." A thereby pulls the strings of B, "fixing and controlling" the way in which B is understood. The "programming" concept is a necessary condition for a symbol to also be a metaphor. Many potent symbols are not metaphors because the conceptions they evoke are not programmed by the meaning of the symbol itself. An example of such a symbol, given by Polanyi and Prosch, is a national flag. A flag can act as a powerful symbol of national identity, but it does not act as a program, and cannot do so, since the flag qua piece of cloth has no meaning of its own with which to do the programming.[72]

But what of the cloth of Oliver North's uniform? It is clearly meaningful, for it is cut to contain a man, and designed to outfit him for combat. This is what, in fact, makes the uniform such a powerful symbol of heroism. Is Oliver North's uniform, then, a metaphor? The uniform is not a metaphor because what it symbolizes is not conceived in the manner of a uniform: the uniform does not program the conception. This does not mean that the uniform cannot *give rise* to metaphor, but it does so indirectly through the image it creates, rather than through itself. Oliver North is not seen "as a uniform," but he is seen in terms of the heroic image the uniform conceives. The image of heroism the uniform creates thus acts not only as a symbol but as a metaphor, because the hero concept is used to program our understanding of North: North is seen as a hero. Symbols that are not directly metaphors, then, can evoke images which can act as symbols which are metaphors.

Metaphor Is Genetic

Metaphor is invasive. It invades the concept it is to program and molds it to its own form. It is in this sense that metaphor can be said to be genetic. Metaphor acts like DNA in that it transports and implants as a code for the development of concept B a set of characteristics which are A-like; the conception of B then develops in the manner of A. Information conforming to A-like characteristics is used in programming the concept. The metaphor hooks onto and transports to B admissible A-like phenomena, and leaves the rest behind.

Coherence in Metaphor

In Black's language, the "system of associated commonplaces" of A is used to conceptualize B, and its components translate from A to B coherently.[73] So, when we say that the king's anger "flared up," related aspects of "flaring" help explain what the king's anger did. Flaring implies a sudden, uncontrolled, eruption. It is a flame that does the flaring. Flames are hot and—when out of control—dangerous. When the king's anger "flares up," we fear the consequences of his *hot* temper; if we get in the way of it, we are likely to be *burned.*

Lakoff and Johnson note that the pictures formed by metaphors fit together coherently, in the sense of making a good general fit, rather than sticking to rigid rules of *consistency* in image formation. Thus, the "Love is a Journey" metaphor may involve a mix of modes by which love travels. The lovers might declare: "We're at a *crossroads*," or "It's been a *long, bumpy road*" (implying road travel). If love has been traveling by road, then you might expect a failing relationship to end up a dead-end street. But it may equally well founder on the rocks (sea voyage), or come off the rails (train trip).[74] These concepts combine to create a coherent picture of what love is doing, if one that is not consistent in the sense of formal logic.

Metaphor Clarifies

Metaphor performs a clarifying function, naming, fixing and structuring "what might otherwise be vaguely troubling situations" according to Schön. "The movement is usually from a more concrete and readily graspable image 'over onto' what is perhaps more vague, more problematic, or more strange. What is familiar is used to explain what is strange."[75] A flame is a tangible way to appreciate the abstract concept of a temper. The strange vagaries of love are brought down to earth when explicated in terms of travel by road, rail or sea.

Many abstract human emotions are understood in terms of physical orientational metaphors.[76] Happy is up, for example, as in "I'm feeling *up*" or "My spirits *rose.*" Sad is down, as in "I'm *depressed*" or "He is really *low* these days" High status is also up—"She'll *rise* to the top," for example, as are good things in general—"Things are looking up," for instance. Virtue is up, as in "She is an upstanding citizen." Depravity is down—"That was a *low* trick," for example.

The basis for metaphor is experiential and cultural. That flames are hot and can cause damage is part of everyone's experience. Explaining the basis of "Happy is up," Lakoff and Johnson point out that a drooping posture typically goes along with sadness, an erect posture with a positive emotional state. Understanding the experiential basis is important when things don't seem to fit. Being "up in the air" is certainly less desirable than having your feet "on the ground," a seeming reversal of the up/good, down/bad relationship. But, Lakoff and Johnson show, we are dealing with a different metaphor with a different experiential basis: "With physical objects, if you can grasp something and hold it in your hands, you can look it over carefully and get a reasonably good understanding of it. It's easier to grasp something and look at it carefully if it's on the ground than if it's floating through the air (like a leaf or a piece of paper)."[77]

Metaphor Structures Understanding

In using the concrete to pin down the abstract, metaphor creates what Ricoeur calls a "heuristic fiction": there is no actual physical flame present in the king's body; derailed train at the site of a failed romance; or things—presumably with eyes—that are "looking up."[78] But, Ricoeur says, "heuristic fiction is not an innocent pretense. It tends to lose sight of its nature as fiction and take on the dimensions of perceptual belief."

While "comparison" theories (such as one might learn in literature class) suggest that metaphor expands the scope of comprehension, since "two lines of interpretation are opened at the same time and put into tension," metaphor as experienced is generally tacit. One is not aware of a likeness—of a comparison—being made, or of the symbolic projection which is making it: "In this respect metaphor is as literal as any normal statement of fact. The 'as' is not the 'as' of comparison, but the 'as' of appearance . . . a likeness to be seen under a given aspect."[79]

The power of the metaphor to present fiction as reality lies in its transparency: we are put under the metaphor's spell without even knowing that it has invaded our thinking. Freud maintained that when thought and wishes become unconscious, they gain greater intensity and generality. "In a similar way, certain metaphors of our culture, as they go underground, intensify and become more generalized."[80]

Lakoff and Johnson demonstrate the power of such a metaphorical concept to structure "what we do and how we understand what we are doing when we argue." The metaphor is "Argument is War," and is illustrated by several examples:

Your claims are *indefensible.*
He *attacked every weak point* in my argument.
His criticisms were *right on target.*
He *shot down* all of my arguments.[81]

We do not simply talk of arguments in terms of war, they maintain, but actually win or lose arguments. But:

> It is not that arguments are a subspecies of war. Arguments and wars are different kinds of things—verbal discourse and armed conflict—and the actions performed are different kinds of actions. But "Argument" is partially structured, understood, performed, and talked about in terms of "War." The concept is metaphorically structured, and, consequently, the language is metaphorically structured.
>
> Moreover, this is the *ordinary* way of having an argument and talking about one. The normal way for us to talk about attacking a position is to use the words "attack a position." Our conventional ways of talking about arguments presuppose a metaphor we are hardly ever conscious of.[82]

Michael Reddy has elaborated another fascinating example of the covert power of metaphor.[83] The "conduit metaphor," he says, silently structures our very understanding of thought. When we talk of trying to *get* our *thoughts across* better we are seeing a thought as an object to be moved across space, as if through a conduit, he says. Communicating is sending, and sentences are containers in which ideas are sent. You might, therefore have to *put* each *concept into words* very carefully or *pack* more *thoughts into* fewer *words.* Your words may, nonetheless, still seem *hollow* and *carry* little meaning.

According to Reddy, "This model of communication objectifies meaning in a misleading and dehumanizing fashion. It influences me to talk about thoughts as if they had the same kind of external, intersubjective reality as lamps and tables."[84] The implication is that a thought may be sent across space independently of the person conceiving that thought. Just as a given model of lamp is the same whether produced in one plant or another, it doesn't matter who sends a particular thought, or what the context of that thought is: a "given" thought is the same, wherever it comes from. The identity of the receiver will have no impact on the reception of a thought, just as a ball caught by one person is the same ball when caught by someone else. "You'll find better ideas in the library" implies that ideas are in the library for the taking, rather than attainable only through the thought and interpretation of the seeker of ideas. If we have a lack of good ideas, the metaphor therefore tells us to build more libraries with larger collections of books—for that is where ideas reside—rather than to educate people better in the conception of ideas.

The conduit metaphor is rooted in the tradition of rationalism, which maintains that there is such a thing as absolute truth and that metaphor can and should be avoided to allow that truth to speak plainly. As Schön points out, believing that a clear nonmetaphorical way of speaking existed "suited the rationalist temper of the eighteenth century, committed as it was to the view that things are inherently intelligible and that theories, if they are based on reason, can convey that intelligibility by the use of clear language made up of 'the ordinary words.'"[85]

This belief is fundamental to our culture, in which the received paradigm of science stresses the search for objective knowledge and the avoidance of bias through the exclusion of self. There is one truth, the paradigm suggests, and it is to be found if one looks hard enough. The truth is assumed to remain constant however and by whomever it is appreciated; that assumption is a convenient protective device that allows science to proceed unselfconsciously. The conception of truth of those under the influence of the conduit metaphor is structured by that metaphor. And the metaphor causes those under its influence to deny that metaphor structures truth!

Generative Metaphor in Social Policy

Schön characterizes the dominant paradigm of social policy as one of "problem solving."[86] There are social problems, this paradigm says, just as there are mathematical problems, and—just as there is a solution to a given mathematical problem—there is a solution to be found to a given social problem. "In opposition to this view," he says, "I have been persuaded that the essential difficulties in social policy have more to do with *problem setting* than with problem solving, more to do with ways in which we frame the purposes to be achieved than with the selection of optimal means for achieving them."

Schön develops the concept of "generative metaphor" to account for the framing of social policy. Problem settings, he says, are mediated by the "stories" people tell about troublesome situations—stories in which they describe what is wrong and what needs fixing. When these problem-setting stories are analyzed, it becomes apparent that the framing of problems often depends upon metaphors underlying the stories. Each story "constructs its view of social reality through a complementary process of *naming and framing*. Things are selected for attention and named in such a way as to fit the frame constructed for the situation." In this way, a few salient features and relations are selected "from what would otherwise be an overwhelmingly complex reality. They give these elements a coherent organization, and they describe what is wrong with the present situation in such a way as to set the direction for its future transformation."[87]

As Colin Turbayne points out, metaphor is not merely "pretend," but "intend," and, through the process of "naming and framing," the stories make the "'normative leap' from data to recommendations, from 'is' to 'ought.'" These stories typically "execute the normative leap in such a way as to make it seem graceful, compelling, even obvious . . . This sense of the obviousness of what is wrong and what needs fixing is the hallmark of generative metaphor in the field of social policy."[88]

In illustration, Schön quotes from Justice Douglas's opinion on the constitutionality of the Federal Urban Renewal Program in the District of Columbia:

> The experts concluded that if the community were to be healthy, if it were not to revert again to a blighted or slum area, as though possessed of a congenital disease, the area must be planned as a whole. It was not enough, they believed, to remove existing buildings that were unsanitary or unsightly. It was important to redesign the whole area so as to eliminate the conditions that cause slums—the overcrowding of dwellings, the lack of parks, the lack of adequate streets and alleys, the absence of recreational areas, the lack of light and air, the presence of outmoded street plans.[89]

The underlying generative metaphor here is of an irreversibly diseased organ. If an organ is diseased beyond repair, it must be surgically removed and replaced to enable the body to become healthy again. In a similar vein, the metaphor tells us, a blighted area has to be torn down and rebuilt from scratch to allow the community to recover. Just as it is obvious that if a patient is wheeled into the operating room with the symptoms of an inflamed appendix, the surgeon had better not waste time withholding the scalpel, so with a sick city the prescription is to excise that which is malignant.

Schön demonstrates that the same conditions can be transplanted to a radically different story, and thereby result in different prescriptive outcomes. The other generative metaphor is of the "natural community." This metaphor focuses on the people in the community, rather than on its physical structure. It sees that the inner city is often a place of comfort and belonging for low-income residents. The task, then,

> is not to redesign and rebuild these communities, much less to destroy buildings and dislocate residents, but to reinforce and rehabilitate them, drawing on the forces for "unslumming" that are already inherent in them ...
>
> What is wrong is that the natural community, with its homelike stability and its informal networks of mutual support, is threatened with destruction—indeed, by the very prophylaxis undertaken in the name of "urban renewal." We should think twice about "dislocating people from the local areas;" "natural communities" should be preserved.[90]

"In our ideas about disease and about natural community," Schön says, there is already an evaluation—a sense of the good which is to be sought and the evil which is to be avoided. "When we see A as B, we carry over to A the evaluation implicit in B."

Schön emphasizes that the basis from which the different metaphors are drawn is not merely experiential, but experiential in the setting of a given culture:

> Each of these generative metaphors derives its normative force from certain purposes and values, certain normative images, which have long been powerful in our culture. We abhor disease and strive for health. Indeed, popular culture seems often to identify the good life with the healthy life and to make progress synonymous with the eradication of disease ... We also have a strong affinity for the "natural" and a deep distrust for the "artificial." The idea of Nature, with its Romantic origins in the writings of Rousseau and its deeper sources in pantheism, still works its magical appeal.[91]

The existence of different metaphors "makes it dramatically apparent that we are dealing not with 'reality,' but with various senses of a reality." As Lawrence Hinman comments in his essay on "Nietzsche, Metaphor and Truth": "What counts as truth depends on the game we are playing ... Truth is only revealed within the context of particular games, and no specific game has a claim to ontological priority."[92]

The sense of obviousness of a generative metaphor's prescription, Schön says, depends on the metaphor remaining tacit.[93] "Often we are unaware of the metaphors that shape our perception and understanding of social situations," he says. When we are unaware of metaphor, we are likely to fall "victim" (Turbayne's language) to it, to be "used" by it.[94] Surfacing the tacitly understood metaphor lays open to critical review the assumptions underlying policy decisions. "That which seems obvious to the unreflecting mind may upon reflection seem utterly mistaken."[95] It is only when the metaphor is surfaced that the "tensions" central to comparison theories become apparent and that we can see how we have been seeing. Then the metaphor through whose filter we have been seeing lays itself open to criticism. Doing this requires that we construct it "through a kind of policy-analytic literary criticism, from the givens of the problem-setting stories we tell."[96]

Schön distinguishes between "surface" and "deep" metaphors. The surface language in which a story is told may offer clues to the generative metaphors which set the problem of the story. Thus (to use another example from Lakoff and Johnson), faced with an energy crisis, President Carter declared "the moral equivalent of war." The deep metaphor is "energy crisis as war" and it generated concepts of an "enemy," a "threat to national security," which required "setting targets," "reorganizing priorities," "establishing a new chain of command," "plotting new strategy," "gathering intelligence," "marshaling forces," "imposing sanctions," and "calling for sacrifices."

The surface language need not, however, offer such direct evidence, even though a particular metaphor may be in force at a deeper level:

> The deep metaphor, in this sense, is the metaphor which accounts for centrally important features of the story—which makes it understandable that certain elements of the situation are included in the story while others are omitted; that certain assumptions are taken as true although there is evidence that would appear to disconfirm them; and, especially, that the normative conclusions are found to follow so obviously from the facts. Given a problem-setting story, we must construct the deep metaphor which is generative of it. In making such a construction, we interpret the story. We give it a "reading" in a sense very much like the one employed in literary criticism. And our interpretation is, to a very considerable extent, testable against the givens of the story.[97]

Myth

We fear what we do not know. Myth provides a convenient way of explaining away the mysterious and of providing intuitively attractive answers to complex problems when the only apparent alternative is to be faced with a troubling void. "The reason that men accept myths can be found in their beliefs that myths do describe *the actual state of the world*. This is the explanatory function of a myth."[98]

Cassirer comments that "The real substratum of myth is not a substratum of thought but of feeling . . . Its view of life is a synthetic not an analytic one." There are no shades of gray in a mythical conception, according to Cassirer. All objects are polarized into "benignant or malignant, friendly or inimical, familiar or uncanny, alluring and fascinating or repellent and threatening." And, in a mythical conception, there is no distinction between "image and object."[99]

The mythical synthesis is derived from the world as experienced in a particular culture, mediated by the symbolic forms discussed above. The mass of symbols which direct us to particular conceptions of their objects; myriad of images which form lasting impressions on our consciousness; and metaphors which unconsciously structure our understanding, come together synthetically to create myth. As in individual generative metaphors the structuring of phenomena is coherent, so in myth the arrangement of constituent metaphors is such as to fit together and create a coherent and therefore compelling synthesis.

Earl MacCormac likens the larger pattern formed to a

> visual picture in which there may be geometrical or proportional relationships among the parts, but these relationships are not primary, for what gives order and harmony to the picture is the sense of coherence that we have whenever we view it. Our familiarity with various patterns of explanation and our confidence that they do cohere enables us to accept such accounts as legitimate modes of understanding.[100]

Once a myth is established, de Bono says, it becomes a self-reinforcing way of looking at the world. Thus, those who believe the myth of extrasensory perception reject the findings of "suspicious people who set out to check the results [and] never seem to be able to reproduce the results of those who are in favor of the idea."[101] The myth incorporates the grounds for rejection. "The explanation suggested is that suspicious people are not in the right frame of mind to get results, that one must be confident and relaxed for extrasensory perception to go through. This makes the phenomenon insusceptible of critical investigation."

The beliefs of people subject to the myth are thereby strengthened, and their critics frustrated. As de Bono suggests, myth is like a language. "Once one accepts and understands the language it makes sense. From outside it may not." Orthodox Christians will accept the virgin birth of Jesus as historically true and dismiss the Buddhist tale of the descent of the Buddha's spirit into the maternal womb in the form of a baby elephant as false and fanciful, Wheelwright says. "Positivists will dismiss both narratives alike." Positivists would, further, claim that they are not subject to such myths— or to any myths, for that matter. The myth of Positivism would have it that myths can be avoided, and that only by avoiding them can "objective knowledge" be reached. The myth, though they would deny it, creates a firm platform on which positivists can operate. The unquestioning belief in it by positivists provides a shared understanding, which allows their work to progress unhindered by thoughts as to its tenuous assumptions. As Turbayne points out, for those who conceive and transmit myth, "there is no make-belief, only belief."[102]

"Myths cannot be eliminated from public policy nor from any area of social life," writes de Neufville.

> One myth can only be replaced by another, as myths are essential to guiding social action. For interaction or collective decisions to be possible, events and actions must have some common significance within a social group. This meaning cannot be deduced from the "facts" of a situation, which in themselves say nothing. But neither can meanings and logic be explained every time an issue arises, or getting anything done would be impossible. So a smile or nod of the head is accepted as a friendly greeting. Cutting an agency's budget is a sign of public disapproval. Though the significance of the action is seldom verbalized, the shared meanings permit cooperative action and policy agreement. These meanings are what motivate people, because they connect to deeply held values.[103]

The major myth to be investigated in what follows will be that rail transit can alleviate the transportation problems of Los Angeles. But, as will be seen, a myth is a "complex of stories," and these stories each create mini-myths to feed the major one.[104] Hypotheses as to the mechanisms by which the various myths function will be formulated on the basis of the symbolic forms—symbol, image, metaphor—studied above, and the viability of these hypotheses will be tested against the content of the stories the myths tell.

Testing the Theory

Testing for the presence of the symbols, images, and metaphors which make up the myth of rail in Los Angeles was done by interpreting of the extensive set of interviews obtained for this project, transcripts of political meetings, and material from media sources. With a large sample—103 transportation actors—interviewed in Los Angeles County, responses can be taken as highly representative for that location. The remainder of the 209 interviews, conducted in other West Coast communities considering or implementing rail transit and in Washington, D.C. provide valuable additional evidence.

The loosely structured questionnaire employed was used to guide discussion through key issues, but respondents were allowed freedom to drift away from specific questions to explore their particular interests and perceptions: the important point was to reveal what was uppermost on the interviewees' minds. While structured questionnaires used in more rigid ways—where every respondent is made to answer every one of the same set of questions—may provide a stronger basis for statistical analysis, such approaches risk asking the wrong questions without a basis for correction and often fail to provide adequate access to the contextual information needed to explore the meanings of the responses given. Allowing people to tell stories provides a basis for studying the coherences which hold those stories together, and in this way the assumptions behind them can be revealed.

Myths come to life in storytelling, in which accounts are given of the respondents' experience in the everyday world, and of how the perspectives governing these accounts are applied to envisage the ideal world of tomorrow. Common sense plays an essential role, and the evidence of the eyes dominates that of the inquiring mind. Myth, as we shall see, is like a language, making sense to those who are versed in that tongue, but not to those for whom it is foreign. A complex of symbolic understandings—rooted in symbolism, imagery, and metaphor—come together to create mythical beliefs. While myths may seem "illogical" to those outside their influence, for those under their control, the understandings build together in logical ways (the "logic" operating within the assumptions of the mythical world), which lend coherence and credibility to the myth. This form of understanding can be quite compelling and create powerful, stable beliefs.

The symbolic and metaphorical understandings of which myths are made can often be complex, and therefore difficult to unravel. As Lakoff and Johnson show, there are often many metaphors that partially structure a single concept.[105] In addition, when employing one metaphorical concept, we often use other concepts which are themselves understood in metaphorical terms. The task of unraveling meaning thereby becomes one of a literary criticism operating at several levels: it requires the isolation and interpretation of the different symbolic concepts at play; and it demands an explanation of how they come together coherently to structure the understandings they create.

Validity here depends on the functioning of a "logic" according to the rules of the theory under test. To test if a symbol is structuring understanding we first need to formulate a theory of the meaning inherent in the symbol, and then establish if those meanings are structuring understanding in the texts. The programming function of a metaphor serves as its "logical" apparatus. To establish the operation of metaphor A, we need to prove that it is acting as a model of our understanding of B. How is B A-like?, we should ask. What is the "heuristic fiction" that is wedding B to A? How is it performing a clarifying function? Are abstract ideas being conceived in concrete terms, and are those terms generated by A? Evaluation requires charting out the assumptions, expectations, and "associated commonplaces" of A to see if they are being mapped uncritically onto B.[106] The presence or absence of a coherent pattern in this mapping function is established to test if the metaphor is structuring understanding on a "deep" level, rather than merely making itself evident in surface language. When flames flare they can go out of control, become dangerous, and burn things in their path. These associations come together coherently to provide a "logic" which structures understanding of a king's "flaring temper," and similar logical mappings will be sought in the examples to be studied below.

Once the internal logic of each symbol, image, or metaphor is established, a similar process takes place at the mythical metalevel in which it is the logic by which the set of these processes come together which is tested. Does it make sense for a particular symbol to give rise to a particular image, and does that image help shape the way a particular metaphorical understanding takes place? Does the system of symbols, images, and metaphors come together coherently to build myth? Just as in a trial where the defendant pleads not-guilty, we can never be absolutely sure of guilt, we can only be sure "beyond a reasonable doubt" that a metaphor is covertly structuring thought. Sure enough to convict the metaphor. And, if there is a lack of coherence when the assumptions of the projective model A are tested against the allegedly projected story, B, we can, equally well, acquit the metaphor.

Clues to Mythology

The accounts in chapters 2 and 6 leave impressions of deep meanings and powerful emotions associated with transportation technologies and of how they might color decision processes. Take Henry Huntington. Not one to reach for the slide rule or cost-benefit analysis, he was a "romantic and imaginative soul," who "found an outlet for his feelings in organizing and directing railroads." Huntington's railroads did spur the development of Los Angeles. But they did more than that: They developed a series of associations of growth and renewal in Angelenos' minds, to linger on well beyond the Red Cars' heyday. If both Long Beach and Watts came to life through rail developments, and the "original nucleus" of several new communities "was the Pacific Electric station itself," conceptions of Red Cars will hold meaningful memories of these places long after the actual cars have departed into history.

The great sense of celebration at the "christening" of the new subway terminal in 1925, preceded by "the greatest luncheon in the history of the Los Angeles Chamber of Commerce and by an impressive parade" shows how much meaning was by then invested in the idea of the train, and what it stood for in terms of development of the community. Snell's conspiracy theory is heavily dependent on evoking this meaning with its heady imagery of the "good old days" of the Red Cars. Snell's depiction of large "gas-guzzling" cars makes for a broad-brush portrait of symbols of evil that transcends the fact that people obtained cars out of choice because they liked the accessibility automobiles could provide.

In our political account, we have seen the contrasting urban ideals associated with highway and rapid transit development: neither was advocated on the basis of analytical investigation, but because the highway and the automobile—directed at the whim of the driver—suggested freedom and decentralization, while the railroad symbolized urban centralization.

Whether the railroad could actually make the city more concentrated—and the contrasting massive scale of highway development against which it would be competing meant that it couldn't effectively do so—is less important than the perception that it could. And so, central-city interests lobbied for rail, while those in dispersed communities worried that with rail transit "local shoppers would travel to Los Angeles to buy a spool of thread," taking away from the economic vigor of their commercial base.

Rapid transit plans dating from Kelker, De Leuw in 1925 show the power of metaphor to render the abstract in concrete terms: abstract ideas of particular forms of social and economic development become symbolized by bold lines drawn on a map, whether or not the solidity of those lines can translate the city into the reality they represent. Alan Altshuler was to remark similarly decades later on the newborn positive perceptions of transit in the 1970s that "this is not to say that transit was an effective way of serving all these objectives, simply that it was widely believed to be so."[107]

As we observe Proposition A and the Long Beach line come to reality, we see Baxter Ward's countywide proposal and Kenneth Hahn's far-reaching Proposition A map coming to suggest that rail would serve the dispersed lifestyle with which it could not formerly be associated; we hear talk of the "untimely death" of the Red Cars, see politicians looking enviously toward the new light rail system in San Diego. We hear buses referred to as "dirty, smelly and definitely not rapid," and that "we need look no further than our clogged freeway system to see the need for transportation alternatives." All these elements have had a necessary place in giving a historical and political account of the return of light rail to Long Beach. We now need to look systematically and microscopically at how beliefs central to the transportation policy process are formed, and at how they can come together to form a myth that to those under its influence constitutes reality.

CHAPTER 8

The Reality of Imagery

> Imagine. Being able to zip from downtown L.A. to anywhere in the
> southland and doing it without ever getting on a freeway or in a car.
>
> —Marcia Brandwynne, Newscaster, KTTV-TV

WHY DID MOST DECISION MAKERS interviewed for this study
think rail projects would be good for their constituents when analytical evidence is so sharply to the contrary? The reason, to unfold from the tangle
of disparate threads to be traversed below, emerges that decision makers,
along with the public as a whole—all of us who are not specialists—operate naturally on a quite separate conceptual basis from that of analytical or
reflective rationality, one which is governed by myth. The exploration of
myth begins with an examination of the role of imagery, the raw "data" of
mythology. It is important to understand how the freeway is perceived in
Los Angeles, and how this relates to the imagery of bus or rail services
which might be provided. These alternatives not only carry powerful associations, but the way such images are structured according to the typical
way we refer to a journey accentuates the polarization of the images of bus
and rail services.

Images of the Freeway

Driving the freeway, writes David Brodsly, "is absolutely central to the
experience of living in Los Angeles, and any anthropologist studying our
city would head for the nearest onramp, for nowhere else would he or she
observe such large-scale public activity."[1] The freeway is a symbol of Los
Angeles, and it symbolizes much of what Los Angeles is about.

> The Los Angeles freeway is a silent monument not only to the history of the
> region's spatial organization, but to the history of its values as well. . . . Los
> Angeles' appeal lay in its being the first major city that was not quite a city,
> that is, not a crowded industrial metropolis. It was a garden city of backyards
> and quiet streets, a sprawling small town magnified a thousandfold and set

among palms and orange trees and under a sunny sky. When the city began drowning in the sheer popularity of this vision, the freeway was offered as a lifeline. The L.A. freeway makes manifest in concrete the city's determination to keep its dream alive.[2]

Brodsly quotes Joan Didion, who calls the freeway experience "the only secular communion Los Angeles has."[3] Agreeing, he adds: "Every time we merge with traffic we join our community in a wordless creed: belief in individual freedom, in a technological liberation from place and circumstance, in a democracy of personal mobility. When we are stuck in rush-hour traffic the freeway's greatest frustration is that it belies its promise."[4]

Brodsly talks of the positives of the freeway of life, compared to riding the New York subway. There is no missing a departure, waiting to transfer, or walking more than a short distance. Outside rush hour, the freeway is a pleasureground of graceful curves and banked interchanges which give a sense of speed without danger. A car driving well on an uncrowded freeway with a good song on the radio can produce a distinctly urban moment of joy, according to Brodsly.

> While it may appear farfetched to compare a peak-hour commute with a stroll down a country road, the freeway has a certain quality that makes driving it the nearest equivalent to such an experience the average Angeleno is likely to have during a typical day. For here, rather than in Griffith Park or along the beach, one receives a daily guarantee of privacy. Safe from all direct communication with other individuals, on the freeway one is alone in the world. You can smoke, manipulate the radio dial at will, sing off key, belch, fart, or pick your nose. A car on a freeway is more private than one's home.[5]

Despite all this, Brodsly says, freeways may be associated with all the ills of a modern metropolis, whether air and noise pollution, congestion, or the loss of the landscape to concrete. "When country singer Jerry Jeff Walker sings of getting off the L.A. freeway, he longs to leave every perceived evil of city life behind."[6]

The interview transcripts yielded almost uniformly negative associations with freeways: places where people are forced to spend a large portion of their lives simply for daily functioning, but which were associated with disease and even death. Experience is a principal source of these impressions and of conceptions of possible alternatives. Jacki Bacharach, LACTC commissioner and chair at the time, listed one of her main sources of information as her "experience in traveling in Los Angeles and in other cities." Burke Roche, then deputy to Supervisor Hahn, talked about his experiences with the freeways: "I don't understand how people can day after day come in bumper-to-bumper on the freeways," he said. One of the reasons he moved home "was so that I wouldn't have to travel the Santa Monica Freeway." Alternate Commissioner Ted Pierce complained of the hemmed-in, claustrophobic feeling when people are "driving down the freeway and they're seeing all the cars, and they're going on: 'there must be another way to get in here.'" Just as in a court of law there is little statement more compelling than "I saw it with my own eyes," the evidence from experience is compelling.

The experience is a painful one. Howard Mull, staff to Supervisor Deane Dana, referred to the "grief and hassle of driving into town." Alternate Commissioner Bob White pointed to the "stress factor involved and the irritability of driving. Oh Jesus, you get into a traffic jam where somebody has a little problem—have you ever thought about when you're in a situation where you can't get out of it, that sometimes you panic a little bit, your

Figure 8-1. Bumper to Bumper. (Photo courtesy of author)

car gets hot, you lose a fan belt and you're out in the middle, you gotta telephone and yer goina be two hours late. I tell'ya."

Mayor Kell of Long Beach described the drive into downtown Los Angeles as "getting myself all stirred up emotionally for an hour and tired by the time I get to work and, at the end of a long day, I got to think of fighting that traffic bumper-to-bumper and getting my nerves all unraveled coming home." Bumper-to-bumper suggests an invasion of private space, a feeling of confinement and stress, quite the antithesis of Brodsly's world where the driver is free to happily fart and pick his nose. It is hardly surprising that, as we shall see, those who wish to return to the dream call for a move toward "balance," while those for whom the dream is smashed demand a viable "alternative" to the hellish daily freeway commute. As Alternate Commissioner King said, "We're starting from a given of three million people trying to kill one another on the freeways, and we're trying to relieve that." Let us now examine in depth why it is thought that the train rather than the bus can bring such relief.

The Imagined Benefits of Buses and Trains

Metonymic Representation and Imagery

"The ballot said 'rapid.' It didn't say mass, it said 'rapid,'" said Long Beach Councilman Wallace Edgerton, a detractor from the light-rail system. His problem with the planned service was that it would be too slow. To him, speed was all-important. Edgerton was focusing on one part of a total journey—the time spent on a train. As will be seen below, many of those interviewed for this project focused on the rail component of a trip. Other necessary elements of any journey—getting to and from the train stations—were given little or no attention. This is an example of the operation of metonymy, a symbolic function in which one entity (in this case the trip on the principal vehicle) is used to refer to another which is related to it (the trip as a whole). In doing so, the former structures the understanding of the latter. In this case, the metonymy we are considering is an example of synechdoche: taking the part for the whole.

We will here examine the theory that transportation systems are perceived in partial terms, and that those partial terms structure our understanding of transportation systems in general. As we shall see, this operates on two levels: the metonymic understanding of a trip first unduly focuses attention on that part of the trip spent on the principal vehicle. Secondly, we will pursue the possibility that images of certain physical aspects of transportation systems provide vivid—and misleading—impressions of the functioning of the whole. Lakoff and Johnson claim that "we typically conceptualize the non-physical *in terms* of the physical—that is, we conceptualize the less clearly delineated in terms of the more clearly delineated" and we will examine the extent to which perception of physical vehicular performance can dominate more abstract—and cogent—conceptions of trip quality, such as the ability to provide direct service between dispersed origins and destinations.[7]

There are two perspectives from which perceptions of technical performance need to be examined, although they are not wholly separable, since the one feeds the other. The first concerns how the image of the technology causes information on performance to be structured. This will be detailed here. The second concerns the ways in which the image of a technology symbolizes technological virtuosity—or sexiness—and discussion of this will be deferred until chapter 10.

The speed of transit has long been a focus of attention. It is stressed throughout the 1925 Kelker, De Leuw proposal, which calls for rapid transit lines to "furnish facilities for high speed train service and make possible the transportation of large numbers of people over great distances in short periods of time."[8] The 1948 report on *Rail Rapid Transit—Now!* likewise demanded speed, since "speed is a requisite for a successful rapid transit system."[9] In both documents, the need for speed is emphasized at the expense of considering what ends that speed is to serve. Neither document considered the possibility that a technology with lower vehicular speed, but better ability to distribute people to where they actually wanted to go, might serve passengers better by providing a faster and more convenient total trip.

The majority of those interviewed for this project understood the total trip in terms of the time spent on the principal vehicle, and underlined the importance of the speed of that vehicle. When Supervisor Kenneth Hahn stressed the need for the "mass movement of the people quickly and efficiently, without any stopping at any signal," he was talking about the time they would be on the train. Dan Caufield, then Long Beach light rail project manager for the LACTC, confidently declared that "this will beat the freeway on opening day." The comparison is strictly between time spent on the train and time spent on the freeway—not between the total journey by car as against the total trip time, including getting to and from stations, when using the train. Ted Pierce was particularly taken with the speed of the train. Rail will make for a quick way of getting out of the congested downtown, he said. "They can just go over and get on a light rail car. I mean, they're—whoosh—gone. . . . With a rail, you know, unless there's a wreck or a stall, it's straight on through."

Most importantly, the description of the journey is couched entirely in terms of time to be spent on the light rail service. This is a common way to talk about a trip. Someone going from Boston to Los Angeles will say "I'm flying from Boston to L.A.," not, "I'm driving to the airport, parking my car, getting a shuttle to the terminal, waiting in line at the ticket counter, going to the gate, waiting around there, flying to Los Angeles, getting a rental car and driving to my final destination." If asked how long the trip takes, the general answer is "five hours"—the actual flight time—rather than the total time to get from home to final destination. The journey is understood as a five-hour trip, even though it takes longer. We note that this form of trip representation is typical of our culture but, as the example of the Ojibwa representation given in chapter 7 shows, it is not the only one. While Ojibwa typically refer to the embarkation point to refer to a whole journey, we tend to refer to the principal means of transportation.

In this way, Ted Pierce is structuring his understanding of an urban trip in Los Angeles solely in terms of the principal technology in use, rather than on the actual journey. All you have to do is "just go over and get on a light rail car," and "*whoosh*," you'll be taken where you want to go. In the excerpt from the newscast which began this chapter, the fiction is even more dramatic: the train will apparently "zip" you just where you want, and problems of actually getting where you are going from the rail station vanish.

Baxter Ward was asked about problems of getting to and from rail stations. Studies had found that people preferred to travel directly where they were going in one vehicle to using means of transportation requiring transfers, he was told. "I think if you had something that just went *whoosh*, you would recognize that getting out and changing vehicles was no consequence at all," he replied. "If I were on the Ventura Freeway—or you—driving, and you saw a train go by at 65 mph, filled with smiling air-cooled faces, tomorrow you're going to take the train," Ward said.

Ward uses a metonymy-within-a-metonymy here. His logic: If you're stuck in traffic, and you see a bunch of happy people streaking past, you will learn from their example and take the train. The "smiling air-cooled faces" metonymically represent the people on the train as a whole and—within the overall metonymy introduced above—represent the people's experience of the trip as a whole. Never mind that they may not be smiling quite so much as they park their cars or wait for their shuttle bus getting to and from the train. The image of the high-speed smiling faces on the train going *whoosh* represents the travel experience as a whole, and it makes the train seem like a desirable form of transportation.

Those interviewed generally acknowledged the need for "feeder buses"—particularly within the understanding of the "natural order" metaphor to be discussed in chapter 9, although relatively little attention was given to problems of getting to and from stations. The possibility that rail might be convenient for the line-haul but inconvenient in terms of a total journey thus failed to be given consideration.

Rail has its own right-of-way and is seen as being exempt from the congestion on the roads. Freedom to move at speed is thus seen as an important advantage of rail. Interestingly, Long Beach Councilwoman Eunice Sato, an opponent of the Long Beach project, criticized light rail in terms of its *lower* speed compared to other rail operations: "If it's going to be a slow train to China, I say who's going to ride it? It has to be speedier for people to get out of their cars to take that transportation system." Sato, then, was also focused on the time spent on the principal vehicle alone, rather than on the problem of the *total* journeys people actually have to make.

Alternate Commissioner Roy Donley would also have preferred a faster system than light rail: "People, say out in Thousand Oaks or Agoura or Westlake Village could get into downtown Los Angeles in fifteen minutes on a very high-speed train . . . One reason I favor trains over buses is high speed. If you can get downtown in fifteen minutes instead of forty-five minutes, that's a big attraction." The fifteen minutes may, in fact, be part of a total trip of an hour or longer—if transfers to and from buses are needed, for example—but only the fifteen minutes on the train itself is seen by Donley as being relevant to the computation of actual travel time. The fifteen minutes to him represents the total journey. The fifteen minutes to a traveler of the future would, however, only represent a part of a whole. The rest of that whole might not be revealed by the imagination, but it would be quite evident in reality. And, as shown by numerous studies cited in chapter 3, it would be the whole travel time that the traveler on the actual, rather than imaginary, trip would take into account in deciding whether to travel by rail.

While the speed of the train was lauded, the bus was typically written off as necessarily "slow." "I think people are looking for modern rapid transportation," said Craig Lawson of Mayor Bradley's office. "The bus does not save you time . . . I think the light rail system is going to be faster than traveling by bus or by car." As Los Angeles Councilwoman Flores put it, "If you go the hours people normally go to work, you're going to run into traffic and buses get stuck in traffic. Light rail will not have those problems."

"Buses are much slower, they don't have nearly as great a speed of the train," said Baxter Ward, whose 1976 proposal promised the "fastest" tran-

Figure 8-2. Inside the first Blue Line train to Long Beach. (Photo courtesy of author)

sit system in the world. And just before the LACTC took the vote on whether to proceed with the Long Beach light rail, Supervisor Hahn talked of the "express, fast bullet-type train" which could be provided. The very term "rapid" in "rail rapid transit" is suggestive of the high speed of the mode and, in our metonymic understanding, gets erroneously projected to the speed of the entire trip. As a result, the performance of the train is over-rated, while that of the bus is underrated.

Peter Hall remarks in his chapter on BART in *Great Planning Disasters*, that the original mistake lay in how the problem was perceived. The 1956 consultants' report shows that line-haul speed was thought far more important to commuters than feeder time, despite a lack of evidence. They were clearly wrong, Hall says, as shown by studies the world over which prove that time spent waiting and transferring between vehicles is felt far more onerous than time in motion, even in congested traffic.[10]

Given this evidence—and other findings detailed in chapter 3—one might well be surprised that the same "mistake" was being made for a project of the 1980s. But, once the nature of our metonymic understanding of a journey is revealed, it is not surprising at all. In our culture we tend to naturally represent the whole journey by the time spent on the principal vehicle alone. Perhaps the job of education, then, is not so much to tell people that they are "wrong," but to make them aware of the symbolic reductions which lead to that error. Perhaps, only then, will they be equipped to change their minds.

Trains Are Said to Be More Comfortable than Buses

Bob Robenhymer of San Diego's MTDB was one of many to expound the view that "there's the inherent attractiveness of rail. That has an edge over the bus system." "The light rail will be a step above the bus in the sense that it will be more comfortable," said Long Beach state Representative Dave Elder.

During Assemblyman Bruce Young's influential hearings in Long Beach, Adriana Gianturco, then Caltrans director, drew attention to the "comfort and convenience" of trains. "I believe most people enjoy riding trains, that's certainly been the experience we've seen in the last few years with our Amtrak trains. Trains are attractive and they can be quite comfortable, allowing people to move about freely while the vehicle is in motion."[11]

Long Beach Mayor Ernie Kell sees the train as something that is altogether pleasant and relaxing: "I could sit and enjoy a newspaper or read a technical manual or even a book as opposed to getting myself all stirred up emotionally for an hour and tired by the time I get to work. . . . That would be reason enough for me to change my mind and take it, sit back and relax and have a nice cup of coffee out of the thermos or something, enjoy the trip and just watch the scenery go by." "I think it is true," said Steve Dotterer of the City of Portland staff, "that trains are more comfortable to ride on." While he acknowledges that rail critics don't accept this claim, he said that rail's greater comfort is "real for most people, it's only unreal to people who are sitting in offices and analyzing things."

While the train's imagery is utopian—symbolizing all that is good in transportation—the imagery of the bus appears to symbolize all that is bad in the world. Buses, said LACTC Commissioner Marcia Mednick, are seen as being "noisy and dirty and slow." Baxter Ward put it more strongly:

> People don't like buses. People just hate buses. They have to sit in the damned sun and they got to sit and take all the fumes from the cars and the diesel Mercedes, and the diesel buses that aren't theirs, and wait until their bus comes along, get in, crowd, lurch, be abused by the operator, and just drag red light to red light or whatever the situation is, until they finally get to their destination.

Jacki Bacharach was also convinced that a train was "much more comfortable than a bus; I just think it's going to be more attractive to ride in them than in a bus, and I think studies have shown that too, that people accept rail more than they will accept bus."

As seen in chapter 3, studies have shown that people rate the comfort of an urban trip as unimportant compared to trip time and cost. There is no reason, furthermore, why buses should not be made more comfortable. Golden Gate Transit in the Bay Area, for example, uses luxury coaches to serve a largely high-income professional clientele commuting from Marin County to San Francisco. We noted, furthermore, in chapter 2 that in the 1930s buses were often introduced into suburban markets because it was thought that their "luxury" would attract a clientele that would not travel by streetcar. The evidence relating to buses, nonetheless came from conditions decision makers observed in the 1980s. And if buses did not appear to them to be comfortable, they inferred that buses are necessarily uncomfortable. Trains therefore clearly seemed preferable.

Trains Are Said to Be More Secure than Buses

Part of a transportation mode's "comfort level is an understanding of where it goes," said Lee Hultgren, director of transportation at SANDAG. A basic human need is for simplicity, certainty, and security. Los Angeles Councilwoman Joan Flores sees an important advantage of rail to be its lack of "complication." In contrast, "The bus makes a circuitous ride. It obviously has to go someplace else to pick up other passengers so that you get a full load."

The train is seen as being more psychologically reassuring because it is on a fixed track. As San Diego County Supervisor Brian Bilbray put it:

> The psychological impact of having rail is massive. People know that if I stand by this rail, something's going to come by. I'm reassured by the existence of that rail. That rail is shiny. That means something travels over it periodically. I'm assured that I'm at the right place and something will come pick me up if I'm here. You don't have that with a bus. You have cars going by, you feel left out, you're not assured; you don't have the security of being assured that something's coming out.

Not only is there the assurance that something will come by. There is also the knowledge of where you are going. As Jacki Bacharach explained:

> Distances here are so far apart; you get on the wrong bus and you're really scared that you're going to be 50 miles from your home before you figure it out. I think a rail system, where people know it's going to go from here to there, and it's not going to go any different because I know it, is going to be a little bit easier to get used to and accept. They're very easy to use. Obviously, you don't even need to speak the same language to get on these things and use them. If you get on the wrong bus, you don't know if you're going to end up in somebody's neighborhood instead of downtown Los Angeles.

"The rail has a physical presence and understandability," stressed G. B. Arrington of Tri-Met, Portland. "You know you get it at station A anytime you want, and you know where it's going and you know how frequently it's coming by," said Mark Wiley, of RT, Sacramento. "They certainly wouldn't take a bus to go there [downtown Portland] because it's uncomfortable and kind of an unknown quantity, foreign for them to take the bus," according to METRO, Portland's Andy Katugno.

LACTC Commissioner Christine Reed complained that bus systems were especially difficult for tourists: "It's difficult to understand, and then there's a multiplicity of systems and you have to transfer between them sometimes . . . and the general public doesn't even know anything about it because the general public by and large doesn't ride it, so you can't just ask people on the street how to use the system." As Alternate Commissioner Bob White concluded, "It's so much *simpler* to park your car and get on the light rail and stay off the highway" [italics used throughout the quotations from interviews are my emphasis]. There is no evidence that regular users of buses find the task daunting: they get to know which bus to take, where it stops, and what time it leaves. But the image of the fixed track—and a knowledge of where it goes and where it ends—lends rail systems an aura of simplicity and certainty.

Nonregular users may find bus information systems inadequate. Most Los Angeles bus stops (other than those installed under the latest Rapid Bus program) do not have schedule information, or even maps of bus destinations. An appropriate response might be to improve these facilities. But to decision makers, the more obvious solution seems to be to install the more intuitively reassuring rail service.

There Is Said to Be Less Crime on Trains than on Buses

"There is a reluctance by a large segment of the population to ride the bus because of the safety aspect," said Alternate Commissioner Barna Szabo. "One of the emphases on the light rail is it's going to be a safe ride."

Bob White has experience from the bus system in his own City of Norwalk: "Buses are notorious for breaking down and cutting seats, and teenagers on the buses," he said. "You have to be a cop, you have to be a chauffeur, you have to be a bus driver, the problems of discipline would be much greater on a bus. We've experienced that right here in our City of Norwalk. We are continually thinking of ideas now to teach our drivers how to run their bus and keep them from getting torn up."

Alternate Commissioner Walter King had similar concerns:

The crime on the buses is fantastic. The decorum of the people on the buses . . . The people themselves have lost their morals, the home is broken down, the church is broken down and the schools are broken down. We don't have discipline any more. You tell the guy to sit down and he won't do it, and that's your problem . . . The security. And the robberies, and out our way I wouldn't be seen on a bus.

Wouldn't the same problems exist on trains, King was asked?

No, you're isolated with a bus. When we build this system, we're going to have coverage you don't have on a bus. We're building in millions of dollars worth of things, it's going to make it safe. . . . You may have to pay a price. We've got a bunch of animals out there that we can't design and build around. We've got to build for the masses and hope the masses will work and will use it, and I think that's where we're coming from.

It Is Claimed That the Bus Is for Criminals, but the Train Is Middle-Class

The very image of the train—as something "modern" and clean—feeds the image of the train as something safe and crime-free. Said Baxter Ward, "I don't know that undesirables get on trains any more than they do on buses. I doubt they would on a brand new transit line." Roy Donley would agree: "There's a lot of crime on the buses. There may be on the trains, too. But I think criminals are more likely to savage buses than they are trains."

Craig Lawson of Mayor Bradley's office went even further:

I have a theory that if you treat people well, then they act accordingly. If you take them and you put them on a nice-looking car with a conductor, that is clean and that's comfortable and air-conditioned, then they're going to act well. . . . So I think the way the system is being designed as a high-quality system; it's a system that's going to be efficient and comfortable, and the people will ride it, and that no matter who gets on there is going to be there as a commuter, not as someone who is going to vandalize the place.

The metaphor is similar to the 1960s view of urban renewal: tear down the physical blight and replace it. If the buses are blighted with crime, then remove the buses, replace them with something bright and modern, and the problem will apparently go away. Not only that, but the people will be reformed and crime will vanish. There is no hard evidence for this, of course. But imagery provides the evidence that counts in maintaining such beliefs.

Because of All of the Above, It Is Argued That Middle-Class People Will
Use Trains

Commissioner Marcia Mednick was one of several interviewees to point out that:

> The rail will attract a certain type of ridership that will not necessarily get on the bus. Because this is a newer type of technology, it can come in with a whole new type of image. . . . I think you'll get a whole range of ridership on the lines. You'll get—I hate to call it quote "your low-income type of person," your transit-dependent because of economics or the people that are traditionally transit-dependent—but you'll also get riders who have not been transit-dependent in the past, who elect to use the service if it's clean, if it's fast, if it gets them where they want to go.

Mednick then showed a rare degree of awareness of the role of imagery: "Part of it is an image. And people have certain images in their mind as far as buses being noisy and dirty and slow and things like that. *Which of course is not always the case.* I'm not saying that's the case, but that's the image to many people. The trains, because it's a new technology, it's a whole new system, can come in and be sold differently." Mednick understands that imagery may not reflect the type of service which *might* be provided. She is nonetheless convinced that the bus image would act as *deterrent* to certain groups of potential rail riders, even though the evidence provided by the substantial white-collar ridership on the El Monte Busway shows this to be untrue. Despite her awareness of the potentially misleading nature of imagery, she advocates rail because of the effect on ridership she wrongly attributes to the image of the train as against that of the bus.

The bus is something only for those without choice, said Craig Lawson of Mayor Bradley's office. "For those people who are wedded to their cars and never ride the bus, they perceive it as being something that they don't use, so they don't like to talk about it." Added Long Beach Councilwoman Hall: "Well, I'm told that studies show . . . that the transit dependent will use whatever's available, but that those who have an alternative and have an ability to choose will, if they have a choice, would more likely ride in a rail system than on a bus." As we saw in chapter 3, studies actually show quite the reverse: that people decide how to travel on the basis of the total trip time and cost, and that little else matters. This goes against the grain of the image, however.

Relieving and Causing Congestion

"I personally hate buses," said Howard Mull of Los Angeles County Supervisor Michael Antonovich's office:

> not for the sake of hating buses, but whenever there is a gridlock in downtown Los Angeles, for instance, you can attribute 90% of that gridlock to the buses, because they're taking up space at intersections at signal changes, what have you. They're sitting there blocking traffic two and three deep . . . They have unlimited turns, they can make a left turn anywhere they want to, and so consequently they're the major cause of gridlocks.

During the Assembly Committee hearings in Long Beach, Assemblyman Nolan Frizelle also characterized the bus as the enemy: "I am personally dead set against the business of the big humungous bus on our local city streets," he said.[12] The bus is seen as more of a contributor to problems of congestion than as part of the answer to them. "We can't just depend on buses," said Supervisor Hahn's deputy, Burke Roche, "because one of the great problems, and you see it every rush hour in the evening, the street is congested with buses, causing most of the congestion."

"Buses, buses, buses. It sounds terrific. It just doesn't work that way. There's not the space on the concrete to run the buses in volumes sufficient enough to accommodate them," said Dan Roberts, of the office of Congressman Mineta (San José). Long Beach Councilwoman and RTD Board Member Jan Hall put it more graphically. "We run 900 buses a day on Wilshire Boulevard," she said. "We can't run any more. The traffic is such that it would do us no good. We would be parking and basically the most efficient system would become entering by the back door of the bus, getting off on the front door and entering the next bus on the back door and walking your way on buses the length of Wilshire."

Debbie George of Supervisor Deane Dana's office amplified the inability to expand bus services further:

> I think the buses are having a hard time getting through. We've double-decked our buses now, should we triple-deck them? Should we make them even wider? Should we widen the road to have four lanes of buses going up and down? . . . We have buses that are not only double-decked, but they are two buses long, you need roughly ninety feet of parking when one of those buses stops; is what they want just a sea of buses, which in essence would be a train?

Buses, from this perspective, are something desirable to remove: "If you take a bus off the freeway system, you're removing a vehicle, a good sized vehicle at that," said Long Beach Mayor Ernie Kell.

By transferring people to trains, Alternate Commissioner Bob White thought, as much as 10 percent of traffic could be taken off the freeways. Baxter Ward went so far as to claim that "Most businessmen would abandon the freeways and use the trains because it would be so remarkably fast . . . If you take only 15 percent off the Ventura Freeway, it's that last 15 percent that are keeping you down to 12 mph, instead of 20. So you'll speed up everything." Debbie George was equally optimistic claiming that "it [the light rail] would ease traffic off the Harbor [Freeway], it'll ease traffic off the 7, and it'll ultimately ease traffic off the 5."

Figure 8-3. Peak-hour buses have a notable presence in downtown Los Angeles streets.
(Photo courtesy of author)

Rick Richmond, LACTC executive director at the time, said, however, that "I don't think we've ever contended that" light rail would immediately reduce traffic on the freeways. There was a lack of consensus on the commission, furthermore, that rail would necessarily help sort out congestion. Alternate Commissioner Blake Sanborn, for example said, "To some folks the ability to transfer the congestion from the freeways to the trolley line was important in making their initial decision. In terms of benefit, I think that's a secondary item."

Buses Are Polluters, Trains Are Not

As KABC-TV newscaster Paul Moyer declared: "Los Angeles Metropolitan area has severe air pollution-transportation problems. Don't have to tell you that; you know. Yet we don't have any real rapid transit." The buses in Los Angeles did not count as "real rapid transit." What Moyer was calling for is rail.[13] As Long Beach State Assemblyman Dave Elder said, "I believe that the light rail is clean: the reason that I'm for light rail deals with the environmental consequences in the southwest air basin." "Trains are quiet and they do not pollute the air," said Adriana Gianturco at the state assembly hearings in Long Beach.[14] Assemblyman Frizelle echoed her feelings, recalling that "One of the great beauties of the Red Car system was that it did not put diesel smoke in the air."

The sight of a bus exuding exhaust leads to an image of the bus as something unclean, an untouchable. According to Ted Pierce: "Those people that wanted rail, and they're driving down the freeway and they're seeing all the cars and they're going oh, there must be another way to get in here, but I hate those buses, because they stink up everything." Alternate Commissioner Roy Donley concurred: "They spew out a bunch of smoke every time they accelerate, they're called stink pots by the people," he said. "They're terrible," said Baxter Ward. "They're smoggy and fume-ridden."

Alternate Commissioner White provided a particularly vivid account of his image of the bus, tying it directly to his experience driving behind a bus on the highway.

> Did you know they planned on having a bus diamond lane and a deal for buses to go on this one lane? But I think that's kinda impractical, those buses are smog buggies, they produce a lot of smog, they use diesel, they're not as comfortable . . . I personally think the buses stir up so much—they're most

of them burning diesel. And have you ever followed the bus, I mean that stink of that diesel? I tellya, I think that's polluting our city right now. A lotta people think it's the automobile—and that's gotta be part of it—but those big buses sure don't help. . . . Not the buses. I'm against the buses all the way.

The image of "the bus as enemy" is common to the accounts of both White and Ward. To both of them, the bus is a part of the automotive illness, not a cure for it. To Baxter Ward, the bus is also a source of actual human illness: "The smog content of the basin is enormous," Ward said "and the most damaging type of smog is that created by diesel vehicles. It is the intention of those now designing the transit services that we add more of these vehicles, with Park-and-Rides out to Pomona, Disneyland and every place else, we'll have more smog and more illness created just to accommodate the view that people don't want to ride trains. To Hell with it!" Alternate Commissioner Roy Donley, when asked why not have busways, rather than railways, said: "because I think that you're not doing a thing towards solving pollution problems." "Considering the pollution of California, it would be counter-productive, it would be foolish to talk about having more buses which is going to create the pollution," Compton Mayor Tucker said.

Long Beach councilman—and light rail critic—Edgerton was not, however, similarly swayed by the image of the dirty bus. "A bus is more polluting than an automobile, but the numbers of people that a bus can take out of automobiles will be a net loss of pollution," he said. We here have an example of symbolic understandings being mediated by prior beliefs. Critics are less likely to be swayed by images which influence supporters.

Capacity: Trains Come Bigger than Buses

Because trains come in larger sizes than buses, it is thought that they can carry more passengers. It seems obvious that not only can they do this, but that they *will* do this. The Rapid Transit Action Group *Rail Rapid Transit—Now!* proposal confidently declared that:

> Automobiles on 3 lanes of a freeway will move 7,000 persons per hour in one direction figuring 1.7 persons per car. Buses operating at 20-second intervals will move about 10,000 persons per hour. A rail line in the center of that freeway inserted at a fractional additional cost, will move 30,000 people per hour. . . .

Rails separated from all other traffic are a must when a city becomes as large as Los Angeles and its sister communities.[15]

The image of trains whisking the masses quickly and efficiently through tunnels is especially potent. Said Jacki Bacharach, "How do you get a bridge across where BART goes under? How many more buses would you stick on that to carry what BART is carrying today?" To Compton's Mayor Tucker, the high capacity of rail was a given fact: "You can haul many more people, everybody knows that all over the world, with trolleys, with light rail, than you can with buses." As a former Red Car driver, R. L. Bacchus, wrote in a letter to Supervisor Kenneth Hahn: "Even with the old Red Cars, when I was a motorman, I pulled a four car train, with four-hundred and eighty passengers, from Los Angeles to Long Beach in thirty-six minutes. Can you imagine what could be done with modern equipment?"[16]

All these cases focus on equipment capacity, not where people need to be taken. The assumption that rails are a "*must*" when a city "becomes as large as Los Angeles" is misleading because it ignores the nature of travel within that city, which can make rail inappropriate to even the largest of cities. In Los Angeles County, complete trips are not concentrated in high-density corridors in which people can easily walk to and from rail stations. Whatever the physical capacity of a rail system, a bus system—which is able to branch out and serve a far larger number of desired origins and destinations—might have the ability to serve considerably more people than rail. But we tend not only to focus on the perceived performance characteristics of vehicles in evaluating the modes of transportation of which they are a part, but to focus on the principal vehicle used alone: in our metonymic way of understanding a trip, access and distribution problems remain invisible.

Efficiency—The Driver Image

The *Los Angeles Times* reported on October 20, 1985 that one of the most frequently made arguments for the rail line was that it would be cheaper to run since a single train driver can haul five times as many passengers as his equivalent on a bus. As Ted Pierce said, "you can put three hundred, four hundred people in at one shot and just move them out of town." Baxter Ward went even higher: "One motorman can carry seven hundred people on his train and it would take ten bus drivers to do the same. So, in terms of labor, you're much better off with the rail lines."[17] The July 6, 1980 *Los Angeles Times* topped even this estimate: "A bus can carry only about 70 passengers per driver, while streetcars can be strung together with one operator for 1,100 passengers."

The picture of train drivers propelling far more people along than their colleagues on the buses was one of the most widespread—and to those under its spell—compelling images among those interviewed, in media reports and in other documentation.

Other examples of the imagery in operation are given in Table 8-1.

Not only are fewer personnel apparently required, but Commissioner Zimmerman went so far as to suggest at the March 24, 1982 LACTC meeting that the job was less skilled than bus driving, and so would attract lower salaries:

> Neither LRT or CST [Cable Suspended Transit] will require the quality of operator that is required to drive a bus. There is no traffic to dodge, no fares to collect, no information to be given, no steering wheel to turn, no signal to be given, no route to remember. One only has to be able to push a lever to make it go and another lever to make it stop. Senior citizens and college students can be perfectly qualified and should be employable at minimum wage with minimum wage fringe benefits.[18]

LACTC deputy director Paul Taylor suspected, however, that: "The introduction [of the Long Beach line] will probably increase [the RTD's] operating budget by 2% . . . Probably the operating cost per marginal passenger is greater than if you did it by bus up to a point, and I don't know where that point is." He said, nonetheless, that he was expecting to get beyond that point.

Other arguments in favor of rail operation included the lower costs of maintaining an electric versus a diesel fuel based operation; because electric power is seen as being "cleaner" than diesel, electric equipment appears cheaper to maintain. Rail vehicles also enjoy a longer life span than bus vehicles.

The imagery of the efficiency of rail demonstrated here is very powerful—and very misleading. The cost of drivers (requiring salaries at least as high as those of bus drivers) is only one of many costs and, as discussed in chapter 3, capital-intensive rail systems are burdened with many costs which bus systems do not face, as has quickly become apparent in the op-

Table 8-1. Comments on Driver Efficiency

*You've going to put more people in an elongated car and you're not going to
have as many operators. One man can handle 300 people.*
—Walter King, Alternate LACTC Commissioner

*One driver: three cars. You can carry 5, 10 times the number of people with
one driver. . . You have the bus driver, so to speak, carrying one bus with 40
to 70 people, and you have an operator with three trains connected
together, so effectively if you did a straight across the board analysis of one
bus driver per people as opposed to one train operator per people, you get a
significant reduction in that personnel cost, which is a significant cost of
running a system.*
—Manuel Perez, LACTC, Rail Construction Committee

*You can sit twice as much on the light rail vehicle. So you have two light
rail vehicles. With the motorman and a cop on the other, you're still
doubling your productivity on the bus, which would require four drivers for
200 people.*
—Richard Stanger, LACTC staff

*By the mid 1970s [San Diego] community leaders realized that buses
simply couldn't haul as many people as efficiently as the rail car.*
—Anna Chavez, KABC-TV, Los Angeles, July 30, 1980

*Within the transit mode, rail is cheaper to operate than the other major
alternative, bus, and this is because rail systems do not require a lot of
labor.*

*It has been estimated that the operating costs of a rail transit system are
only one-third of that required by a bus system transporting a comparable
number of riders. As a result, 80 percent to 90 percent of a rail system's
operating cost can easily be recovered from the farebox as opposed to the 20
percent to 40 percent of farebox recovery on a typical bus system.*
—Adriana Gianturco, Director, Caltrans, Hearings, California
    Assembly Transportation Committee, Long Beach, August 14, 1981

*The state Transportation Department recently completed a feasibility study
of the Long Beach route, the only one that it has surveyed. According to
Adriana Gianturco, state transportation director, the line could handle 15,000
commuters daily, with rush-hour schedules of 15-minute intervals, for
about one-third of what it costs to move the same number of people by bus.*
—"Hip, Hip, Hurray," Editorial [in support of rail],
    Los Angeles Times, October 15, 1981

> *Preliminary estimates put the cost of building light rail to Long Beach at less than two hundred million dollars.*
>
> *Trollies are also inexpensive to operate. In San Diego, for example, as many as four hundred fifty people can be carried on one three car train. Each train requires only one operator. The trains make the sixteen-mile journey from downtown to the border in forty two minutes. Before the trolley, it would have taken at least four buses, with four operators, one hour and ten minutes to carry the same load.*
> —Gene Gleason, KABC-TV, Los Angeles, December 31, 1981
>
> *We made the decision from a transit perspective based solely on operating cost, and we thought and continue to believe that light rail is more efficient than buses because you have one operator in a two-car train and you can carry more people.*
> —G. B. Arrington, Tri-Met, Portland
>
> *Characteristically from a pure, real gut level feeling about it, certainly it should be more cost-effective on the operating cost side because of the fact that you only have a single operator for however many people you're pulling.*
> —Jim Pierson, San Jose
>
> *As do most transit operators, RT [Sacramento] expects their rail service to have lower operating costs per passenger than bus, mostly because of its lower ratio of drivers to carrying capacity, and therefore to have a higher farebox-return ratio.*
> —Johnston and Sperling[19]
>
> *We will be carrying more people with the same level of expenditures.*
> —Mike Wiley, Asst. to General Manager, RT, Sacramento

erating budget of the Long Beach line. Not only are items like right-of-way and station and fare equipment maintenance costly, but feeder buses to bring passengers to and from rail stations have to be paid for, and the cost of these buses must be included when comparing rail operations to the cost of providing direct one-bus bus service. This cost is generally overlooked. There is a trade-off, furthermore, between the quality of service of a large (rail) vehicle on a small number of fixed routes and the flexibility of a smaller (bus) vehicle serving a larger number of neighborhoods directly and more frequently. These—invisible and rather abstract—complexities are not readily perceived, as compared to the commanding image of the speeding train efficiently transporting hordes of commuters to work.

Shaping the City

The association of rail with forces for centralization, and of highways with the culture of dispersion has played a central part in the political account given in chapter 6. Yet, ultimately, rail became more acceptable to dispersed communities when the Proposition A map suggested it could serve them as well. In the process immediately leading to Proposition A and the choice for implementation of the Long Beach light rail line, issues of city form did emerge from time to time—Assemblyman Elder referred during Assemblyman Young's 1981 Assembly Transportation Committee Hearings in Long Beach to the value of light rail "in helping to accelerate the already fast moving redevelopment of downtown Long Beach"—but did not come into the foreground. Attention focused on the transportation crisis afflicting L.A.'s freeways, more than on issues of urban development. This emphasis is reflected in interviews, too, although the potential of rail to bring revitalization to Watts and Compton (to be discussed in chapter 11) was important, as was the ability of rail to generate pride in a city and give identity to its core (see chapter 10).

Rail was seen by some of those interviewed as a tool for guiding growth and, in particular, for bringing order to the chaos of a dispersed megalopolis. As Jerome Premo, former LACTC executive director said, "Now, if we have a bunch of Adam Smiths whose view is that people should do anything they want whenever they want and let's not worry about it, then fine, but I happen to think that there's some logic to planned and coordinated growth."

The rail system is symbolic both of concentrated development and of a form of order: it provides service to a limited number of foci and, it can therefore be argued, will tend to encourage development at those points. As we saw in chapter 2, this was certainly the effect of Henry Huntington's Red Car system. Some of those interviewed thought the new light rail could also have an impact, if a less significant one. Commissioner Barna Szabo saw light rail as a way of bringing the urban core to life:

> It relates very well to certain land use patterns, and can also enhance certain land-use developments. . . .

Let's suppose you have about a 10 block to 20 block major street, on which there are offices, restaurants, a hotel and other related service facilities. Given the cost of parking these days in urban areas, and the inconvenience of getting into a parking space, getting out of a parking space, getting into a new parking space, a person who wants to leave for lunch, for example, can get out of his office, get on the light rail, go six blocks, and go to a restaurant or go to a store, pick up what he wants and come back and so forth without a major expense. . . .

And that type of concentration is what then helps you create and enhance and make functional an urban core, which is a retail service core. And then you can link it to nearby residences. And I think in this Long Beach line, for example, you can create sub-areas, like going through Long Beach, Long Beach Boulevard could become the new North-South commercial center for the city, the back onto the city that light rail can serve very well. Cars cannot, because cars would allow you to go all over the place.

In Szabo's vision, the restricted mobility provided by rail is actually an advantage, because he sees it as having the potential to concentrate certain development, which he says cannot be done with cars. Despite a lack of evidence that rail could have such an impact in an autopolis of the 1990s, or that people would actually wish to be bound by the restrictions Szabo sees as advantageous, Szabo's understanding of light rail's functioning—which was formed from experience of rail in Canada and in Europe—leads him to believe it can bring order to the metropolitan area of Los Angeles.

Gerald Leonard, former aide to Baxter Ward, thought likewise, this time citing the flexibility of buses—rather than cars—as problematic, not advantageous:

There is a belief and I think empiric evidence to prove it, that rail has an attractive quality that the flexibility of the bus doesn't possess. If you build a rail line and it attracts increased property values at least along the station areas, if not the whole line, so it's an inducement to the community which ultimately is a self-fulfilling prophesy that it attracts density and it feeds on itself. The rail line needs density, and it attracts density.

Ted Pierce also had a vision of the concentrated city: "To me it's less expensive to go ahead and build one area and have everybody within walking distance of that area, because then it costs—the transportation nightmare of getting people back and forth between those areas is just compounded."

All these quotations focus on the creation of order out of chaos and link that vision to the construction of rail facilities, believing that rail can have that impact. The very restrictions to mobility imposed by rail—rigidly fixed links of steel-symbolize control and order, and do so to an extent that the competition of an existing sophisticated road system is ignored along with evidence that recent rail systems have shown little power to shape development. With rail having the ability to transport only a tiny proportion of total demand, how could it be expected to compete with the freeways in organizing development?

Los Angeles Councilman Ernani Bernardi sat away from the mainstream view of the desired city form, attributing congestion to having "too many people to concentrate in too few areas, instead of dispersing the population as well as the workforce . . . I support the concept of dispersing the activities and limiting the activities in one area and making the activities more available in more areas of the city and more areas of the state . . . My improvements would be to limit the growth on congested areas." Bernardi saw focusing on downtown as making people "prisoners, they have no choice: their choice is you come downtown, or else. . . . I don't know why people want to be herded like sheep, and that's what we're talking about doing here. This was so nice in 1940 when I first came out here."

To those who want concentrated urban forms, rail appears to have much to commend itself as an aid to desirable development, despite the lack of proof that the introduction of rail will indeed induce such development. Bernardi, on the other hand, came at the problem from a different frame of assumptions—he favored dispersed development—and he was against introducing the train to Los Angeles. These arguments reflect those heard time and again during the political development recounted in chapter 6; while they were not central to the arguments for and against rail, it is clearly still understood that the train can have a role in shaping urban form, despite the auto-dominance in Los Angeles.

Imagery Sells

Everyday imagery and understandings create impressions of what constitutes good policy, and what does not. According to our normal metonymic understanding, we tend to think of a journey in terms of the time spent on the principal vehicle making the trip. Access and distribution problems sink into the background: the main issue becomes one of vehicular speed. Trains appear to be more comfortable as well as faster than buses. They seem more secure and less prone to crime. They have a middle-class image. Buses do not. Underlying many of the statements is a subtle racism suggesting that "undesirable" people to be found on buses will not contaminate trains. As Walter King said, bus users were a "bunch of animals."

Buses are seen as a cause of congestion and of pollution: a symptom of L.A.'s urban woes, not a solution. Trains were not universally said to be the answer to congestion, but they were seen as clean and pollution-free, and able to travel on a congestion-free right-of-way. Perhaps the most widespread impression was that trains cost less to operate than buses because they required only one driver to carry the equivalent load of two or more buses. This misconception, resulting as it does from an obvious "common sense" logic, is intuitively appealing, if wrong, because it ignores the complexities of actual operations.

Most disturbingly, false conclusions about the alleged advantages of rail over bus service found their way into the conclusions of decision makers, not only in Los Angeles but in other West Coast cities adopting rail systems, too. They appeared, furthermore, in TV and newspaper stories and editorials on transportation, providing conduits for a most compelling—if unintended—misinformation of the public as a whole.

We see in the next chapter how the images identified here do not exist by themselves, but fit into a structure of metaphorical understandings. The images of buses are bad, we've seen, while those of trains are good. But one response to bad bus service might be to ask how it might be made better. That was to happen to a limited extent as the twenty-first century got underway with the introduction of the Rapid Bus program, a second-best response to the financial impossibility of continuing with the planned rail program even if propelled into existence with the type of imagery which had glorified rail (see chapter 10). The metaphors underlying thinking about transportation, as we shall see, tell us why in the absence of such drastic circumstances, rail reigns supreme.

CHAPTER 9

Integrating Metaphors

In Greek "transport" is metaphora which also means "metaphor." The train constitutes one extended metaphor conveying an inexhaustible supply of lesser ones. It is a metaphor-transporter or a metaphor of metaphors. A three-dimensional mobile metaphor for metaphor itself.

—Tiresias

IMAGES ASSOCIATED WITH TRAINS and buses tell us that the former eclipses the latter at solving transportation problems. These images do not, however, exist in isolation, but are integrated by a series of metaphorical understandings. Together, these understandings set the definition of the problem to be solved. They create a coherent picture of why one possible solution would succeed, and the other would fail. The images—while performing an evaluative function themselves—define the characteristics of the modes of transportation to be evaluated by the metaphors. The output of the images becomes the input to the metaphors. The metaphors then provide a basis for evaluation.

Figure 9-1. Destination boards read Los Angeles (see p. 302). (Photo courtesy of author)

The metaphors we shall be exploring are often rooted in organic understandings: of organisms alive and vital; and diseased and unwell. Underlying these metaphors are further understandings of physical systems which are used to structure solutions to complex social problems.

Organic Metaphors for Transportation

If you have ever fumed in a blue haze on a clogged freeway at rush hour, watched with dismay as the gas pump meter clicked towards a figure resembling the national debt, or stood at the bus stop and seen overloaded buses lumber by without taking you aboard. . . . If, like "Network's" Howard Beale, you are mad as hell and don't want to take it anymore, we have good news for you: You don't have to.

Proposition A on the Nov. 4 ballot offers the seven million traffic-cramped, transit-starved citizens of Los Angeles County an opportunity to regain their dwindling mobility while winning a measure of independence from the automobile.

—Editorial, *Los Angeles Herald-Examiner*, October 21, 1980

The human body is our most ready source of everyday experience, and we project our experience and understanding of it into many domains. Writing on strategic planning in London, the urban planner Douglas Hart describes, for example, how urban planners came to conceive of that city as an organic whole: there was perceived to be some "natural order" under which the city could function healthily.[1] The planner's task was to identify how the city had deviated from that natural order and then to take corrective action through which that order could be restored. Hart quotes Patrick Abercrombie, author of the influential "County of London Plan," who, in 1933, wrote that "Town and Country Planning seeks to proffer a guiding hand to the trend of *natural evolution* as a result of careful study of the place itself and its external relationships. The result is to be more than a piece of skillful engineering or satisfactory hygiene or successful economics; it should be a *social organism* and a work of art" (my emphasis).[2]

A basic source of metaphorical understanding in transportation appears to be organic and, specifically, projected from our experience of the human body. But there are other concepts at work, too. Some derive from the body concepts; others are determined by how the body itself is metaphorically understood (and we shall see that the "body-as-machine" metaphor is powerful here); others still come from quite separate metaphorical sources of structuring, but mesh together to create coherence in meaning.

Fundamental metaphors at work here conceive of a broken-down transportation system as a body afflicted with disease. There appear to be two separate metaphors, but with a fuzzy, gray boundary between them. Most striking is the metaphor of a body in balance—or fallen out of balance. When all of the body functions according to the natural ordering of its parts and the relationships and flows between them, it is balanced. When it departs from the equilibrium nature has designed for it, the body—or transportation system—becomes sick. An alternative account emerging from interviews sees the need for an "escape valve." While the balance metaphor suggests that a cure is possible—enabling a return to balance— the escape valve metaphor sees organs as incurably diseased and advocates bypass surgery to allow circulation to continue through an alternative set of arteries.

The Circulation Metaphor

Both the "balance" and "escape valve" metaphors come to life through a conception of circulation, a metaphor which is at once organic—the circulation of traffic is metaphorically akin to the circulation of blood through the human body—and mechanical: both blood and traffic flows are seen as substances being pumped through a system of interconnecting tubes. When the transportation system breaks down, it is as if afflicted with a disease which impedes the natural flow of traffic.

Restoring Circulation in London

Historical evidence from transportation planning in London exposes the circulation metaphor in action in another context and helps illustrate its principles.

As far back as 1945, one author wrote that "Transport in all its forms is a system of arteries and veins through which the blood stream of the city carries oxygen to the brain and nourishment to every part of the organism."[3] In our everyday language we talk of "arterial" and "circulator" roads, and it follows from the metaphor that if these vital channels get blocked—and flow constricted or even brought to a halt—the affected organs suffer malnutrition, weaken, decay, and can even die. There are a number of logical responses under the metaphor, but they all focus on the restoration of free flow.

The most obvious response under the circulation metaphor is to construct a new system in which free flow can once more prevail and decaying organs be thereby revitalized. Thus, Abercrombie maintained that "the essence of good planning is to *canalize* main streams of traffic" (my emphasis). His response to the congestion which threatened to constrict London's life was the creation of a new hierarchical "circulatory system." "Corrective surgery was required," writes Hart, "and would necessitate the insertion of artificial channels, or canals, to drain away traffic from areas where it was both unnecessary and unwanted."[4]

The minor vessels serving the extremities of the body cannot carry major flows of blood: they would quickly become clogged and fail under the undue pressure applied by an attempt to force too much blood through them. Larger arteries—from the aorta down—serve instead as the higher-level channels of a hierarchical system which distributes blood to all parts of the body. The hierarchy serves to keep major flows from congesting parts of the body where they are not needed, allowing flows appropriate to each part of the metabolism to reach the organs which are to be nourished. The problem in London was seen as one of lack of structure: with major channels in a state of disrepair, the city's lifeblood was being forced through minor arteries, congesting fingers and toes.

We note that Abercrombie's thinking mirrors that of the *Major Traffic Street Plan for Los Angeles*, which saw congestion as a function of "unscientific" street width and design, and "improper" use of existing street spaces.[5] That plan had sought to produce a "balanced scheme for handling a tremendous traffic flow" by establishing different classes of roadways for different traffic needs as a way of avoiding the "promiscuous mixture of different types of traffic," which the authors said caused congestion. The London example is used in more detail here, primarily because links to the circulation metaphor are explicitly spelled out, but also to show that the metaphor is far from localized in its influence.

A three-tier hierarchy was devised to provide a solution for London's problems: arterial roads for fast, through traffic; sub-arterial roads for local traffic; and local roads for access to particular destinations—the fingers and the toes. Through traffic—streaking by on segregated arterials—would avoid lower-order roads which could then perform local functions without getting congested. With each type of road carrying a free-flowing level of traffic within the limits of its design capacity, the parts of the system would come together to form a whole road system in harmonious balance.

The circulation metaphor generated a new solution to transcend the inadequacies of the old arrangement. In every sense, it had obvious appeal: More capacity was clearly needed because the existing system was clogged; but, while this added capacity was needed to enable the city to continue its vital functions, it also had to improve environmental conditions for individual localities. By ordering the flows of traffic, the road hierarchy appeared to serve both aims, promoting movement while appearing to keep neighborhoods free from through traffic.

Public protests in London led to cancellation of most planned road building. If the circulation solution offered to protect neighborhoods from high-volume traffic, it ignored the walls of noise, pollution, and visual obstruction which would be built by placing major arterials at the borders of those neighborhoods. The emphasis on canalization of traffic gave priority, furthermore, to the free flow of *vehicles*: it ignored the possibility that the movement of people might be eased by controlling the circulation of vehicles, perhaps by emphasizing public rather than private transportation. It also bypassed equity issues: it promoted circulation by vehicles without asking *who* would be circulating in those vehicles or who would pay for that circulation in both monetary and environmental terms.

The Body As Machine

Our everyday understanding of the human blood circulation system is, in fact, a mechanical one: the body is seen as a machine. In the above account, there is a clear mechanical understanding: circulation is seen as the pumping of fluids through tubes, whether to maintain flows of traffic or blood. This comes from what the philosopher Mark Johnson identifies as the "Body is a Machine" metaphor. Entailments of this metaphor include:

The body consists of distinct, though interconnected parts.
It is a functional unity of assembly serving various purposes.
It requires an energy source of force to get it operating.
Breakdown consists in the malfunctioning of parts.

Repair (treatment) may involve replacement, mending, alteration of parts, and so forth.

The parts of the functioning unity are not themselves self-adapting.

[partial listing][6]

To this may be added the implication that there *is* an equilibrium state in which the machine functions efficiently: the job of repair is to restore that equilibrium. Sometimes, however, the machine breaks down beyond repair.

Illustrated here is the concept of *chaining* of metaphors: it is not simply that one concept is seen in terms of another, but that the latter concept is itself metaphorically understood. So, the metaphorical understanding put forward here is one of a road system understood in terms of a blood circulation system which is itself understood in terms of a machine.

Expectations of the Circulation Metaphor

Before examining evidence from Southern California, let us see what we would expect from a circulation metaphor in operation there. First, we would expect the transportation system talked of as an interconnected series of tubes through which a flow of traffic is pumped. Just as the human circulation system connects all organs of the body, the transportation system would be expected to connect all points of the city. As in a blood circulation system the principal organs are connected by the largest-volume arteries—with the greatest flows through the heart—we would expect a need to be seen for the greatest capacity between major nodes, and with a "heart" to the system at the principal urban core.

In the same way as the human blood circulation system is seen as a system of tubes with a pumphouse at the center, and its job is seen as keeping the flow of blood moving, we would expect a transportation system understood under the metaphor to be seen as a system around which traffic is to be pumped. Note that the blood of a human circulation system is a homogeneous substance which has nutritional value to whatever part of the body it is delivered. Under the circulation metaphor, we would therefore expect traffic to be seen as a homogeneous substance. Furthermore, under the "body is a machine" metaphor, the emphasis would be on keeping traffic flowing, rather than on analyzing the composition of the traffic or on studying the needs of the *people* who make up the traffic. Due to the homogeneity of the flow, what happens to *particular* units of the flow upon arrival would not be of interest.

Just as blood provides nutrition wherever it goes, the circulation metaphor would suggest that any communities through which a flow travels could expect to derive economic benefits therefrom. Since the source of nutrition is homogeneous, and since its nutritive value is established, merely to supply that flow would be to supply benefits. By implication, in the same way that a bodily organ deprived of blood becomes diseased and dysfunctional, deprivation of traffic flow would be expected to have a depressing effect on a community. Conversely, restoring flow to a depressed community would be expected to be conducive to that community's revitalization. To keep the circulation system healthy requires the maintenance of free flow through its tubes. If any arteries become constricted or blocked, the prescription under the circulation metaphor would be to take whatever action is necessary to restore free flow.

The Circulation Metaphor in Southern California

As in the human circulatory system, this traffic, deprived of its main artery, has to find relief in the veins of city streets. The major vein between Alhambra and Pasadena is Fremont Avenue. A four-lane street in Alhambra, it narrows to a two-lane winding road in South Pasadena.
— J. Albert Curran, *Pasadena Star News*, February 24, 1980

Organic metaphors are alive and well in the transportation planning world of Southern California, and evidence suggests that, taken together as a coherent metaphorical system, they not only program understandings of what the problems are, but they also help shape the policies developed to respond to those problems.

We have three questions to address if we are to identify the metaphorical entailments of a diseased organic system: First, what is the disease afflicting the transportation system? Second, what is causing the disease? And, third, what is the cure to the disease? The disease, we shall see, is one of a constricted circulation system. The system is overloaded and blocked; free flow cannot take place anymore. The disease is due to a lack of *balance* in the transportation system. And the cure involves either restoring that balance, thereby enabling a state of free flow to return or—if that is not possible—providing an "escape valve" through which the excess pressure on the system can be released.

The Los Angeles metropolitan area is seen as a series of points connected by tubes through which movement occurs. People are conveyed through these tubes as blood through arteries and—by extension through the "body is a machine" metaphor—as crude oil through a pipeline. The overriding impression is of a body paralyzed by disease. The flow of the circulation system is being restricted but it is assumed a surgical—in other words a mechanical approach—can provide a cure. A complex social problem is thereby metaphorically reduced to one of engineering. Decision makers are akin to the doctor who always wants to dive in and "cut it out," rather than the physician who calls for attending to the whole body and its relations.

Jacki Bacharach, who was then chair of the LACTC, talked of a system under stress: "We are at capacity or over just now," she said: "The traffic. We just saw yesterday at the Commission a presentation on the highway system; and, by the year 2000 we're going to be choked. . . . I think we need a basic arterial system of high speed transit, and then infill that with other modes." Bacharach's language, in common with that of many others, suggests a metaphor of disease afflicting a system of circulation; it suggests, further-more, a disease which may not only be diagnosed, but which has a cure. And she goes on to prescribe the cure: provide an arterial system where free flow can be maintained, then balance it with other modes of transportation.

How do people think about the freeway system and its attendant problems in Southern California? Everywhere there are parallels with human condi-tions. Congestion of an organ is "an abnormal accumulation of blood in its vessels, by which its functions are disordered."[7] Vehicles are blood cells in the transportation setting, and they build up to such an unhealthy level as to dis-rupt free flow. Long Beach Councilman Edgerton talks of "our clogged-up environment," while, according to Long Beach Councilwoman Eunice Sato, "the freeways are clogged now. . . . private automobiles are clogging the free-ways in commuting to and from work." As we saw in the previous chapter, buses ("clogging streets" according to RTD Board member Marvin Holen) are widely understood as not only a part of congested traffic but as a cause of congestion for other traffic, too. Buses are seen as a part of the disease, not as a potential cure. The circulation metaphor leads us to believe that we need to unclog streets, and the image of the bus, as further interpreted through the circulation metaphor, tells us why the bus cannot do that.

Congestion is seen as a disease which makes one suffer: "People who try to move suffer severe congestion," said Long Beach light rail project manager Dan Caufield. Some see the effects of congestion as not just clogging or choking, but more violently in terms of strangulation. "They're strangling the county," said Baxter Ward of cars on freeways. "When I first came out here twenty years ago or thirty years ago, you could just speed along the Ventura Freeway; now you can't even move. It's getting worse every day." Ward's former aide, Gerald Leonard, commented similarly that "as planning technicians, for years we have all been saying that we're strangling ourselves in the freeways."

Some other examples are given in Table 9-1. The Gene Gleason television presentation is particularly dramatic, suggesting as it does the disease resulting from loss of the Red Cars, and death if trains are not brought back.

Choking or strangulation are not, of course, consistent with afflictions which might affect a blood circulation system. As noted in chapter 7, however, metaphorical systems do not invariably involve complete mappings from one set of understandings to another, but rather collect a series of entailments that come together to form a coherent picture. Choking and strangling involve restricting a flow through a tube—which is indeed what happens when an artery becomes clogged. So, although they affect different bodily organs, these metaphorical understandings are consistent with the circulation metaphor, if not with the blood circulation system itself.

Associated with the idea of congestion or clogging, choking, or strangulation, is the conception of a system with a finite capacity. It is as if an abnormally large quantity of blood is being forced through someone's arteries which fail under the stress or as if a constricted windpipe were unable to accommodate a flow of oxygen sufficient for life. As Alternate Commissioner Walter King put it, "I think there's no question, our congestion, if you drive the freeways around here and we hear all the experts for many years we've been told that we're at capacity." Furthermore, said Alternate Commissioner Ted Pierce, "we're almost out to capacity as far as the amount of buses the downtown can handle."

There remain more entailments of the circulation metaphor to be tested against the evidence of the interviews conducted in Southern California.

Table 9-1. Strangulation by Freeway

An officer of a major petroleum company may seem an unlikely advocate for mass transit. But if we don't build a subway in Los Angeles, our city is going to strangle on automobiles and buses.
—Rodney Rood, Vice-President, Atlantic Richfield Company.
Testimony to Senate Appropriations Committee on Transportation, April 27, 1983.

A recent energy shortfall strangled the country.
—Presenter, KJOI-FM

Unless we get a rapid rail system started, we're going to reach a point where we'll start to strangle on our own problems and the ability to pick up people. We're passing up people right now and things can only get worse.
—Thomas Newsom, General Manager, RTD, KABC-TV, Los Angeles, August 11, 1981

The car gave us freedom. We could take the freeway anytime. So we ripped up hundreds of miles of streetcar and inner-urban train tracks. This is the result: freeways that can no longer be expanded. Congestion, pollution, frustration. L.A. has a fine bus system, but it too is over-worked and stuck in the same traffic jams. The city which rejected rapid transit three times on the ballot since 1968 now finds itself threatened with strangulation if we don't build it.
—Gene Gleason, presenter, KABC-TV, Los Angeles, August 12, 1981

This is best done, however, after introducing a further central metaphor. To understand and further extend the idea of disease and cure as applied to transportation in Southern California, we need to expose and appreciate the potency of one of the most powerful metaphors to emerge from the interviews and documents examined: that of balance.

The Balance Metaphor

Proposition A will mean a more balanced transportation system for all citizens of Los Angeles County.
—Supervisor Kenneth Hahn, press release following California Supreme Court approval of Proposition A

"The structure of balance is one of the key threads that holds our physical experience together as a relatively coherent and meaningful whole," writes Mark Johnson. "Balance metaphorically interpreted also holds together several aspects of our understanding of our world."[8] The meaning of balance, Johnson says, comes from experience of our body. In our daily lives we expect there is a bodily equilibrium, although it is invisible once attained. The stomach does not draw attention until we get hungry or it gets too acidic. We only know that the bladder exists when it is full. The environmental temperature only attracts our attention when it is too hot or too cold. The act of physical balance when walking is intricate, but we are unaware of it until we stumble and fall.

It is only when these things happen that we see ourselves as being "out of balance." Being "out of balance" means "too much" or "not enough," so that "the normal, healthy organization of forces, processes, and elements is upset." Our response to the loss of balance is to add or subtract what is necessary to restore it. We eat to dispel hunger; we take an antacid to quiet our stomach's rumbling complaints. We relieve ourselves when the bladder is full, and turn the thermostat up or down or add or subtract clothes when it is too cold or too hot. And if we stumble, we quickly act to right ourselves and restore a state of physical equilibrium. "As you stumble, and fall, balance becomes conspicuous by its absence. You right yourself by rising back to your typical upright posture. That is, you reestablish a prior distribution of forces and weight relative to an imaginary vertical axis. You are balancing out, once more, the relevant physical forces." As Johnson explains, "In every case, balance involves a symmetrical (or proportional) arrangement of forces around a point or axis." By definition, then, we can respond to losing balance by physically readjusting these forces to bring about its restoration. We know that there is such a thing as a state of equilibrium; that it is attainable; and that it is healthy for us to reach it.[9]

Balance as experienced physically is fundamental to our existence; it is a basic assumption of everyday life. As such, the concept readily extends itself metaphorically into other areas of our lives. Johnson cites Lakoff and Kovecses to show the concept of physical balance as metaphorically translated to psychological balance.[10] Emotions like anger, these authors claim, are experienced on a model of hot fluid within a container (usually closed):

> Emotions can *simmer, well up, overflow, boil over, erupt,* and *explode* when the pressure builds up. In such cases an equilibrium must be reestablished. One can *express, release,* or *let out* the emotions *(blow off steam)* to lessen the

strain. One can try to *repress, suppress, hold in,* or *put a lid on* one's emotions, but they will not thereby disappear—the forces are still present within the system. In short, we tend to seek a temporary homeostasis where we are emotionally *balanced, stable,* and *on an even keel. . . .*

On the other side of the scale of emotional balance, there can be too little emotional pressure. One can be *drained, emotionally bankrupt,* or *exhausted.* The result is lethargy, dullness, lack of energy, and absence of motivation. In such cases we try to *pump up* our emotions, to stir them up, to recharge them, to generate some emotional energy.[11]

In our lives, Johnson says, we seek to balance intellectual, physical, social, religious, and moral activity. "If too much *weight* is put on one activity or enterprise, to the exclusion of others, the individual is unbalanced. Furthermore, financial, marital, political, or sexual problems can *weigh on our minds*, throwing us out of balance." As Johnson points out, this reflects not merely how we talk about the effect of our problems, but our *experience* of them, and so how we conceive a cure. "As Jaime Carbonell has pointed out, the notion of *weight* is intimately related to the structure of the BALANCE metaphor. We experience power and force in terms of weight and mass (which is equated with weight in common understanding)."[12]

Johnson shows how we trade off weights in establishing *"The balance of rational argument."* We *pile up evidence, amass facts,* and build up a *weighty argument.* People judging what we have to say will *weigh* our argument's merits. If two arguments *carry equal weight,* we try to *tip the scale* in our favor by *adding* further evidence. The type of metaphorical balancing at work here is well-represented by a mechanical balance scale with two pans—as in the popular image of "The scales of justice."

Schön's account of the metaphorical entailments of a balance scale with two pans provides an interpretative tool for analyzing the presence of the balance metaphor in conceptions of transportation problems and remedies in Southern California.[13] Schön notes in particular that objects come to the weighing process ready to be weighed: "Objects are brought to the scales. They do not have to be invented in order to be weighed. In a sense, they are given for the weighing process; from the point of view of the weighing they are assumed. The issue is not how they came to be, but how much they weigh in comparison to one another." In a process where the balance metaphor were operating, we would expect actions "to be treated as given for evaluation. Problems of invention or formulation would be ignored."

We should therefore see if people were more concerned with deciding whether to take a certain predefined action, than with thinking about what actions they might possibly take. In the course of weighing on a balance scale, objects do not change. So, Schön says, "we would expect a theory of deciding based on a displaced theory of weighing to treat objects of decisions as unchanging."[14] The advantages and disadvantages of different given objects might therefore be discussed, but not the possibility of reformulating the objects themselves.

Schön concludes that "because of the very structure of a balance scale, weighing is always a comparison of two things or sets of things."[15] We would expect to see an evaluation process operating under this metaphor to perform trade-offs between two opposing options or sets of options. All that is at stake is adding or subtracting particular substances—as in filling our stomach or emptying our bladder, there is a predefined response to the problem which it seems obvious will result in its resolution.

Balancing Roads and Rails

The term "balance" crops up very frequently in the interviews and elsewhere, giving surface evidence of the metaphor operating at a deeper level below. We see politicians of opposing viewpoints in agreement on the need for "balance." In a letter to the editor of the *Los Angeles Times*, the liberal Supervisor Kenneth Hahn said he had recently visited San Diego, whose example he wished to emulate: "By employing San Diego's can-do attitude we can undo the wrong that was done by narrow industrial interests and bring a balanced transportation system back to Los Angeles."[16] Mike Lewis, former RTD Board chair and then deputy to former conservative Supervisor Pete Schabarum, declared that "Pete's been an advocate of what he calls 'balanced transportation.'" Alternate Commissioner Walter King wants Los Angeles to have the "balanced transportation" of Paris. Richard Stanger of LACTC staff talks of light rail helping to "balance transportation subregions." And, reported the *Los Angeles Times*, John C. Cushman III—developer of ARCO Plaza, Crocker Center, and other major projects—"argues for a 'balanced transportation system,' including road improvements, better bus service in some areas, as well as Metro Rail."[17]

The concept of a necessary balance between different modes of transportation is not a new one. According to the 1948 report *Rail Rapid Transit—Now!*:

> There are three ways to move people daily in a community—by auto, by bus, and by rail. The group is convinced that a combination of all three is necessary. Autos are too expensive for most people. Both autos and buses congest the streets. Rails separated from all other traffic are a must when a city becomes as large as Los Angeles and its sister communities.[18]

In 1966, California state Senator Randolph Collier, known as the "father of the freeways," came out with a similar sentiment, declaring that: "I want you to know that I support rapid transit as part of an integrated, balanced transportation system—a balance that seems to be lacking at the present time. . . . A natural partnership between rail and rubber waits to be put to work to help solve the enormous problem of moving people in metropolitan areas."[19] County Supervisor Deane Dana therefore fell naturally in line with these historically established understandings when he spoke at State Assembly hearings on light rail transit in Southern California:

> Until the late 1940s, it [the Pacific Electric] provided our citizens along with our expanding highway network with a good balanced transportation system. . . .
> We now have to keep pace with the future and we require a more balanced system. Streets and highways alone cannot provide a reasonable level of service to keep pace with even the most conservative population and development projections in this area.[20]

Steve Siegel of Portland Development Corporation echoed this point. If it hadn't been for the Banfield light rail project, "we would have had a freeway, which probably would have caused a major imbalance in our transportation system."

There seem to be two ideas of balance when it comes to transportation. If a transportation system operates under conditions of free flow, it is in balance. If it is overloaded, it falls out of balance. Second, if the components of that transportation system—say road and rail—are in the wrong proportions, they are out of balance. The two understandings are connected: if a road system loses its internal balance by being overloaded, that balance can be restored by transferring the load to a new rail system.

The current problem is, indeed, mainly characterized in terms of the *overloading* of the existing road system. It is often referred to in terms of weight. "The traffic right now is *unbearable*," said Debbie George, aide to Supervisor Deane Dana (italics in all quotations from interviews are my emphasis). "We need something that gets people off the roadways," declared LACTC Commissioner Jacki Bacharach, while Long Beach Mayor Ernie Kell praised light rail because "it's going to take some of the *load off* of the freeway system."

If there is an overload on the freeways, at the other end of the balance scale there is too little weight being put on "mass transit" or "rapid transit." As Jerome Premo, former executive director of the County Transportation Commission, saw it:

> I think in a historical sense, the tragedy of transportation development in Los Angeles isn't necessarily the freeways, but how it was an issue of using those old transit right of ways for freeways to the exclusion of transit. So there was a tradition of exclusion in the decision-making process—it was an either/or. I think the expectation in California in the mid-70s was that there could be some thinking about balance.

In Premo's statement, we see the idea that there are two distinct entities— freeways and transit—to be balanced. But, as we shall see, it is not simply that roads are to be balanced with transit in general, but with *rail* transit in particular. As deputy to Supervisor Kenneth Hahn, Burke Roche, said "buses cause congestion on the streets, and the light rail system we would hope would not." Buses must therefore be put on the roads side of the balance scale for weighing—they are a part of the problem: an extra burden which it is up to rail to relieve. As Hahn said in a release of March 24, 1982: "Every other major metropolitan area in the nation and the world has a balanced mix of rubber and rail transit. Only in Los Angeles have we tried to get by with only automobile and bus transportation and for this we have to pay a steep price in pollution, in hour-plus commuting times, and in the necessity for every family to own two cars." Kenneth Hahn is nonetheless a supporter of the area bus system, so long as it is balanced with rail: "You have to have *two* forms," he said in an interview: "The rail is not the substitute for the bus, Jonathan. You have to have buses. I'm a strong believer. And you have to have mass transit, too: rail."

Alternate Commissioner Walter King, while defending rail for Southern California, conceded that new articulated buses were being tried out in San Francisco: "But they also have a balance. They have the heavy rail, they have the light rail, and then the buses, and then the electrified." King evokes a "natural order" metaphor here, which is both consistent with the balance metaphor and provides additional implications.[21] In King's mind the bus and the train each have their places in the *natural order* of things. There is a desirable equilibrium balance between them at which the transportation system as a whole works harmoniously. If this order is disturbed, the system will be knocked out of balance: "I don't want to be limited by cars, I don't want to be limited by bus, I don't want to be limited by rail; I want them all in their place." As Dan Roberts, of the office of Congressman Mineta (San Jose) said: "The trick is to balance them [different modes] off, a desire to plan that puts each mode where it needs to be." "It's like an orchestra," said Congressman Jim Bates (San Diego). "You've got the violins, and the trumpets and the horns, the cellos, you know." San Diego Councilman Ed Struiksma also used the musical metaphor, calling for light rail to work "in harmony in an overall system."

Just as we bring items ready to be weighed to a balance scale—and they undergo no change in the process—the discussions above center on balancing predefined technologies, not on changing ways in which those technologies might be used, let alone in considering changes beyond the scope of transportation technology choice or outside the realm of transportation itself. In calling for a return to balance, there is little talk of innovation which might, for example, have improved the operation of buses, freeways, or both. Interviewees talk of buses and trains as givens, as things which come standardized out of a box to be put into operation. In the same way that you can't make cellos sound like horns, it is thought that you can't give buses the supposed advantages of rail-like characteristics. The train is thereby seen as a necessary part of a "balanced" system, excluding the possibility that rail service might not be appropriate for all cities.

There is another important implication to the balance metaphor: If a system is out of balance because there's too much loaded on the road side relative to the rapid transit side of the scale, there are two options: add weight to the transit side; or remove it from the road side. Adding to the rail side avoids *forcing* anyone off the road side, while the metaphor tells us that conditions will be improved for those who remain on the roads. As Assemblyman Richard Katz said in a *Los Angeles Times* interview: "The rail line could siphon off 5% of the daily commuter traffic from the nearby

Ventura Freeway. . . . Even if the light rail doesn't go directly through the community and you don't ride it, you'd benefit. . . . If you were among the remaining 95% on the Ventura Freeway, you'd be doing 45 m.p.h. where it used to be 3½ m.p.h."[22]

This shows the balance and circulation metaphors linked together. "Siphoning" implies relieving pressure in one tube by adding another tube to share the flow. Behind the concept is the idea that people are like a homogeneous liquid that will flow equally through one pipe as through another. This is not true because road and rail are not similar "pipes," and travelers will continue to choose road travel in the many situations where rail is not a competitive substitute. More fundamentally, the concept falsely assumes that transportation systems are like closed systems of liquid. If rail did draw riders from the roads, the initial reduced road travel times would attract more cars onto the system which would return to an equilibrium level of congestion. "Balance" would not be restored.

The idea of balance is central to our existence; without it, we could not even stand or walk. If balance is associated with good health, it is quite natural to think that for a sick system to be made well again, it must be returned to balance. The dangers of uncritically applying this understanding to transportation planning emerge when the metaphor is surfaced and we see that such a concept of balanced transportation can be no more than a fiction. The new Los Angeles rail system will probably make no visible difference to road loadings. While the comforting goal of balance is never achieved, the vast expenditures on the rail system take away opportunities for the more productive use of scarce resources.

The balance metaphor serves a basic function of channeling thought: far from inviting reflective thought, it makes it seem unnecessary by providing a solution of obvious appeal. It enables decision makers to see the remedy to the transportation malaise in Los Angeles merely in terms of adding or subtracting certain given technologies. Although they are at best only reacting to symptoms of the perceived transportation disease, they feel as if they are on to a real cure.

Backbones and Feeders—Part of a Naturally Ordered and Balanced System

"Buses should be feeders, they shouldn't be the *backbone* of any system," said Baxter Ward. The idea of a "backbone" supporting a network—though not the same as the concept of a major artery at the core of a circulation system—has similar implications. In both cases, there are smaller structures with which the backbone—or principal artery—are symbiotically linked. The understanding came up many times. Some further examples are given in Table 9-2.

There was a general acknowledgment, in the interviews, that were conducted, of the need to have buses "feeding" the rail system, a view underpinned by the circulation, balance, and natural order metaphors. First, according to the circulation metaphor, the total transportation system consists of a hierarchy of tubes through which traffic is pumped. Higher order arterials provide for the major flows, while smaller vessels distribute the

---

Table 9-2. Rail as Backbone

It's the backbone of the system; it's not the replacement for the whole system. And you can't expect it to operate in isolation. Just like you'd never expect to have thousands and thousands of miles of freeway in a metropolitan area. . . . If you were to try to do urban highway travel with just an arterial system, you wouldn't be able to serve it; you need some freeways to serve as the backbone of the system.
  —Andy Katugno, METRO, Portland

Well, I think the trolley system can be the backbone of the public transportation system, just as the freeways are in the street and highway sense.
  —Tom Larwin, General Manager, MTDB, San Diego,
    KABC-TV, Los Angeles, August 2, 1982

Their [Sacramento's] concept is to use the light rail system to handle the heavy mainline traffic in the future, and then to reroute the buses so that they feed into it. And they think the total cost of that will be much cheaper than if they were to go to an all-bus system expanded in the future. And I think the same would apply here.
  —Adriana Gianturco, Caltrans Director, Assembly Transportation
    Committee Hearings, Long Beach, August 14, 1981

---

traffic to and from its final destination. In this instance, the rail lines are seen as providing the "arterial" connections, while bus lines serve for collection and distribution. The images of high speed and high capacity associated with rail suggest that rail should carry large volumes over major distances, while buses should provide for local distribution.

This formulation is a function of the natural order metaphor as well as of the balance metaphor. Not only are rail and bus seen as operating in the appropriate relative quantities when they are in balance, but each functions according to its assigned role in the "natural order": rail to carry large volumes at high speeds; and buses to reach out into neighborhoods from train stations. As Richard Stanger of LACTC staff saw it: "It's not really a question between bus and rail. It's a question of trying to combine the benefits of each mode well. . . . So you use the area service of the bus and the line haul capabilities of the rail, and the task is to provide a good, convenient transferring and try and figure out where to put the rail." If that's done, "I think it's very flexible if we tap in with the buses," said Debbie George of Supervisor Dana's office.

Commissioner Marcia Mednick also stressed putting each type of technology in its place. She was asked why people would be willing to ride on feeder buses, given the assertion that they didn't like to ride on buses in general: "Because the bus is merely a means of getting to the system; it's not their means of getting to where they're going. The buses, you're just talking about just taking people a relatively short distance here from wherever they live to the station. I think it'll fly."

Associated with the concept of feeding is rearranging bus systems to link up with the rail. Ted Pierce said: "Our other transportation systems will have to be adjusted so that if the rail is going east-west, then the other will have to go north-south to develop the artery, as feeder systems." As Jim Sims of the LACTC staff explained: "We want to focus the local bus systems onto the rail systems, wherever possible. It means, where there are parallel routes which can be moved from a parallel surface street or parallel freeway onto the rail system, what we will do is eliminate the parallel bus route. Second approach is that we would focus local transit to the station locations to the extent that that's possible."

In other words, direct bus services to downtown would be replaced with shorter lines feeding the rail stations, removing competition and supplying the trains with passengers at the same time. Richard Stanger has a quite revealing metaphor for this concept. He calls it "force-feeding," which he says has worked in Atlanta, and which he says should have been implemented in the Bay Area, where AC Transit bus service continues to duplicate BART, providing direct service to San Francisco from a large number of neighborhoods somewhat distant from BART services, rather than only taking passengers to the BART station: "Any rational distribution of transit in the Bay Area would have used AC Transit to feed the BART stations and you would have used BART to pump people across the Bay." But would that be better for the passenger than a direct bus service? Stanger replied: "Well, if we do it right, he won't have a choice. Is it right to give him a choice?"

There is an element of violence to the "force-feed" metaphor and the suggestion that people would do otherwise if given the choice. There are two ways the metaphor can be considered. In one sense, it represents "transit as food," something needed for survival: if people won't feed themselves, we'll force them to eat. From another angle it is the train—a terrifying anthropomorphic monster—itself which is being fed, commuters being directed forcibly—though its doors—into its jaws, waiting to swallow them up and clamp shut.

Compton Councilman Maxcy Filer, a minority community activist and one of the most fervent of light rail critics, laughed when he heard of the idea of force feeding: "That is the solution. That's the solution. That's a good thought, I never thought of it that way, but I shall remember that from now on. I shall remember that. They must ride the train [laughter]. That's all right, I'm telling yer" [laughter].

The Addiction Metaphor: Trains as Antidote to Gas Addiction

"Dependency of foreign oil must be broken—immediately," declared an editorial in *Coast Media* Newspapers of October 23, 1980. "One sure way is to finance and construct a rail rapid transit system," the editorial continued. We see here the "addiction" metaphor in operation.

While the understandings of metaphor we have just studied suggest that the problem is that the transportation system—or its component parts—are out of balance, the "addiction" metaphor is more personal, suggesting that, just as when we become addicted to a narcotic drug our body chemistry goes out of balance, an "addiction" to gasoline is causing us to be diseased and the resultant behavior to lead to the transportation system in turn being "unbalanced." Nowhere does this understanding come through more clearly than in television news reporting for KNXT-TV, Los Angeles. On April 21, 1980, with talk of another transit proposition in the air, presenter Marcia Brandwynne detailed the problem as follows: "What gas has made us is addicts. We depend on gasoline much like a heroin addict depends on a fix. Now that gas is harder to get and costing more, we are starting withdrawal symptoms."

The camera focuses on an "addict" to prove the point: "I gotta have it, I use this in my business." We quickly home in on another "junkie." "My gas bill has just gone up tremendously high, you know seems as though I'm working just for gas." And just in case viewers have yet to get the point, we move to a third "addict": "I guess I am hooked on gasoline, because it's a necessity, I just have to have it."

In line with treating the problem as one of substance abuse, the TV show calls on a psychiatrist to make an analysis: "What will happen when it's taken away? It'll be a shaker-upper. They will be in a sort of transportation shock. I suspect that some people will succumb, they won't be able to overcome the idea that their movements are constricted." "Withdrawal symptoms" are referred to as when a human body is in shock. Brandwynne now returns to confirm to viewers that: "In Southern California our dependence is staggering."

The series continued on April 28, when anchor Connie Chung opened by telling viewers that: "Tonight Marcia Brandwynne is here to tell us how we might have avoided getting hooked." The answer, Brandwynne says, lay in the Red Car system (shown in operation for viewers), "a system that flourished in a Los Angeles of yesterday." Baxter Ward now appears on the screen to declare: "Life in this County will come to a standstill, economically, socially, recreationally, you know, in all forms, if we don't have transportation."

Brandwynne returns to tell us that: "It didn't have to be this way and here's the reason: it was the greatest mass transit system in the world, and we had it right here. . . . It was called the Pacific Electric." We now pan to Bill Meyers, a rail historian who is seen in the Red Car he owns: "The Pacific Electric was a very efficient system. Even a big car like the one we're sitting in this afternoon was far more fuel-efficient than any passenger motor vehicle, even a bus, today, but with only 50 people in the car, it's 26 times more efficient than a modern passenger automobile."

Brandwynne then resurrects the false conspiracy theory of the Red Cars decline, asking:

> Who killed Big Red? There's no easy answer, but it was a slow and painful murder, with many accomplices. . . . In 1949, General Motors was convicted of criminally conspiring to replace electric transportation with diesel buses in 40 American cities, Los Angeles was one of them.
>
> But although GM made hundreds of millions of dollars by this scheme, it was fined the sum of $5,000, and that didn't stop them. By 1955, 88% of the nation's electric streetcar network had been eliminated, and in Los Angeles, all that was left was the red car run to Long Beach, and that died in 1961.
>
> And so the seeds of our addiction to the automobile and to gasoline were born.

And the antidote to the addiction is to bring back rail. As Kenneth Hahn declared in a press release dated January 10, 1985: "We should set this project [Long Beach light rail] as the number one priority so we can begin to reduce our dependence on the freeways and smog-producing automobiles."

Although the drug addiction metaphor provides an easy way to understand a complex problem, it leads those under its influence to faulty conclusions. An addiction is bad, something to which healthy individuals do not succumb. Few who are not addicted to heroin would see anything favorable about it. The metaphorical understanding puts gasoline consumption and automobile usage into the same category. As someone may be driven to heroin by the removal of normal life opportunities, so the public is seen as driven to the road by the elimination of Red Car rail service. As we saw in chapter 2, this is untrue. The automobile was not something people were forced to use; people stopped using trains because they found road travel offered more convenience and freedom.

Once again, the metaphor is fed by images of road and rail: pictures of high-performance Red Cars paint a picture of well-being, while snarled-up roads seem like an evil to which we have involuntarily fallen prey. The addiction metaphor paints a misleading picture both of why road travel grew and rail travel declined, and in casting the automobile as the villain and the train as the savior, points to a simple—rail—solution that will not work precisely because the automotive life is for most people the way of choice, not of desperate compulsion.

The Escape Valve Metaphor

> Los Angeles needs to develop that alternative transportation system.
> —Los Angeles Mayor Tom Bradley,
> *On The Move*, Vol. II, No. 4, LACTC, July 1980

> As the freeways become more clogged and more clogged, this system [light rail] may look rapid in the very near future.
> —Long Beach Mayor Ernie Kell,
> *Los Angeles Times*, March 21, 1985

An alternative to the balance metaphor in good currency was that of the "escape valve." Out of thirty Los Angeles area interviews examined in detail for the presence of either metaphor, the statements of fourteen respondents displayed characteristics consistent with the escape valve metaphor, while the answers of twelve were consistent with the balance metaphor. In only four cases were the comments of respondents in line with both metaphors, suggesting some fundamental difference between them.

The escape valve metaphor sees the need for an alternative to an ir-reparably diseased road system. The idea of an alternative need not be inconsistent with the concept of balance. Burke Roche, deputy to Supervisor Kenneth Hahn, uses the term "alternative" consistently with the balance metaphor, for example, when he complains of the lack of an alternative to the: "freeways bumper-to-bumper. . . . And I do think if you provide a decent system, that they're not going to be coming in the same bumper-to-bumper." In other words, the new "alternative" system will allow the old one to return to balance.

Yet, if balance implies a state of equilibrium on both sides of the balance scale, the concept of escape valve does not necessarily imply balance. Instead, it can suggest the need for some new and supplemental form of transportation because the primary system is incurably diseased and—if not already defunct—on the way to death. It is as if a tube of the heart were clogged-up beyond repair, and the only action—if free flow is to be restored—is to provide a bypass. A system in balance implies the desirability of having different entities in balancing proportions. A desire for an alternative, on the other hand, need only arise when the existing system shows signs of failure. In this context, an alternative is a secondbest, a necessity born out of crisis rather than a favored choice.

When Jacki Bacharach talks of being "choked" by traffic, she expresses a common feeling that the existing transportation system is overburdened beyond repair, necessitating some new system to provide relief. "We have to have another alternative," she said. The bus is not seen by her to be an alternative because buses travel on roads, and roads are incurably diseased: "I think we need something besides bus—we need something that gets people off the roadways. I think we need another system. We need a new right-of-way. I'm stuck on that. We can't use existing rights-of-way. We have to find new alternatives."

Commissioner Marcia Mednick aired the same concern: "We need alternatives to the freeways and the bus system, which is what the light rail is." Kenneth Hahn called for supplementing "our freeways [which are] filled to capacity. . . . with a rail transit system or risk choking our local economy in an eternal traffic snarl."[23] Manuel Perez, a citizen member of the LACTC Rail Construction Committee, also demanded alternatives to the freeways which were "incapable of moving the people currently." He made no claims, however, that the light rail would relieve congestion, bringing the road system back into balance. Light rail would not be improving on the existing situation, "just coping with the detriment. It will be less bad, rather than being more good," he said. Rail, in his eyes, is an overflow device, preventing an already established disease from leading to complete paralysis, rather than eliminating it altogether. Or, as a *Los Angeles Times* editorial entitled "Yes on A" put it: "The long and short of Proposition A on next Tuesday's ballot for Los Angeles County is that it is an economic health-insurance program with very low premiums."[24]

As Bob Robenhymer of MTDB, San Diego said: "I guess it's the philosophical question of waiting until there is a dire situation, or trying to short-circuit that and provide an alternative before the things get real bad." Rail, in this light, does not even have to perform well compared to today's conditions: it just needs to appear good compared to an expected dire automotive future. This was clear in the statement of Long Beach councilwoman and RTD Board member Jan Hall, that light rail: "won't beat the car there today, but I predict there will be a time when in fact that forty minute trip will beat that car. That's when they'll start to use it. And when they do that, they relieve congestion on that San Diego freeway. . . . [For] the first five years it will not be competitive, and therefore there will be no reason to get on it if you have an alternative." But, once congestion crosses a certain threshold, light rail will act as an escape valve, and provide relief: "And I think that because they are seeing this overcrowding that the realization that we have to have a system that is an alternative is coming to the people."

The image of free-flowing light rail is compelling. The concept of rail being separate and so inherently immune from the diseases of the road system is so powerful that rail becomes defensible even when there is no claim to its higher speed than buses, as in this example from a KABC-TV news broadcast by Bill Press: "Light rail may not be the perfect system. It may not even run as fast as express buses. But at least, at last, it provides a real alternative to the dead-end streets we call freeways. Just think, tomorrow morning, choked in traffic, Los Angelenos will know, for the first time, there's relief on the way. Now there's a reason to celebrate."[25]

But perhaps the most astonishing justification for light rail as escape valve came from executive assistant to the general manager of RT, Sacramento, Mike Wiley. Some people had suggested that light rail could lose Sacramento's transit system riders because the need to make transfers would make trips longer than on existing direct express bus services. "One of the problems that we have here is that we've designed too effective of an express [bus] service," he replied. The bus was even at times faster than driving, he said, while

> if we started the rail operation today, some trips would take longer, in the neighborhood of ten minutes. But, the difference is there are no expansions to the freeway system planned, that our buses are traveling. We're experiencing more and more congestion on those freeways. The buses are in the same congestion as the automobiles. There are no plans to put any kind of express bus lane, carpool lanes, or anything of that nature that would allow those buses to travel at a higher rate of speed than the automobiles. Over time and in a very short period of time, those buses are going to be trapped in congestion and approaching stop and go for a lot of their time on the freeways.
>
> The rail line doesn't experience that type of problem. The rail line is an exclusive right of way. It is not trapped in congestion like the bus is.

Interestingly, Long Beach Councilman Edgerton opposed light rail precisely because he did *not* see it as an alternative. A rapid transit system, he said, "demonstrates the capability in our clogged-up environment to offer the driver of an automobile an alternative. And in suggesting that a World War II streetcar is an alternative to the automobile is, I think, ludicrous on the face of it, because it will not get the person there any faster."

The Evolution Metaphor

A further metaphor leading to a belief in the need for rail transportation in Los Angeles is constructed in terms of "evolution." Transportation can be seen as evolving through a series of set stages as the city-species advances through history, coming to life, facing death, and either—through adaptation—arising like a phoenix reborn, or facing extinction.

Commissioner Ted Pierce explained it like this:

> The city streets became congested. Then back in the '50s—you know '40s and '50s—they developed a freeway system that gave people the freedom of their automobile. They were mobile, and if they wanted to leave at four o'clock or three o'clock or two o'clock or whatever time, seven o'clock, they had the ability to just get in their car and go, not home, but some other place.
>
> Then you got to the point where the freeways started getting, you know, really congested, so they brought in the buses, and they did the park-and-ride lots, and they did various bus operations and developed the bus transit system for downtown Los Angeles, you know, for the commuter situation. And then the next progression is that, because the congestion in the downtown area is just going to get worse, then you have to then go to the next step, and that would be light rail or—depending on the corridor and the amount of density in that corridor—then going to a heavy rail system that will stay on, that will last for years.

"It was a *natural* that the bus came on the scene for a while. . . . but we've now saturated that and you can't handle any more buses on Wilshire," said Alternate Commissioner Walter King. "You could put more buses out there. We've tried it. You'll have to check with the bus people on that, but you'll find that the volume is such that we have to have a high-capacity mover, and that's how you got to the rail. It's just an evolution, and not the evolution that they teach in the schools."

This was a theme common to many interviews. RTD Board member Carmen Estrada was another to define a series of set stages in development:

"Well, I think that the bus system can carry only a certain number of people, can't it, per hour and per day?" she said. "We have, I think, the El Monte Busway, and it carries I guess twenty-five to forty thousand people a day, and then at that point it sort of maxes out. It can't carry any more is my understanding. So then it's necessary to *move one up*, and I imagine that would be perhaps light rail, but light rail carries about that many, doesn't it?" Peter Ireland, staff to Supervisor Deane Dana, talked of having "outgrown" the bus system, while Craig Lawson of Mayor Bradley's office advocated rail since "we're so far behind where we should be in developing transit in Los Angeles." Bob White felt likewise: "Everybody's doing it all over the world, and I think it's like keeping up with the world, and doing what the rest of the world is doing." To White's mind, "The buses are *obsolete.*"

Common to these views is an understanding of a fixed path of technological progress from the limits of highways to the allegedly greater capacity of rail. In focusing on physical capacity, the evolution metaphor sidesteps the question of where people actually need to be transported. In this sense, it is coexistent with the circulation metaphor which sees transportation as the propulsion of a homogeneous liquid through a series of tubes. The needs of growth have to be addressed, the metaphor says, by evolving towards larger-capacity tubes. The question of the Los Angeles urban form as well as peoples' actual day-to-day interaction patterns and the preferences which these suggest are ignored in the belief that technological evolution alone can be a successful vehicle of progress.

A Connected System

At the October 31, 1985 groundbreaking, Los Angeles Mayor Tom Bradley declared that the Long Beach light rail would be "a part of the great connection when Metro-Rail is built, when the lines along the freeways are built, when the rest of this light rail system connects the entire county of Los Angeles."[26] The circulation metaphor implies the existence of a closed system of tubes, through which liquids (blood or, metaphorically, people) circulate to provide sustenance to all parts of the body (city). Clearly, by implication, all parts of the body must be connected or linked if they are to be nourished and sustain the body as a whole. The end result of such connection is a whole transportation system of far greater significance than the sum of its parts.

"Connection" was seen as centrally important. According to Peter Ireland, for example, "one of the objectives of the light rail system is to link the various centers or nodes together." Implicit (and sometimes explicit) in the statements of many of those interviewed was the linking of major centers with major arteries. Several of those interviewed, for example, pointed to the importance of linking the "two largest cities in the county" (Marcia Mednick); "it links the two largest cities in the county" (Ted Pierce); it "will link the two major population and job centers, namely Long Beach and Los Angeles. . . . of overriding importance is to link those two job centers" (Los Angeles Councilwoman Pat Russell). "The Long Beach to Los Angeles proposal is a good starting point because it connects two major population centers," said Deane Dana.[27]

Perhaps particularly pregnant about the above statements is their definition of Los Angeles as a point: the terminal site in Los Angeles is seen *as* Los Angeles. As further evidence of the prevalence of this understanding, destination boards on Blue Line light rail trains headed for downtown L.A. read "Los Angeles," (Figure 9-1) as if one were arriving at the heartthrob of the metropolis instead of at one point—albeit a significant one—in the vast mass of centers, subcenters, and all the sprawling infill that occupies the spaces in-between to create the real Los Angeles.

There is, in fact, relatively little demand for travel between downtown Los Angeles and downtown Long Beach. One comment on the Long Beach line Draft Environmental Impact Report criticized the assumption "that a substantial number of trips made by residents in the corridor can be captured by a fixed rail system and that a major focus of these trips (at least the home-to-work trip) is the Long Beach and Los Angeles downtowns. We question this assumption."[28]

The lack of demand between the two end points was acknowledged in some interviews. Alternate Commissioner Ted Pierce's comment later in his interview that most of the expected light rail use would be on the middle sections of the route was particularly revealing. Yet, the significance of the "link" between the two major hubs was stressed, despite the knowledge of data suggesting that only a small proportion of travelers would go from one end of the line to the other. Such an otherwise strange understanding would be expected with the circulation metaphor operating: the concept of "connecting" the two principal cities with major arteries is as powerful and necessary under the circulation metaphor as providing major connections from the heart to the principal organs of the human body.

At the same time that the metaphor stresses the importance of connecting the two downtowns, it allows for the statement that this particular link need not be successful *of itself* to justify its construction as part of an *interconnected* larger *system*. Many of those interviewed did find the Los Angeles–Long Beach line justifiable as an entity unto itself, including LACTC Executive Director Rick Richmond: "Building a system without that line is going to be pretty difficult because you're not going to get the connections that you need to make," he said. "Nonetheless, I think the line unto itself is also a valid investment."

Many of those who supported the Long Beach line of itself nonetheless attested to its greater significance as part of a complete *system*. And some, who had doubts about the Long Beach line alone, found it necessary and beneficial in terms of the overall light rail network. Jacki Bacharach was one who supported the Long Beach line alone. Yet, she stressed the importance of "connectability to the rest of the county. It's going to be a *system* just like the freeways, so it will get you, you know, where you want to go all over the county.

Others also latched onto the freeway parallel, particularly two members of the LACTC staff, Richard Stanger and Jim Sims. Stanger pointed out that: "In 1937 they built the Pasadena Freeway and it went downtown about eight miles to South Pasadena where all the rich people lived, and you say, is this really going to serve the regional interest? Well, then you add a little bit here, a little bit there. Then there was a big surge of about twenty years of construction, and now you cannot picture L.A. without a freeway system." And, said Sims: "We're talking about that being the first line in a system, an overall system. So I think it's unfair to focus too closely on a single line. What if you had done a detailed cost analysis in 1941, when we opened the Pasadena Freeway? What would it show compared to an analysis of the system?"

These arguments in favor of the system recurred again and again. "You have to start someplace," said Los Angeles Councilwoman Joan Flores. "You can't get the system without the components." "If you said, let's make Hollywood bus and Santa Ana rail and then something else bus, then you'd have no connectability," said Jacki Bacharach. "I think the strength of what we're going to have is going to be a system that's going to connect." "If you were to call out a segment of rail, whether it's subway or light rail or overhead or underground, whatever," said Peter Ireland, "calling out only a single segment of any system in the world and ask the question: how would you assess the benefits of that one small segment, it would be more difficult to say because it means people are going to go from Point A to Point B. If you overlay it with the complete system, certainly it's a little easier."

At some point you have to start, said RTD Board member Carmen Estrada, "and at this point that Long Beach to Los Angeles light rail is part of this overall system." "There were a lot of people disagreed that it should be the first line," said Craig Lawson of Mayor Bradley's office. "However, you have to start somewhere. . . . you have to look at L.A.–Long Beach as part of the regional system." Christine Reed was among those who questioned the strength of the Long Beach line by itself but who nonetheless supported it as part of a system: "Yuh, I do support it. It is the first line in a much larger system, and so, we can't build the whole system at once, we have to do it incrementally, and this is the first line that we—the Commission—decided to do."

Walter King put it more strongly:

I don't like you talking about the Long Beach. When you're looking at a report—and I want to get it into your recorder—you cannot, you cannot look at just the Long Beach line. Because Proposition A is 150 miles, and you must keep that perspective. I would be the first to say if all we're going to do is Los Angeles–Long Beach, let's go home. Let's don't take another dime for that, because that's only a little piece of the chain.

Commissioner Barna Szabo—who defended light rail largely on grounds of the development and improvements to the urban environment he said it would bring, rather than for its transportation benefits—however, questioned that light rail would function on a regional basis:

I don't view light rail as a connection between Long Beach and L.A. I would not take the train to go from Long Beach to L.A. But I will take the train to go from Ocean Boulevard to 20th St. [in Long Beach]. Or from the courthouse in Compton two streets down to the shopping center. It's cheaper, more efficient, more convenient for local service.

As a commuter rail, it's a bloody failure. It's slow, and it's not efficient—you take a bus or you'd take your car. It just happens to run between L.A. and Long Beach. I don't anticipate people driving their car, parking their car, getting on the light rail to go as slowly as their car would have to go up to L.A. You take your car. But if you're already in Long Beach, you take the light rail. Why hassle with your car? If you look at light rail as your answer to interurban commuting, you've got the wrong stuff.

Compton Councilman Maxcy Filer, who was against light rail altogether, did not go for the system argument at all:

Can Los Angeles be subwayed, if there is such a word? I'm not sure. I'm not sure that Los Angeles could. We've had this grid system as far as the buses are concerned, and we get that going and everything. I'm not sure we will ever have a so-called rapid transit system in Los Angeles. It's just too spread out. . . . How can I say I'm going from Watts to the Valley, and it's going to take me two or three hours to get there, and work every day—some do it.

Those who support the Long Beach light rail because of the "system" concept are sold by the idea of interconnectivity—the logic that a completed system will allow the passenger to get from anywhere to everywhere. It seems obvious under the influence of the circulation metaphor that a closed system of tubes, touching every major center, is needed to circulate lifeblood throughout the region. In this context, the Long Beach light rail is playing a pioneering role, like the first freeway. Unfortunately, however, the Long Beach light rail is quite unlike the first freeway. The freeway constituted a technological revolution which provided a degree of accessibility the "Red Car" network—a previous revolution—could not deliver, and did provide for one-vehicle travel from any point to any point. Light rail does not, however, provide improved accessibility to most points—and the length of time needed to complete cross-regional trips makes it unlikely that people will transfer to do so. The inability to serve actual origins and destinations directly is also a deterrent to light rail travel, although this is not perceived by those who visualize the region as a series of points, rather than as areas that are spread out.

The Logic of Metaphor

A series of metaphors coherently comes together to suggest a rail solution to what is seen as an automotive problem. The circulation metaphor, with its entailments of a system of interconnected arteries focused on the heart of the city, explains the desirability of linking major cities with major arteries, even though there is little travel between them, and even though "downtown"—though a significant business district—is not at the "heart" of the city. With freeways seen as "choked" or the city "strangled," the metaphor prescribes a new system—isolated from contamination—through which free flow can prevail.

Like blood, traffic is seen as a homogeneous substance to be "pumped" around the system: the actual origins or destinations of particular units of traffic—or, more importantly, of the people that constitute that traffic—are not seen as being of great importance. While the metaphor is guiding thought, no consideration is given to the possibility that transportation might be improved by *regulating* flow—charging tolls, encouraging the use of high-occupancy vehicles, or implementing other forms of road management—because the goal is simple "free flow." It is this picture—which highlights certain features compatible with the metaphor, excludes others, and makes solutions consistent with the metaphor appear obvious—which suggests that the metaphor is operating at a deep level.

Common to the "balance" and "escape valve" metaphors, both of which nest into the circulation metaphor, is the notion of providing a separate system in which free flow can be maintained. The balance metaphor operates on the basis of equilibrating certain predefined technologies, the desirability of which—or lack thereof—is determined by the symbolism and imagery associated with each technology, as studied in chapter 8 and to be extended in chapter 10. The metaphor tells us how the predefined technologies need to be assembled to create a state of equilibrium. The "natural order" metaphor tells us how technologies are to be functionally allocated so that each serves in the "correct" place in the natural order. The apparent oddity of interviewees who malign buses but approve of their use as feeders to rail transit is explained when it can be seen that the feeder function puts buses in their correct place in the envisaged natural order: it is not that no buses are required, but that they are seen to cause imbalance when they try to perform the line-haul function of rail. Just as a cello performs a different function from a horn in an orchestra and the one cannot readily be substituted for the other, the train and bus are needed together in a certain balanced combination: too many buses, like too many horns, may lead to a lack of harmony.

This impression is false. Buses can perform the line-haul function quite efficiently, and more so than rail services when a dispersed population is to be served. Yet when roads are seen as "diseased" and "out-of-balance," buses—which use roads—do not seem to be a viable option. The "addiction" metaphor generates further false impressions by leading those under its influence to believe that the craving for the road is part of an addiction resulting from the destruction of Red Car services. This is untrue, since the market showed public preference for road travel well before the Red Cars went out of service; rail travel, furthermore, cannot act as an "antidote," since it cannot reach many of the destinations of the roads or do so nearly as efficiently.

While the balance metaphor assumes an equilibrium is attainable, the escape valve metaphor does no such thing. Under the balance metaphor, rail is part of a natural ordering of transportation systems and is desired as part of an ideal balance. Under the escape valve metaphor, however, rail is an alternative needed only because of the inadequacies of the road system: it is not natural, but an artificial device to cope with the excess pressure the "natural" road system cannot withstand. While the metaphors are different, of greater importance is the same conclusion to which both metaphors point: that a rail system is required, whether to return free flow to the entire system or to provide it only on the new rail part of it. Both metaphors fit coherently with the circulation metaphor; both fit coherently with the symbolism and imagery of freeways, buses, and trains and, indeed use that symbolism and imagery as the subject matter that the metaphor interprets and endows with its particular meaning. A further metaphor—of evolution—which charts the development of transportation technology on a fixed path of progress, also fits with the above: if the evolutionary step up to rail has not occurred in Los Angeles, then, of course, the system is out of balance. If there are no speeding trains in a city which has evolved beyond the capabilities of the roads, the metaphor says, it is no surprise that L.A. grinds to a halt.

Above all, there is a "logic" shown in the understandings promoted by these metaphors: if circulation is blocked, relief must be provided; if the system is "out of balance," the weights assigned the different technologies must be changed to bring it into balance; if the city is behind in the evolutionary chain, it must grow with the times.

If we have seen that the images suggesting the benefits and disadvantages of each technology are powerful, there are other symbolic processes, connected with human themes such as sex, romance, pride, and a fascination with power, which have little to do with the transportation characteristics of trains or buses, but which also play a vital role in structuring thought, and which mesh coherently with the processes studied so far to show rail as the ideal outcome. These are the subjects of the next chapter.

CHAPTER 10

Technological Sex Symbols on Steel Rails

God, who Made the Man

I hear the whistle sounding,
The moving air I feel;
The train goes by me, bounding
O'er throbbing threads of steel.

My mind it does bewilder
These wondrous things to scan;
Awed, not by man, the builder,
By God, who made the man.

—Cy Warman, *Tales of an Engineer*

Trains: All that is vigorous and go-ahead in the dreamer
—Tom Chetwynd, *Dictionary for Dreamers*

Trains are sexy, buses are not
—Christine Reed, LACTC

The Larger Meanings Enshrined in Technology

We are thinking beings, but we transcend thought in dream. We have memories of what life was like in the past; concerns about how it is today; and hopes for what it might be in the future. With needs for experiencing the spiritual, we do not restrict religion to the church or temple: we seek out the Godlike wherever we can, and find far more than merely functional meanings in the artifacts we encounter in our everyday lives. Technology can come to symbolize good or evil; a particular style of life; power; or impotence. It can take on anthropomorphic qualities, and become the subject of love or hatred. As it may seem to come to life, it can die and—as a loved one in death—be mourned.

In chapter 8 we assembled a series of images which show why rail seems to offer more transportation benefits than buses. In chapter 9 we explored metaphorical understandings which use those images as inputs and demonstrate that action should lie in developing rail rather than improving bus services. Here we investigate symbolic understandings which may have little or nothing to do with transportation benefits but which can powerfully give the impression that one mode of transportation is more attractive than another.

The Train as a Toy

Childhood memories are replete with trains. Trains are the heroes that today's adults want plucked from the storybooks to have as the playthings of real life. "This impossible dream is steadily coming closer to reality," wrote Noel T. Braymer in an op-ed for the Long Beach publication *Uncle Jam International.* "I'm talking about the little Trolley that might well make it."[1] Jim Pierson of the San Jose system pointed out that "There are a lot of neat little gadgets and things with light rail or rail in general. It is like a toy." "And I always say," said Richard Stanger of LACTC staff,

> When was the last time you heard a kid ask for a bus for Christmas? Kids don't ask for buses for Christmas, they ask for trains for Christmas. And we're just older children.
> We like to play war games. And when we grow up we fight wars. We like to ride our bicycles. When we grow up we buy motorcycles. We like to do all these kinds of things and then we grow up and do it, so what's wrong with liking to play with trains and then *wanting to ride trains?*

The little trolley, fighting against the wicked trucks and cars, has a personality. In Portland, the system is referred to as "Max," and talked about as of a brave little engine or even a precocious child. According to Ted Spence of Oregon Department of Transportation: "The pride in Max is incredible . . . People won't get on a bus and go downtown. But they'll get on Max and *play on the darn thing all day long* . . . There was a lot of enthusiasm for something new and it's shiny and like a toy. People like trains. And I can get enthused about the thing."

Stanger's claim is interesting because it legitimizes bringing toy trains to full scale by claiming that people will *ride* trains because of their friendly childhood associations, not just because they make for the best way of getting to a given final destination. Spence appears to provide evidence that this is happening. But he is referencing high weekend levels of recreational ridership. Overall ridership has not only been below initial forecasts but, considering that many riders were already using the public transport bus, Tri-Met (1998), Portland, has estimated that only twelve thousand of the twenty-seven thousand average weekday riders on light rail in 1996 were "new system rides attributable to light rail." There is no evidence, furthermore, that the toylike qualities of trains attract a long-term commuter ridership more concerned with the speed and cost of getting to work than with the cuteness of the technology. Perhaps Brock Adams, then U.S. Secretary of Transportation, captured the reality best when in testimony before the House Transportation subcommittee in April 1979, he said, "They are talking about riding the trains, but a great many of the people in the community would like to have it there; they do not ride it."[2]

We need to look at why it is that we like having toy trains, and why we want to turn them into real trains: a close look shows us that trains have associations of excitement, sex, love, and even religious significance that buses do not possess.

The first light on the roof outside; very early morning. The leaves on all the trees tremble with a soft awakening to any breeze the dawn may offer. And then, far off, around a curve of silver track, comes the trolley, balanced on four small steel-blue wheels, and it is painted the color of tangerines. Epaulets of shimmery brass cover it, and pipings of gold; and its chrome bell bings if the ancient motorman taps it with a wrinkled shoe. The numerals on the trolley's front and sides are bright as lemons. Within, its seats prickle with cool green moss. Something like a buggy whip flings up from its roof to brush the spider thread high in the passing trees from which it takes its juice. From every window blows an incense, the all-pervasive blue and secret smell of summer storms and lightning . . .

"Last ride," said Mr. Tridden, eyes on the high electric wire ahead. "No more trolley. Bus starts to run tomorrow . . .".

"Last day?" asked Douglas, stunned. "They can't do that . . . ! "Why," said Douglas, "no matter how you look at it, a bus ain't a trolley. Don't make the same kind of noise. Don't have tracks or wires, don't throw sparks, don't pour sand on the tracks, don't have the same colors, don't have a bell, don't let down a step like an accordion . . ."

> *Bing! went the soft bell under Mr. Tridden's foot and they soared back over
> sun-abandoned, withered flower meadows, through the woods, towards a
> town that seemed to crush the sides of the trolley with bricks and asphalt and
> wood when Mr. Tridden stopped to let the children out in shady streets . . .*
>
> *"School busses!" Charlie walked to the curb. "Won't even give us a chance
> to be late to school. Come get you at your front door. Never be late again in
> all our lives. Think of that nightmare, Doug, just think it all over."*
> —Ray Bradbury, *Dandelion Wine*

Nostalgic Associations

"You have to understand the history of the P.E." said Daniel Caufield,
Project Manager of the Long Beach line:

> What it meant to people. They started dating, met their wife. They went to
> movies. It was part of the fabric, it was part of their life . . .
>
> The point is that the public is fixated with rail as a mode. People don't
> know [Peter] Gordon [the rail critic]. They know rail. They go to another
> city: they see trains. They remember the P.E. and they could pay a dime. I re-
> fuse to get into that sort of an argument [is rail good or bad?] because that's
> the wrong point.
>
> While the academics are arguing, the people are voting. Point is, why
> would people vote on it? The sense of loss that people feel, that the P.E. has
> gone. They say rail: that's mass transit. I'm not going to say right or wrong.
> That's the world we live in.

Nostalgic associations with trains came through loud and clear. Take the
news spot of KNBC-TV, Los Angeles on July 5, 1985, when presenter Nick
Clooney announced that: "Light Rail is coming, and Channel Four's Saul
Halpert says it might just equal the *glory days* [my emphasis] of the Red
Cars when some said we had the best rapid transit system in the world." A
clip shows Red Cars in action. Next Supervisor Hahn comes on the screen:
"I remember those Red Car lines. I rode 'em. And it was the best rail transportation in America, and the fastest." Baxter Ward also has boyhood
memories, in his case from Seattle: "I don't want to keep saying how good
everything was when I was a kid, but there was a trolley every minute."[3]

> Orange County Supervisor Ralph Clark testified before the Assembly Trans-
> portation Committee Hearings in Long Beach:
> I think Mr. Chairman that we are all in agreement that it was a shame that

we *lost* [my emphasis] the valuable Red Car system. As a boy growing up in Hollywood I can attest to that. I can speak personally about that important part of my life, because we had no automobile, and we depended on the big Red Cars to take us everywhere in the region. . . .

And I'm really happy to see the red trolley back in Southern California [in San Diego]. As I indicated it was something that I had a lot of respect for when I was young and I am glad to see the color.[4]

Like the "Jordan River" symbol examined in chapter 7, the color has attained an independent identity as a symbol. The color represents all that was good about the trolley in Clark's formative years and continues to command his respect whatever the real benefits of the vehicle it is to be painted on. Long Beach Representative Elder remembered riding the Red Cars. "Oh, they were wonderful," he recalled. To Alternate Commissioner Walter King, the Red Cars were one of the "finest systems." To LACTC Rail Construction Committee citizen member and Long Beach architect, Manuel Perez, the Long Beach light rail is the "reestablishment of the Red Car."

Technological Virtuosity—The Trolley as Spaceship

Arnold Pacey writes about the "virtuosity values of technology, the enjoyment of

having mechanical power under one's control, and of being master of an elemental force. The teenage enthusiasm for motorcycles reflects this. Many farmers, it is said, buy larger tractors than they really need, to the detriment of soil structures, because of the pleasure they get from using such powerful machines. Some automobiles are designed to appeal to this impulse; others, of more modest power and pretension, seem designed mainly to enlarge personal capability. . . . The dominance of the automobile in the western way of life is not due to blind imperatives, but to the fact that its usefulness is complemented by these two very considerable satisfactions; and as Florman says, "Technologists, knowing of this desire, were, in a sense 'commissioned' to invent the automobile. Today it is clear that people enjoy the freedom of movement of which they had previously dreamed." In most invention, basic human impulses like these precede the technological development. Dennis Gabor talks about "archetypal human desires" which include the wish to communicate at a distance, to travel fast, to fly.[5]

Pacey says it is desires such as these, rooted in "virtuosity values" and not economic aims, that lead to the urge to travel to the moon and the development of hot-air balloons or fixed-wing gliders.[6] It is the meanings related to power and virtuosity which the train symbolizes that most convincingly seem to focus attention on that technology. There is more to the *"whoosh"* image of chapter 8 than a belief that trains go fast. Talking of trains going *"whoosh"* carries the image of powerful technology, dazzling the senses as it passes by in a blur. As LACTC staff member Richard Stanger said, "I think that trains evoke an appreciation and an image and a sense of fascination and power in our history and in our lives to this day."

Trains, seen as "powerful," "advanced," and "modern," were often conceived of as spaceships. As former state senator, Jim Mills, explained the attraction of the San Francisco Bay Area's BART: "They made it a twenty-first century system, they said, because they wanted to appeal to the public. They wanted to say that this is not something like anybody else has. It's something new and wonderful. It's totally automated. It's space age." And, as San Diego Congressman Jim Bates saw it: "A bus has a stigma that a bright red trolley does not have. There's something modern, exciting about a red trolley moving along at a nice clip, a little more of the space age."

The same image is popularly disseminated on television. Take this report, suggesting the arrival of a magic airborne transport: "A plan to shuttle *earthbound* [my emphasis] travelers between two Southern California cities is getting underway now," the announcer said.[7] San Diego County Supervisor Brian Bilbray sees the positive in this kind of imagery: "you need to have that shtik or that marketing element that beats General Motors and Ford's commercials," he said. In contrast, the bus could not be made attractive, because "You're not going to match a Toyota commercial, you're not going to match the hype that goes into it. You're not going to hype something as new and different and exciting." Bilbray overlooks the fact that bus systems—such as are operated by the San Francisco Bay Area's Golden Gate Transit or on the El Monte Busway in Los Angeles—have been made attractive by providing a high-quality level of service which, in the real world, is of more interest to the commuter than advertising "hype."

While Supervisor Hahn strongly supported the development of rail services as part of a "balanced" system and shared in the excitement about rail of so many other politicians, he was immune from his colleagues' negative view of buses. "I think the bus is the workhorse of public transportation," he said.

It's not glamorous; it's not sexy; it's not exciting. It's more easy for the transportation experts to talk about mass transit and the fancy subway and
monorail or mass transport. . . . or double-decking the freeway with transit
and subways, than it is to say get a bus on time, that's clean, it's comfortable,
it's safe, it'll pick up that person who needs to get to work on time in the
morning and get them back and home that evening.

Hahn's daily dealings with a constituency which depends on bus service,
and his reputation for dealing with the nitty-gritty mechanics of getting
things done in government appear to have attuned him to the needs for bus
service, showing the compelling but not compulsory nature of symbolism.
But, being subject to the "balance" metaphor, Hahn does not need to deride
the bus to support the train.

> *Every time he touches the throttle the swift steed shoots forward as a smart
> roadster responds to the touch of a whip. When the lever is forward and the
> stroke is long, the steam flows in at one end of the cylinder, and pushes the
> piston head to the other end. When this exhausts, another flow of steam
> enters the other end of the cylinder to push the piston back . . .*
>
> *Absence, we are told, makes the heart grow fonder. The pain of parting is all
> forgiven in the joy of meeting; and now, as we begin to swing round the
> smooth curves, all the old-time love for the locomotive comes back to me . . .*
>
> *It was nearly midnight now, and the frost on the rail caused the swift steed
> to slip. When we had reached the speed of a mile a minute, and gone from
> that to sixty-five miles an hour, I thought she would surely be satisfied; but
> every few minutes her feet flew from under her, and the wheels revolved at
> a rate that would carry her through the air a hundred miles an hour. . .*
>
> *It seemed the harder he hit her, the better she steamed.*
> —Cy Warman, Tales of an Engineer

Technological Virtuosity as Sex Appeal

"I think that aesthetic sex appeal is important in everything we do in this
world," said Alternate Commissioner Roy Donley, an architect by training.
It's commonly understood that there's more to a car purchase than mere
functionality. "A guy buys a BMW not because it drives better than a Dodge
Colt. It does drive better, but it doesn't get there a heck of a lot faster at 55

mph," said Ted Spence of Oregon DOT. He said there was difficulty getting support for the bus. "And, you know, it's not very pzazzy frankly." "Try to sell checker cabs," Brian Bilbray said. "They do the job. They're great engineering. But try to sell a checker cab in Southern California: they don't sell. And that's what you're trying to sell with a bus."

Freud referred explicitly to the sexual connections of rail travel:

> Shaking sensations experienced in wagons and railroad trains exert such a fascinating influence on older children that all boys, at least one time in their lives, wish to become conductors and drivers. They are wont to ascribe to railroad activities an extraordinary and mysterious interest, and during the age of phantastic activity (shortly before puberty) they utilize these as a nucleus for exquisite sexual symbolisms. The desire to connect railroad travelling with sexuality apparently originates from the pleasurable character of the sensation of motion.[8]

"Light rail is sexier than bus," said Bob Robenhymer of MTDB, San Diego. "It's sexier than buses," according to Jim Pierson of San Jose. The train is perhaps the only technology to have both sexes—it is both a woman and a penis. Rail historian Bill Meyers, who owns a Red Car, appeared on KNXT-TV, Los Angeles, on April 28, 1980, and described the "exciting experience" of going as a boy to "see the gleaming rails in the middle of the street and *this gigantic steel thing, huge, panting and whining and puffing away like it was alive.*" Another rail fan (interviewed at a rail historical society meeting) told of how "as a kid I'd watch the PCC cars *pull in and out in the middle of the night,*" while his "fondest memories was when an old conductor said to me one time, 'come along with me son, *I'll show you what that train of mine is all about.*'"

Nowhere does the appeal of technological virtuosity have a greater influence than in the thinking of Baxter Ward, and the sexual imagery comes through loud and clear, too. Ward has explained that his fascination with trollies as a boy first got him interested in public transportation.[9] The fascination was clearly with the technology, not the economics. And, as in the Warman engineer's tale above, the train is a seductive woman. "I'm sorry you never saw them," said Ward, talking of rail cars on a commuter service briefly instituted through his influence. "They were stunning. I've got pictures of them. They were knock-outs. . . . Ours was *lush.* We had lovely café cars, and smooth not heavyweight—there were fluted sides."

Buses are "terrible," Ward said. "But it's *thrilling* to be on a train car in a Pullman at night, or a bedroom, and look outside, you know at America passing by. . . . The opportunity for riding from Chatsworth to Disneyland on a train far exceeds in excitement and popularity the dismalness of getting on an RTD bus." Ward was at this point told about a friend in San Francisco who prefers the bus over the train because of the more direct service it provides: "If he wants to ride the bus, let him ride the bus. There are some people who buy brown cars," Ward replied.

Ward's concern for displaying technological virtuosity, rather than with allocating resources to provide *transportation* services, really came through when he talked about the Hollywood Bowl. "That's an *elegant* function, it should be made comfortable. And you should have a train right there. The present interests supporting the current Wilshire subway proposal say don't stop at the Bowl, it's too expensive." Ward was told that the stop would only be used for a few hours a day. "So what? So what? What a *thrill* to get there is twenty minutes and not clog the freeway and cause wrecks and kill people." If efficiency is subservient to "thrill," little thought is given, either, to how the elegant people—presumably in evening dress—are supposed to get home when they emerge from a suburban subway station late at night.

The "thrill" and "elegance" factors of trains come through in documents supporting rail construction. The 1948 report on *Rail Rapid Transit—Now!* is obsessed with the technological marvels of vehicle design and performance, highlighting not only the train's acceleration, but reveling sensuously in all aspects of its virtuous construction:

> The wheels on these modern cars will be built with rubber inserts. They will be relatively noiseless. Throughout the car there will be a lavish use of rubber mountings. The resilient wheels take up road shock. Coil springs and stabilizers will tend to eliminate excessive and sharp swaying. The cars will have improved ventilation systems, and high intensity lighting for ease in reading. The cars will be wider than most of those now in use and the seats will be larger and more comfortable. . . .
> The cars will be streamlined and may be articulated. They can be connected to operate in trains. The cars will be of the low-floor type. Results in the latest experiments in the use of colors to ease rider fatigue will be incorporated in car interiors.[10]

The picture on the report's cover sums up the portrait of a technically conceived transport of delight capable of attracting all manner of riders whatever its destination. A similar picture emerges in the reports prepared for Baxter Ward's "Sunset Coast Line" and "Sunset Ltd." proposals. The report for the "Sunset Coast Line" contains a great deal of technical information, as if presenting the results of a thorough engineering analysis. We are told that the system will have "230 miles of welded-rail, heavy duty track, totally grade-separated, built to carry full trains at 85 mph." Exciting-appearing Monorail service is suggested to provide feeders, creating "A totally-balanced system, built for Main Line speed, plus the convenience of Light Rail extensions and Monorail feeders."[11]

Unheard of in transit history, and unconnected with any notion of economics, Ward proposes that "A small number of domed bilevel cars would be provided on the local service of the Sunset Line to add an attractive quality to the car fleet." Ward also calls for special "Airporter" cars. "The Airporter would utilize a lower seating density with swivel rocker seats similar to those utilized on the Metro Club cars in the New York–Washington service." To top it all, special facilities would be provided for "excursion service. . . . Tourists, weekend travelers, beach visitors, and weekday shoppers will all find the excursion cars to be a happy way to travel." While the "Sunset Coast Line" proposal calls for "automated refreshment service and restrooms," the "Sunset Ltd." provides for trains with a "Beverage Bar" and lounge.[12] Figure 10-1 shows the "Local" and "Interurban" trains for the "Sunset Ltd.," together with the seating plan which would be common to both types of service. Figure 10-2 illustrates the "Airporter" trains and seating plan from the "Sunset Ltd.," and also shows the interior of an "Airporter" for the "Sunset Coast Line."[13]

While both reports give details of train performance and appear to offer the key to a technically engineered paradise, an account of the attractiveness of the system for travel, given the realities of the Southern California spatial economy, is lacking. That this did not seem of concern to Ward perhaps comes through most pointedly in a comment not on trains but on automobiles. "Things aren't as nice as they were," he lamented. "Cars are not the *big things*. Who the hell cares if you can drive a Honda Civic to the Civic Center? What the hell thrill is that? Nothing. But you can drive an Olds 98 to Civic Center or a Town Car, or a Ferrari or something: GREAT!!!"

While elsewhere in his interview Ward complains about the pollution caused by traffic, he admits a preference for exciting large cars. While he

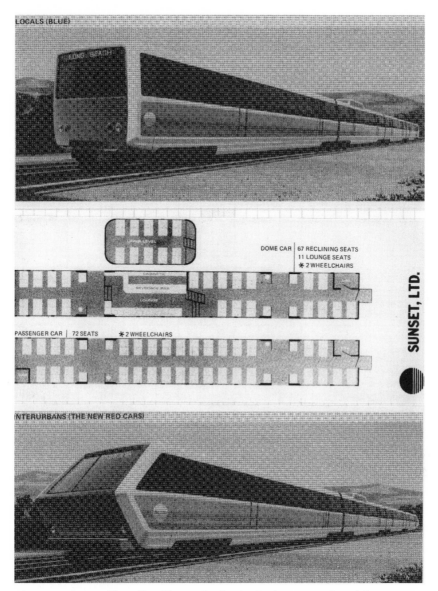

Figure 10-1. Proposed "Local" and "Interurban" trains for the Sunset Ltd., and the seating plan to be shared by both types of service.

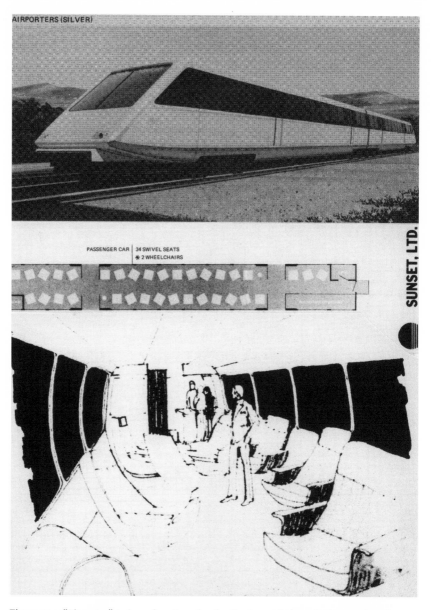

Figure 10-2. "Airporter" train and seating plan for the Sunset Ltd., and Airporter interior for the Sunset Coast Line.

does not advocate getting people to drive larger cars—a demand which would be inconsistent with his image of "car as polluter"—he can promote "exciting" trains without seeing any inconsistency with his professed transportation and environmental priorities. But the way Ward sees the train suggests that its attraction to him is similar to the attraction of a large, stylish car. While on the surface trains might be advocated for transportation purposes, it does not take probing far below the surface to see that sex appeal and technological virtuosity dominates economic considerations in driving advocates' desires.

The Romance of the Rails

To celebrate the San Diego Trolley's third birthday Thursday, gifts will be handed out to morning commuters and evening commuters will be serenaded.

—*Los Angeles Times*, July 24, 1984

Trains are frequently referred to as "she." They are seen as women; perhaps if the obsession with mechanical performance is phallic, the romantic-nostalgic is feminine.

"People have to become enamored with systems," said Alternate Commissioner Roy Donley, implying a connection between the love of a system and the likelihood that people will use it. Rail, said Manuel Perez, is something "you can *relate* to. I don't know how many people get terribly excited about a bus that's running on a freeway." The San Diego light rail, said San Diego Councilman Ed Struiksma, "is something that people can *relate* to, and they relate to it in a favorable light."

In one metaphorical understanding, the current problem is seen as breaking a "love affair" with cars, and the objective to bring a technology that will be at least as lovable, and preferably more so. Marcia Brandwynne on KNXT-TV, showed viewers "a major mass transit system we once had, a system we *loved* almost as much as we *loved* our automobiles, a system that flourished in the Los Angeles of yesterday" (my emphasis throughout this chapter).[14] The October 7, 1985 edition of *Passenger Transport* meanwhile reported that, "*Infatuation* with Auto is Cooling. L.A. Turns to Rail to Alleviate Congestion." As San Diego Congressman Jim Bates put it: "There's a *love affair* with the trolley, which overrides the love affair with the auto."

"There's nothing romantic about a bus," however, according to Long Beach Mayor Ernie Kell. "A bus is a bus is a bus. Who cares?" said METRO Seattle Chairman Bob Neer: "It's just a hunk of equipment that runs back and forth. You've got to really be a nut to have a *love affair* with a bus." Kupferberg helps explain this phenomenon:

> The bus has never attained its place in romance, legend and folklore. The chariot has Ben-Hur, the ship has the Flying Dutchman, the railroad train has Casey Jones. But literature has done little for the bus. . . . The mundane commuter bus, of course, is the most neglected of all the species. Like the commuter himself, it leads a harried life full of jolts, false starts and unexpected detours. About the best that can be said for the average commuter bus is that it beats walking—just barely.[15]

As we learn from Baxter Ward's "Sunset Coast Line" proposal:

> Los Angeles might be having a *love affair* with the auto, but after 30 years, there is not a flicker of feeling for the bus. And their fumes add to our problems with the air.
>
> Even the gasoline shortage and the 25 cent bus fare failed to *affect us emotionally*. There was to be no shotgun wedding with buses.
>
> Because *the heart still returns* to the rails, people still talk about the Big Red Cars, and ask these sensible questions . . .
>
> The Sunset Coast Line, this proposal, brings all that talk and nostalgia and hope together—into an 85 mile an hour Main Line that is guaranteed to *break up* any *love affair*—it'll pry us right out of our autos and take us where we want to go, in style, in comfort, and faster than any law-abiding car.
>
> It'll be a *whole new romance* for commuters, and they'll like themselves in the morning, and in the evening as well. . . .
>
> So we settled on Sunset Coast Line [for the name]. It is the kind of name that pretty much says all there is to say, keeps some nostalgia, while still looking forward to the future. Someday its tracks might be converted to air cushion, with speeds up to 150 miles per hour on certain longer stretches of the Main Line—and the name still will be good.
>
> Of course, by then people might want to shorten the name somewhat, to suggest *feelings of warmth and familiarity*. A name with three or two words could be just too long.
>
> So we suspect it will become known as the Sunset. In fact, it possibly will be called that while it is still going only 85.[16]

When Ward did get a chance to begin his experimental commuter rail service, he settled on February 14 as opening day. "Let's start on Valentine's Day," he had told the Santa Fe railroad: "it's a nice romantic touch, I think.

And we began Valentine's Day." Valentine's Day 1991 was also chosen for the opening of the Blue Line tunnel into downtown Los Angeles. The front page of the LACTC (1991) publication *Metro Moves* headlined "A tunnel just waiting for a train" and a picture of the tunnel was contained in the outline of a heart (Fig. 10-3).

When on October 6, 1980, the *San Diego Union* reported on the public presentation of the first new trolleys the previous day, it referred to the event as the "San Diego Train Unveiling." Here was the first look at the delectable bride, a woman to truly fall in love with:

> Ed Herold, 60, and Eric Sanders, 66, watched the red and white balloons lift a striped parachute canopy to *unveil* the San Diego Trolley's first double-car train on 12th Avenue yesterday and remembered a Monday morning more than 31 years ago when they both rode the San Diego Transit System's last trolley car to the Adams Avenue barn . . .
> Most of the crowd was *transfixed* by the trolley.

The "love affair" metaphor misleads because it assumes that people choose transportation for the joys of the relationship, rather than because they want to travel somewhere. It sees people as traveling by car through infatuation, but it also appreciates the potential to lure them away by supplying a more ardent lover. While people with spare cash might buy a more dashing car than those with less money to spend, the principal reason for buying a car is for the transportation services it can supply. As empirical evidence cited in chapter 3 indicates, another mode of transportation will only be competitive with the automobile if it provides a trip as convenient as by car. Rail cannot do that. By making transportation choice appear to be something as subjective as a romantic relationship, however, the false impression is given that a lovable trolley will pry people out of their cars.

In many ways, the "love affair" metaphor is similar to the "drug addiction" metaphor. Both ignore the core reasons people choose particular modes of transportation. And while one admits that the car is lovable—making it seem understandable why it should be loved—and the other sees it as a dangerous drug, both see a new and healthier relationship as the prescription for change.

METRO MOVES

"A tunnel just waiting for a train"

L.A.'s first underground METRO tunnel officially opens on Valentine's Day. Shown here in an earlier construction stage, the completed tunnel will bring the METRO Blue Line to its new underground rail station in downtown Los Angeles.

Transportation Milestone — The Blue Line Moves Underground

On February 14, 1991, the Los Angeles County Transportation Commission (LACTC), and its subsidiary, the Rail Construction Corporation (RCC), unveils another milestone in Los Angeles County rail transportation history when it opens its first underground Metro rail station — four months ahead of schedule! The tunnel brings the Metro Blue Line into downtown Los Angeles via a six-block underground tunnel — marking the completion of Los Angeles' first rail transit segment a short seven months after its spectacular grand opening last July.

Beginning on February 15th, the day after the official tunnel opening ceremony, Metro Blue Line cars will suddenly dip down and disappear from sight at 12th and Pico Streets in Los Angeles, and passengers will ride — many coming all the way into town from Long Beach — the rest of the way to the intersection of 7th and Flower Streets through an underground tunnel. There, the Metro Blue Line will connect, at a shared station, with the forthcoming Metro Red Line.

The Metro Red Line, meanwhile, is currently under construction. An 18-mile underground heavy rail system that will serve the densely populated regional core of Los Angeles County, starts at Union Station where it links up to commuter rail lines, moves south into downtown Los Angeles (to the 7th Street Metro Station), then west into Hollywood, and finally out to the San Fernando Valley. Phase I of this system — a 4.4 mile portion connecting Union Station to the Wilshire/Alvarado intersection — is scheduled to open in September of 1993. ■

Figure 10-3. Cover of *Metro Moves.*

The Religion of the Rails

Baxter Ward has been called "the 'Rail Messiah,' and he certainly clings to his convictions with an intense quasi-religious fervor," reported Shaffer.[17] Religious belief implies acts of faith: neither fact nor reason is necessary to justify devotion. To California Assembly Transportation Committee Chairman Bruce Young, speaking at Hearings in Long Beach, it was praiseworthy enough that "The people in Long Beach *kept faith longest* as far as the P.E. *[sic]*. They were the last to give up and fought to the bitter end to this city's credit."[18]

Manuel Perez is one of the faithful. He admitted to a "great love of trains . . . Politically, technically, emotionally and spiritually I'm very committed to the light rail." Perez was asked why it was important that he was emotionally or spiritually committed. "Because I *believe* in the system very much," he replied. While Perez admitted to both a love and a belief in trains, Alternate Commissioner Roy Donley claimed that unlike most other Commissioners, he did not "*worship* at the *altar* of light rail." He called instead for a higher-tech system. "These would be high-speed trains, 100, 120 mph, and they would have about a five mile run to gain speed," he said. He called for "*high-powered locomotion*" to "get up to the speed pretty fast." The trains would stop at "megastations," at freeway interchanges where they would interface with "surface transportation. Now, this will include surface buses, taxicabs, private automobiles with park'n ride facilities and also I see this interchange being developed in cooperation with private enterprise with commercial development, possibly even residential, that is hotel-type things and that sort of thing. And a heliport. People can fly in to these things from LAX and other airports."

Perhaps most fascinating is that Donley uses economic arguments when criticizing the light rail program, pointing out that the old streetcar systems—local "Yellow Cars" as well as interurban "Red Cars"—used to cover the city with a dense grid, and that their service would not be replicated by the light rail: "People say: well, we had a beautiful system thirty-five or forty years ago, and we dismantled this system and what a crying shame that was. If the truth were known, it was obsolete at that time." Yet, when overcome with visions of powerful speeding trains and "megastations," these economic arguments are absent. Donley's vision is founded in technological fascination and power. He may not worship the trolley car, but Donley is nonetheless a member of the rail religion, if of a different denomination.

Converting to the True Faith—The Need for Religious Education

If rail is a religion, then its priests must get idolatrous automobile worshippers converted to the true faith. Commissioner Christine Reed talks of "getting *converts* to the idea that rail transportation is in fact a good alternative." And if, as Roy Donley says, "the majority of our commission right now are light rail *believers*"—members of the cult—why should the public, of whom the lay members of the commission are in many ways a subset, not become "initiates" of the faith with a minimal amount of "religious" education? As then LACTC Chair, Jacki Bacharach, explained when asked if people would transfer vehicles to use light rail:

> I also don't think transfers. . . . once people become transit users—are not as onerous as they are to the *uninitiated*, and on a system where you understand the system and get used to riding, you transfer, and it's not a big deal. It's interesting that Southern California has this feeling about itself, because what are we except transplanted Easterners? There's how many people came from New York where they wouldn't have thought to drive their car? So, we say it's an attitude problem, but I think it's because we have a system right now that's so complicated and so unclear that it's really hard for people to understand. We have former rail users, so I don't think there'll be a problem in getting them reinitiated.

If Bacharach sees auto users as lapsed rail religionists who must be returned to the faith, Bob White relies on his coaching experience to show how it can be done: "We're going to have to train the people. Any time that you change and you're innovative and you change transportation or you change your method of coaching, like I was a coach and if I took over a ball club and ran it differently from the guy before, second guessing. We can educate the people to ride."

Alternate Commissioner Ted Pierce gives an example of such education in practice.

> For instance, you take Toronto, has the Blue Line, and whatever they do whenever they're building a system, what they'll do is parallel that system with a bus route. And it would be called the Blue Line, so people get used to riding it, taking feeders to the Blue Line. And then, once they get on the

feeder, they get off and then they transfer onto the bus; takes them into downtown and wherever they're going. Soon as they get the rail completed, they pull the bus: people just go right onto the train.

All these notions share the assumption that people will conform to a desired form of behavior if shown the way. All ignore the realities of consumer choice which dictate that people will continue to travel by other means if the rail system does not prove to be more convenient. Users of the Long Island Railroad commute to work that way because it provides the best service in the context of New York. There is no evidence that they would use a rail service in another situation simply because they have been brought up to know trains and had their latent desire for rail worship reawakened. So long as choice remains, it is unlikely that they would respond to "training." If the public is neither being paid to play a particular game nor derives particular satisfaction from doing so, they will turn to a different sport. The "students" are looking for the most convenient way to complete their journeys, and there is no reason why they should accept instruction which goes against their best interests.

Another troubling aspect of seeing rail in religious terms—and the implied need for the acceptance of religious dogma—is the authoritarian stance it shares with the idea of "force-feeding." Such assumptions clash with notions of freedom of choice.

A Railway Funeral

> The railroad station at Billings is old, clean, wooden, and deathly quiet, except when the long coal trains roll past along Montana Avenue. We had a passenger train that used to come through Billings on the run between Chicago and Seattle. They called it the North Coast Hiawatha. It's *dead* now, been *dead* since October—the victim of a head-on collision with a thundering herd of numbers, statistics cooked up at Amtrak headquarters in Washington to justify *killing off* some trains [my emphasis].
> —Phil Primack, "Death of the Hiawatha," *The Washington Monthly*

Love and religion are combined in the rituals of death. A "Streetcar Called Desire" has a personality, and to "kill" it is murder. Associations of death hang heavily over tales of the ending of the Red Car system. Baxter Ward's study for the "Sunset Coast Line" talked of the Red Cars' removal as of the death of a great author or artist whose works were lost: "The Pacific Electric system did not simply *die*—it was *killed* . . . The *tragedy* in its *passing* is that its fine network was not preserved."[19] And, as KNXT-TV's Marcia Brandwynne asked on April 28, 1980, "So if it was such a great system, the question is who *killed* big red? There's no easy answer, but it was a slow and *painful murder* with many accomplices."

There are many examples of the loss of rail service seen in terms of death. The *Los Angeles Times* reported on the departure of one last train:

> "Railroad stations used to be filled with action. Now they are like tombs," said Joseph Stoddard, who was among 30 railroad buffs who gathered in Pasadena last Saturday to say goodbye to the Desert Wind on its last stop. . . .
> Some Amtrak employees and many of the rail fans who spent the afternoon photographing and *mourning* the *passing* [my emphasis] of the Desert Wind believe that its rerouting signals the same *destiny* for the Southwest Chief, and perhaps oblivion for the two depots.[20]

If, as Craig Lawson of Mayor Bradley's office confided, "People still have a fond spot in their heart for the old Red Car system," it is no wonder they were "forlorn" or felt "upset" at the passing of the Red Cars as at the death of a person, as other interviewees commented. Or that, with the loved one gone, there should be a desire to face death and propagate continued life in that most human of ways: with a new birth, a new romance, a new transport of delight.

The Nature of Evidence—The Role of Experience

Dan Caufield, Long Beach project manager, described how he impressed visitors. He would "spen[d] 8 hours with them. And by the time they get to milepost 21—they go wow!" A tour of the physical works—though it says nothing about the economics—excites an admiring response. In a *Los Angeles Times* interview, Caufield described the Long Beach project as a "real class facility. It's the biggest light rail project in North America, and it's real rapid transit."[21]

People trust evidence they can see; they are skeptical about the abstract and the unknown. Playing on the primacy of direct experience, Baxter Ward came up with an experiment in which he legitimizes the employment of sense impressions for the making of major decisions. To test whether people would prefer buses or trains he would see how they felt about riding them:

> Well, for one thing I would put a person in a bus. I would take an expert from Harvard or MIT or UCLA, USC, make him ride through Watts on a bus, or to East Los Angeles, or to Pomona or whatever the hell, put him on the bus.
> Then I would let him rest up a bit.
> Then I'd budget $500 and give him a first-class train ticket on an AMTRAK Coast Starlight, put him up in a bedroom in the Coast Starlight, and let him ride to Seattle on a train, in just the loveliest comfort possible, and then have—find some means of reminding him, frequently either on that trip or immediately thereafter—the way the bus was.
> And then say to him, your job is not to do a cost analysis or anything else, but which ride did you like the better, A or B? What can you design for public transportation that comes as close to that lovely trip to Seattle as you possibly can make it? What's the best route available? You can start with a set of tracks, and have electric power, and then you're going to have to design seats that are comfortable, passageways that are pleasant, talk areas, service for refreshments if that's appropriate, and all kinds of nice things to make it just a lovely, lovely experience. Colors are important, upholstery, fabrics, softness, cushioning, is it leather? What do you want to make it right? . . .
> Let them have both experiences. One is the gut reaction, which your gut's going to tell you that the train ride was sensational, comfortable, smooth, fast, whatever. . . . Then the second sensation, the reality part, is all right, how do you compromise on the ultimate experience, the Pullman that you can't have, you can't have individual rooms on board a commuter train, but how close to it?

I asked Ward if luxury express buses should be considered. "But they've ridden the bus, and they've canceled it out," he replied. "The first question on the quiz would be which ride did you like better?" To a hard-nosed transportation economist, Ward's test might seem ridiculous and designed to select a fantasy experience rather than an efficient public conveyance. But Ward's system of evaluation is, in fact, quite in line with our normal, everyday, common sense way of reaching conclusions about phenomena we experience. We will now look at some examples of how experiential evidence is used to make inferences.

The Disneyland Effect

If many negative views of buses come from observation of local buses, concepts about what the train might have the potential to do for Los Angeles are reinforced by brief visits to places which already have such systems, along with images of those systems in operation transmitted by the media. This phenomenon is most graphically illustrated by the effect that Disneyland, with its monorail system, has had on monorail advocacy.

Although monorail is nowhere near a viable option for the operation of commuter services, Craig Lawson of Mayor Bradley's office said his office receives letters calling for monorail "all the time. People go to Disneyland and they think that because the monorail works there it can work anywhere in Southern California. If you took a poll right now of people in Los Angeles, they would probably support a monorail over light rail or subway because they're familiar with the monorail and they've seen it in Disneyland. They've ridden on it, it's lots of fun, and it seems to go very fast."

Mike Lewis, former RTD president and deputy to Supervisor Pete Schabarum, has had similar experiences: "I can't tell you how many people call every day and say why don't they just build monorails down the middle of the freeway," he said, agreeing that the Disneyland imagery was powerful: "If a person hasn't traveled, it's the only form of transit they've seen," he added, stating that he himself favored some monorail application. Long Beach Councilwoman—and light rail critic—Eunice Sato wanted a system "more like monorail" than the light rail system proposed. Conservative Supervisor Michael Antonovich also favored monorail: "I went down to Disneyland to see theirs . . . And it seemed something that was economical and nonpolluting. No noise. And it fits above the freeway."

"People were seeing in other cities such as in San Diego that light rail can be very successful and very popular," said Craig Lawson of Mayor Bradley's office, citing this as part of the reason for the rail orientation of people in Los Angeles. Interviews brought a multitude of references to experiences of rail in other cities, and they were used to justify bringing rail transport to Los Angeles. A visitor to a subway system experiences its physical exterior, but not its larger socioeconomic consequences. Ride on BART, for example, and you will marvel at the slick trains and artful stations but won't learn of the communities unserved by the system or of the regressiveness of BART's financing.

The evidence for eyes counted for much among those interviewed. "I see how well it works in Europe," said Debbie George of Supervisor Dana's office. "Experience in traveling in Los Angeles" was a primary source of information for Jacki Bacharach. "Why are people putting in rail all over the United States? Why isn't bus the answer everywhere else?" "Why?" Bacharach was asked. "Because people are riding it. It's being used." The fact that trains are seen being used elsewhere suggests that the train *qua* train is "popular"—irrespective of the context in which it is operating— and that, whether in Washington, D.C. or Los Angeles, it will be equally popular.

Jacki Bacharach and other commissioners decided on an automated system for the Century light rail line, despite concerns by staff:

> "What turned me around was other successful systems," said Commissioner Jacki Bacharach, referring to driverless trolleys used in Vancouver, Canada, London and Lille, France. Bacharach, an influential commissioner who chairs the rail construction committee, and other commission officials have traveled at public expense across Europe, Canada and the Far East in recent years to inspect various rail systems.[22]

There had been criticism of the cost of the Washington, D.C. system, said Burke Roche of Supervisor Hahn's office. But, "I don't see how anybody can criticize it, who's *used* that transportation."

Peter Ireland of Supervisor Dana's office thought it justified for commissioners to travel to other cities to observe their rail systems: "It is probably a more beneficial way to present the information to the commissioners, is for them to see it firsthand," he said. Once more, a high value is put on direct experience, even if it does not convey the complex economic or contextual information needed to properly evaluate whether a rail system could translate successfully to Los Angeles.

Bob White is one of the commissioners to travel abroad to look at trains: "I went to Canada, I went to Edmonton, Montreal, Toronto and visited the plants where they make the light rails," he said. "I've seen the people in Canada *love* the darn thing and those cars fill up, and guess what, Jonathan, when that light rail comes up and stops would you believe, in thirty seconds I think it is, maybe not even that long, they open the doors and you can get on any car and in thirty seconds they're *ready to take off*. And they don't mess around and take all day to move their train." White feels that Los Angeles residents would be equally impressed by the service to be operated in Southern California. "And once they ride it, they will see how *smooth* that it operates, like they did in Canada and I think that it would go."

The sole criterion for evaluation is the physical experience of the system. Asked whether getting to and from the light rail would be a problem in Los Angeles, White replied:

> I'll tellya exactly: We'd do the same thing that they do in Canada. When I was in Montreal and Toronto, the train would pull into the depot, the buses are in the parking lot to take these people to where they live. . . . They do it very well in Canada. I've observed it, and ride on the light rail and taken the buses out like that, and it just works; in fact I think they give them as much as five minutes to get off the train and get on their buses.

On a common sense level the Canadian systems seem quite beguiling, and it is quite understandable why they would appear attractive in Los Angeles. White's travels create vivid images, which combine impressions of speed, which make rail seem an effective form of transportation, and of technological virtuosity and sexiness, which make it alluring. Roy Donley was similarly taken by the rail technologies of other cities: "I see subways in Paris and London and New York and other cities where the damn train is almost up to full speed before it exits the station," he said. And here is Burke Roche on the Calgary system: "The way they get through, down and out of downtown, it seems to be very convenient," he said.

No city's system has proved more convincing for the light rail case in Los Angeles than that of San Diego. At Assemblyman Bruce Young's Long Beach Hearings, rail historian William Meyers had testified that "the same light rail technology that's working so well in San Diego is definitely feasible here," while Assemblyman Young stated that: "And again I think that San Diego has shown us the way."[23] As Commissioner Ed Russ said at the LACTC meeting on March 24, 1982: "In closing, of course, I would like to say, of course, that we took the trip to San Diego, we saw what they did. San Diego is one of the greatest examples in the United States."

Television reports have focused heavily on the shiny technology of San Diego, showing Los Angeles viewers time and again enthralling images of what could be theirs. One Los Angeles station explained that: "By the mid-1970s, [San Diego] community leaders realized that buses simply couldn't haul as many people as efficiently as the rail car . . . They decided to build a light rail system. It was a sleek reincarnation of the old street cars."[24] The shots of trolley cars look impressive.

Comparisons to Los Angeles are readily made. On one slot, the presenter opened: "Mr. Taylor [Deputy Executive Director, LACTC], as you and I speak we can show our friends at home some video tape that is in fact a mass transit system now existing in San Diego and apparently we are looking to—I shouldn't use the word copy but it *works* [my emphasis], I guess—copy that."[25] On February. 26, 1982, KABC once more reported that: "Light rail is already *working* very well elsewhere. . . . Lovingly called the Tijuiana Trolley, it's a favorite model of transit cities can build without counting on dwindling federal dollars. Now, finally, Los Angeles may be ready to catch the train of thought." On another report, an on-location reporter tells us that "The L.A.-to-Long Beach Light Rail system will be similar to this one in San Diego."[26]

The comparison being made is in terms of hardware: the trolley cars will be similar. A flashy looking trolley car is shown working in San Diego ("work" being understood in the simplest technical sense of their motors working), and the image translates into one of a car "working" to move passengers around in Los Angeles, too. Social or economic considerations do not come into play. The image tells a very small part of the story but it is taken as the whole.

Christine Reed remarked that: "You can't help but go other places and marvel at the wonderfulness and efficiency of it." I asked whether that was an appropriate experience to apply to Southern California. Her reply displayed a rare self-awareness, absent in most of the others interviewed. "It probably isn't, but it still makes an impression that you never lose," she said.

> If in fact you worked in Boston, you might not want to ride that transportation day-in and day-out, you know, you might not want to be jostled and standing and all the rest of it all the time, but it is truly wonderful to go and be a visitor in those places, and to have those wonderful transportation systems available to you to use, and I have personally taken advantage of them in Boston, in Washington, D.C., in Japan, in France and in London, and in Holland and in Switzerland, and it's great, you know.

Walter King finds the drawing of inferences from such experiences to be quite legitimate. Rebutting the assertions of USC Professor Peter Gordon that light rail was not appropriate in Los Angeles, he said, "All I'd want this man, whoever says that, to go to Europe. I just got back from Europe and you would have a desert [without rail systems]." He claimed that "All I'm saying is good sound logic. You show me one city in the world that's an important city without a major transportation, without some type of rail service." The others cited above believe that they are being logical, too. They have amassed evidence—albeit evidence of the senses—and used it to deduce conclusions about the desirability of rail in general. The fact that analytical processes are not engaged does not mean that the imagery lacks logic: it provides evidence, and anchors inferential mechanisms in a quite definable and powerful way.

Light rail critic and then Long Beach Councilwoman, Eunice Sato, was one interviewee not sold by the experience of other cities. "Just because other big cities do it and make a mistake doesn't mean we have to follow suit," she said. Los Angeles Councilman Ernani Bernardi was similarly unimpressed by subways elsewhere: "I don't know of any city that's more in gridlock than London; I don't know of any city that's in more gridlock than Paris, and they have very elaborate subway systems. You can't even drive in Manhattan and that has an elaborate subway system." Bernardi felt that it was the concentrated nature of these cities that causes congestion. His different framing of the transportation problem made him immune to images of technological performance alone. He had images too—of a congested London, Paris, or Manhattan—but the images were different because his framing was different.

Experience Is More Powerful than Argument

Some television reports have given time to opponents of rail, but not enough time to allow viewers to form reasonable opinions of the reliability of what they have to say. On one slot, for example, Peter Gordon says, "Well, the evidence is pretty clear these systems don't work," after which the newscaster announces that "The Los Angeles County Transportation Commission is looking into a very different crystal ball," and hands over to Jacki Bacharach, who says, "When people realize that the freeways are gonna be down to twenty miles an hour max in the peak hour. . . . when people realize they can read their newspapers and do a tremendous amount of work on the rail line I think they're gonna take it very happily."[27]

Bacharach has all the televised images of speeding trains and the viewers' experience of congested freeways in her favor. And while fleeting images of only a few seconds can be extremely powerful, Gordon lacks the chance to show that his reasons are stronger than Bacharach's. Both are said, by the reporter, to be looking into crystal balls, with no one view having higher validity.

A Trip to Disneyland—Brazil

Federal Urban Mass Transportation Administrator Robert Petricelli, visiting Seattle in October 1976, announced that funding would not be offered for rail planning in that city, but would be available instead for an all-bus system. Faced with the loss of federal dollars, Seattle's transit agency, METRO, came up with an all-bus alternative to include a downtown bus tunnel with five rail stations served by dual mode diesel/electric buses. The bus system was designed to have a sex appeal missing from regular buses. As Dan Graczyc of METRO staff said in an interview prior to project implementation, the new bus "will be designed by the same guy that designs Ferraris . . . I mean, it will be a pretty slick-looking bus . . . As a matter of fact we even have the option of putting his signature on the buses. It'll be a very class operation, just like riding in a Ferrari."

Even so, the agency was careful to state in the project's environmental impact report that the bus tunnel was designed to be convertible to light rail: "This approach enables the Downtown Seattle Transit Project to address problems now and in the future."[28] As Wes Fryztacki, Executive Director of Puget Sound Council of Governments put it during an interview: "You could say that the downtown bus tunnel is actually a rail tunnel that's being used in the interim by buses . . . Technically we need rail eventually. And that's been concluded and endorsed by our local elected officials. That's the common goal of everybody."

Faced with the impossibility of continuing development of their rail system as planned, the opportunity for developing a technologically virtuous bus approach opened up in Los Angeles as in Seattle. "The magic that we have seen in Curitiba is amazing," declares Los Angeles Mayor Richard Riordan on a documentary video entitled "In Our Lifetime," which recounts the visit of twenty-four Los Angeles civic and transportation leaders to Curitiba, Brazil in January 1999. "The whole reason I took them down there is it's operating as if it's light rail," said Los Angeles architect and prime rapid bus proponent Martha Welborne, who organized the visit and the video.

If the bus can be seen as a virtuoso performer anywhere, it is in Curitiba with its high-capacity bus system complete with platoons of buses operating at high frequency on busways and stopping at rail-like stations where, fare paid, passengers rapidly board and alight. Supervisor Yvonne Brathwaite-Burke says on the video she found the system "very impressive. I was impressed by the buses and its ability to move large numbers of people in a very short period of time for long distances. Busway—we need to do that. We also need to give preemption for bus transportation in order for us to cut down the time of travel." Supervisor Zev Yaroslavsky, full of admiration, described it as a "subway on wheels . . . I don't think there's any question of its applicability to Los Angeles . . . We have a number of wide boulevards that serve some of the most transit-dependent communities of the region. We've got right-of-ways. Let's use the right-of-ways."

Back in Los Angeles, MTA staff members did their best to market the Rapid Bus program, instituted in July 2000 at the same time as the opening of the Metrorail Red Line to North Hollywood, with the sexy imagery associated with rail. The Rapid Buses are red. "My opinion is that red connotes speed," said a principal MTA staffer on the Rapid project. The Rapid logo is slanted "the dot above the 'i' is a comet sign . . . and the word 'Rapid' is another comet shape. There are lots of comet shape ideas—the whole logo is a comet, which connotes speed."

Another member of the staff celebrated the change in perception of the bus: "Before what was dismissed as just a bus, now is seen as a transportation improvement. We are evaluating a variety of articulated vehicles. We would like to bring in new vehicles and add another level of excitement." Figure 10-4 shows one design image for the next generation of Rapid Buses, designed not only to look sleek, virtuoso, and phallic, but to be quite hard to distinguish from a train. Gone is the suggestion of dirt, dependency, and poverty associated with the traditional local bus; arrived is a new world of speed, adventure, and service to which Los Angeles politicians can relate, the very image of "new and different and exciting" which was previously unavailable to the bus.

As with the rail dreams of yesteryear, sense experience—in this case from a Disneyland-type visit to Brazil—delivered the message to politicians that the bus was the way to go, but it is no accident that the bus has been conceived with as trainlike an image as possible. Nor can we ignore the context in which the Rapid Bus program arose: one which precluded development of the planned rail program. The loss of rail created the vac-

Figure 10-4. Design for future development of Rapid Bus system. Note that the bus appears virtually to be a train, and is distinctly phallic in looks. (Reproduction courtesy Suisman Urban Design)

uum which allowed the Rapid program to emerge, and for the sense imagery provided in Curitiba to play a catalytic role.

Bob Nair, former chair of the Seattle METRO Transit Committee, acknowledged the appeal of the city's bus tunnel project, but declared in an interview that "They love the bus system, but they dream about rail." When the opportunity to pursue rail once more arose, a funding proposal was put on the ballot and, on November 5, 1996, 58 percent of Seattle voters approved a proposition to provide $3.9 billion of transit funding, two-thirds for light rail and commuter rail development. A proposed light rail system is in the midst of a stormy review process at the time of writing in fall 2001. Local critics are protesting at the high cost of the light rail system compared to the small number of people it would serve and the loss of the downtown tunnel for buses, with the environmental cost of putting them back on city streets. METRO is intent on implementing the rail system as part of its vision for future transportation, oblivious to flaws which will likely make the system as poor an investment as in most other new light rail cities.

In Los Angeles, a principal MTA manager was careful to present Rapid Bus to the Board as "an interim step until we organized the rail program." For Supervisor Yaroslavsky, on location in Curitiba, it was not that rail was any the less an object to be desired, but that "We can either sit here and wait for fifty years until everyone here has a light rail or a subway, and I think fifty would be optimistic, or you can try to do something in our lifetime."

> *"The Calcutta metro-rail is very nice," Ajit remarked over dinner. "The stations are clean, the trains run on time, and while one is waiting for the train one can listen to the pleasant tunes of Tagore's music played through the public broadcasting system. There are flower pots in the platform and inscriptions from Tagore's poetries on the wall," he added to lure me into the exciting experience of the Calcutta metro. As I listened to the different comments around the dining table that evening, I was struck by the pride that was felt about this new improvement in the city. "Calcutta may be deficient in many ways compared to Bombay or Delhi," Ajit concluded, "but we are leading in at least one aspect: we are the only Indian city with a metro-rail" . . .*
>
> *As we walked down the stairs, I was really surprised. The walls were clean and white washed, with no posters, no graffiti, not even hand marks. The floors had been swept clean and washed, it seemed only a few hours back. Not a single hawker or beggar was inside the station—not even the type one finds in the affluent countries' metros . . . Calcutta's subway was truly clean: even cleaner, it seemed, than some of the city's hospitals I was familiar with . . .*
>
> *. . . it occurred to me that beneath Ajit's reaction was a search for meaning—the meaning that sustains Calcutta's middle class, whose emotions fluctuate regularly between despair and hope . . . To retain one's sense of meaning amidst these despairing trends, hope is essential. And that hope expresses itself in many forms: the pride in a clean and efficient metro-rail, a large stadium which is believed to be Asia's biggest, a multi-storied, air-conditioned supermarket, a new fly-over near the Sealdah Station. These visible "improvements" of the city all add up to an image of modernization: the end product of the middle classes hope, the "India in the 21st century" that Rajiv Ghandhi expounds to motivate people.*
>
> *In questioning the relevance of Calcutta's metro-rail, I had unknowingly undermined this very basis of hope that makes living in Calcutta endurable for Ajit . . . For this group, [the middle class], the make-believe world of the metro-rail is a symbol of hope, much more meaningful than would appear to a visitor of the city.*
> —Bish Sanyal, "The Make-Believe World of the Calcutta Metro-Rail," *The Statesman*

The Train as Symbol of Community Pride—Penis Envy in Los Angeles

"I'm proud that Long Beach will be the first to be served by the new network," said Long Beach Mayor Ernie Kell. As Commissioner Marcia Mednick put it, in more global terms, it "gives a community a feeling of pride to have a new type of service going through it."

Rail was frequently seen as a mark of identity and status symbol. As Lee Hultgren, Director of Transportation for SANDAG commented:

> I think a community is known by a rail system more than it is known by a bus system. The London Underground. Even The Loop in Chicago. It becomes part of the Chamber of Commerce image of the community. You don't see buses representing a community, but you do see the Vienna Underground or London, San Francisco—BART and cable car—are much more prominent symbols of a community.

As Jim Pierson of San Jose said, rail has an "uptown image," an understanding that shows up in other communities too. Johnston and Sperling quote Anne Rudin, longtime light rail transit study committee member and mayor of Sacramento, who saw light rail as "representing good urban life in Sacramento," while the image of Portland was widely seen to be enhanced by the arrival of the train.[29] According to Steve Dotterer of the City of Portland staff, light rail had "captured the imagination of people throughout the state. People come up here to ride the thing." Ted Spence of Oregon DOT felt that the trolley was far more than something to ride on. "It's progress—there's something exciting, something to focus on—what else are you going to focus on? And it's something that's new. There's a lot of monuments in Washington, D.C. And if you were a pragmatic economist, you probably wouldn't have built those things."

Bill Robenhymer of MTDB San Diego staff pointed out that although the "technical analysis didn't come out overwhelmingly in favor of light rail," light rail "could do more for the image of San Diego" than buses. Alternate LACTC Commissioner Barna Szabo contended that the light rail had "improved the prestige" of San Diego. San Diego Supervisor Brian Bilbray agreed: "We're a very cosmopolitan city now," he said. "We're the best of Los Angeles and the best of San Francisco without the bad parts." "The trolley right now is motherhood and apple pie in San Diego," said San Diego Councilman Ed Struiksma. "If you want to do something good, stand next to the trolley and have your picture taken."

The trolley is a Washington monument or a St. Louis Arch. The trolley is symbolic of progress, of a vibrant city, and the symbolism can operate independently of the transportation characteristics of the service provided. As Kenneth Hahn put it: "If you're going to have a *great city* [my emphasis], you have to have rapid rail transit."[30]

Conversely, there "seems to be a feeling," said Aubry Davis, regional UMTA Administrator in Seattle, "it's sort of a macho thing. If you don't have a railroad, you're not a city." As we can see from the remarks of Commissioner Christine Reed, L.A.'s problem was penis envy.

> There was an intense amount of institutional ego over the fact that San Diego had *whipped* a trolley system *out*, kabloom, like that. They just did it. And I mean everybody else was like, oh my God, you know, what an affront that that little city could do that, and here we are—a big county—powerful, two-thirds of the population of the state, blah, blah, blah, blah, and we can't do this.

The "envy" showed up in many places. As *Los Angeles Times* reader Dorinda Humphrey wrote in a letter to the editor: "I have ridden the model for our trolly [sic], the San Diego Trolley, and I find it to be clean, quick and efficient. It is really embarrassing that a small city has showed us up like this."[31]

Speaking of Angelenos as naughty children, KABC-TV, Los Angeles, reported on August 2, 1982—the first anniversary of the San Diego system—that "While Los Angeles fights, San Diego celebrates." Alternate Commissioner Roy Donley joined in the self-flagellation: "Los Angeles is really a screwed-up mess, in my opinion, with respect to solving its transportation needs. We're the only major city in the world that doesn't have a subway system. . . . We've been screwing around with this transportation problem ever since World War II. We're the last major city without any kind of decent transportation system in my opinion." Alternate Commissioner Bob White really told it from the heart. "Let me tellya," he said, "we're about fifty years behind the rest of the world, and if we don't catch up now we're going to be the laughing stock of the world. Japan, England, Sweden, look at the countries right now, England—they have their transportation ten to one better than our country, and we've gotta get going." And, as Walter King added, "I'm *jealous* of what they do with their trains over there, and I hope some day we can get it here. Buses can't do that."

"There's a lot of civic ego in it," said Christine Reed:

There is, there really is. . . . that we are a great city and every other great city has a subway, therefore we must have a subway now. . . . I mean, you go to New York and you ride it and you don't feel that way about it, but there's this notion, and it comes back to the civic ego stuff, it's like all great cities that people talk about: Paris and London. Which are great for other reasons, not their metro rail systems.

But a dented civic ego hurts. While Los Angeles is condemned to continually search for its identity with overpriced Beverly Hills psychoanalysts, a rail system, like a sports stadium, opera house, or monumental arch, shows that a city has "arrived": it is a symbol of success. It is an easy logical leap to conclude that "if only we had rail, we could be successful too." In this respect, Los Angeles is little different from the Calcutta described by Bish Sanyal: in the "screwed-up mess" of Los Angeles, light rail is a symbol of hope, something to be proud of, something to at least momentarily make the larger problems—which remain unsolved—seem to go away.

Light Rail as Cargo Cult

In a New Guinea cargo cult, docks or runways may be constructed to receive the cargo thought to arrive when such facilities are in place, despite the lack of evidence that any cargo-carrying ships or aircraft would actually bring the goods.[32] Cargo cults are symbolic ways of doing something about situations otherwise thought to be unendurable, and have as their central tenet a sudden and complete transformation, a complete change in circumstances and material wealth that would come about "if just the right actions were taken, just the right incantations invoked."[33]

As Raymond Firth points out, "there tends to arise a gap between the results of symbolic action and those of pragmatic or empirical action. It is not a matter of which is real; it is a matter of sorting out the different implications of each." The New Guinea islanders'

> activity may be classed as no less real to them than is our technological construction to us. But the results are of a different order. To expect a material aeroplane to land on a symbolic airstrip, or material rice or calico to come out of a symbolic aeroplane, is a confusion of implications. But such confusion of implications of symbolic action can occur, and out of it can come much dissatisfaction and disturbance of social relations.[34]

Joseph Conforti suggests that such behavior is not limited to traditional societies:

> At least a nascent cargo cult orientation exists within American society (and probably within similar societies as well). It constitutes *an ideology that contains its own logic* and fosters a set of beliefs about how things are accomplished, how things are acquired, which actions are likely to have desirable consequences, and which are not.[my emphasis]
>
> Just as the Pacific Islanders focus on some singular and decisive factor or pattern of action rather than the complex relationship between causal factors and their effects, so too do many Americans. . . .
>
> The ideology presented is perhaps most obvious in the kinds of programs commonly watched by young children: cartoons and adventure series. These programs present very dramatic, oversimplified depictions of life, stressing contention and sudden resolution. The hero, who is all good, is also inherently endowed with all the qualities necessary to success. As the child grows older, these programs give way to a broader array, incorporating sports, popular music, soap opera, situation comedies etc. They also all continually reinforce the idea that things come into existence fully formed; there is little or no process depicted, no growth, no development, no accumulation; there is also neither ambivalence nor compromise, something either is or it is not, one does something or one does not.[35]

As a symbol of hope, the train in Los Angeles is a cargo cult, an answer to an otherwise unendurable situation. Just as the New Guinea islanders entertain the unfounded belief that airstrips to which goods are delivered elsewhere will bring them cargo, Angelenos believe that the rapid transit systems of "successful" cities will make Los Angeles successful, too. As in the case of the islanders, Angelenos focus on a technology of hope, rather than on "the complex relationship between causal factors and their effects." Theirs is an oversimplified world in which the train is the hero and is all good, the road and the bus the bad guys—and all bad: The train is a transport of delight, seen as the key not only to solving Los Angeles' transportation woes, but providing a pleasant experience and giving the community a new pride in its identity.

The Unconverted: Maxcy Filer

Compton Councilman Maxcy Filer, a grassroots community activist, closely connected with and deeply aware of the problems of his constituents, was not party to the myth of rail. "I think what has happened:

Nostalgia is setting in," he said. "I think many of us look at it from the standpoint, we've tried buses, we've tried cars, we're tired of the freeway, well now let's go back to the train. I'm not surprised. I think they might bring the horse and buggy back from a nostalgia standpoint." Was this appropriate he was asked?

> I think that we know what the Red Car did before. And now all at once we all say, well we want to get back on the Red Car, that's where we're really going. Again, the only thing we're doing is painting it silver. And of course we might [make] it go a little faster. I'm not sure that we will, because the Red Car used to get up to I think sixty-five or seventy mph, so it's one of those things you aren't talking about really getting it there that much faster. And the Red Car had the right-of-way, too, most of the time, especially when it was express, it had the right of way. So I think [there]'s a nostalgia aspect that we're going through.

"Was that valid?" I asked.

> No, that isn't a valid way. It is not a valid way. Those that want to ride the Red Car, let them do it on Sunday afternoon, have their pleasure of going to Long Beach, to Los Angeles, from Los Angeles to Long Beach, and say fine, now I've gotten over it, I've enjoyed it, so forget about it. Those that want to ride the old paddle boats, they should go to the Mississippi, ride on one, and don't bring it here and say get rid of the Queen Mary or whatever the steamships are. You just go ride the paddle boat. . . .
>
> I'm not looking at the pretty issue, the glamorous issues or anything of that nature. Will it do the job? If it won't do the job, then why have it? Just doesn't make sense.

Searching for Meaning and Identity

Trains are toys, objects of nostalgic, futuristic, sexual, and romantic desire and of religious worship. Their symbolic power, extending far beyond any functionality as providers of transportation, derives from a rich context of experiences, memories, and historical associations. The 'thrill' and 'elegance' associated with trains goes to the heart of our human quest for the colorful and the pleasurable, rather than the utilitarian and the mundane. Like humans, trains are to be mourned when they pass away: their death symbolizes the passage of a way of life, and the loss of a connection with our history and our very identity. Trains, like objects which become transformed by worship into pagan gods, come to acquire religious identities through their rich, positive, historical associations, and their promise to bring blessings in the future: they become subjects of worship.

The desire for trains in Los Angeles, as in Calcutta, is part of a universal human search for meaning and identity. Associated with the "good life" of other, supposedly successful cities, trains seem to offer Los Angeles the secret of success, too. Although this belief is as absurd as to expect "a material aeroplane to land on a symbolic airstrip" in the New Guinea cargo cult, the collection of symbolic associations manifest in the idea of the train make that belief quite compelling and real.

These associations have little or nothing to do with rail's functionality as a provider of transportation services in Los Angeles, yet they reinforce the images examined in chapter 8 which are directly—if misleadingly—linked to transportation attributes. The apparent technological virtuosity of a train strengthens the image that it is fast. The sex appeal and romantic imagery make it seem more comfortable. The concept of pride in having a train goes along with the idea that the community won't vandalize trains and that they will be safe to ride. These associations all help draw attention to that part of the total trip spent on the train (rather than to access and distribution to and from the train), and perhaps provide clues as to why—in our culture—we metonymically reduce our total trip to that part spent on the line haul.

CHAPTER 11   # Light Rail and the Symbolic Promise of Community Renewal in Watts and Compton

From a conversation with Professor C. West Churchman.

Richmond: What should be our priorities in transportation policy and why?

Churchman: I'm not interested in priorities within transportation.

Richmond: Why?

Churchman: Because . . . the majority of people in this world are still using sticks for energy.

Richmond: So we should not attend to localized problems?

Churchman: No, I just say if you ask me where the priorities should be. The Catholic bishops put out a pastoral letter and they asked that question: "for the world, what's the top priority?" and they said "poverty." Much more so even than nuclear threat.

Richmond: We've moved from transportation to poverty.

Churchman: Do you want me to shut the door and go into the transportation room?

THE RED CARS GAVE WATTS ITS NAME. Later, the Long Beach light rail project—which crosses the depressed city of Compton in addition to passing through Watts (a part of the City of Los Angeles)—came to be seen by many as a way to brighten prospects for a sector of the metropolis where poverty and deprivation continue to rule. The relief the new Blue Line was expected to bring, however, was conceived in symbolic terms, which projected a far rosier outcome than the benefits economic principles dictated the light rail system might actually bring.

To study the attraction of the Blue Line to Watts, we must go beyond transportation issues to understand the larger problems the community faces and appreciate why light rail was seen as a cure. It is only then that light rail's symbolic status as a beacon of hope emerges; only then is its appeal to politicians is seen for what it was: an action both feasible and virtuous in appearance; an action taken at a time when neither easy solutions nor political commitment existed for overcoming the complex social problems at play.

> *Listening to Ajit, I began to realize that underlying our apparent disagreement over the metro-rail were different visions of Calcutta's future. There was an irony in our different visions: Ajit, who lives in Calcutta, seemed less concerned about poverty than about modernization; yet, I, who was now living in one of the most affluent countries of the world, believed that Calcutta's poverty was of such magnitude that efforts to change the city into a vast technological landscape were futile at best. To put it another way, Ajit, who daily confronts Calcutta's poverty, seemed less pragmatic about it and, instead, longed for a "modern" Calcutta. The construction of the metro-rail seemed to him a step in the right direction. I, who have virtually no direct contact with poverty of such dimensions, as in Calcutta, was being more pragmatic in thinking that Calcutta will essentially remain a city of the poor, around whose needs the city should be primarily organized. That would mean more jobs, shelter, transport and health care for the poor; not the kind of development usually associated with an image of modernization.*
> —Bish Sanyal, "The Make-Believe World of the Calcutta Metro-Rail," *The Statesman*

History

In 1883 the completion of the Santa Fe and Southern Pacific railroads launched a wave of land speculation in Los Angeles. One of the communities to appear, along with the railroad, was Tajuata, founded on the right-of-way of the old Los Angeles and San Pedro Railroad. As the McCone Commission on the Los Angeles riots recounts, two of Henry Huntington's Red Car lines—from downtown Los Angeles to Long Beach, and from Santa Monica to Venice—intersected close to Tajuata on land owned by the Watts family.[1] The station at the intersection was named Watts, and soon afterward Watts displaced Tajuata as the name of the community at that location, too.

The railroad brought the first blacks to Watts. Employees of the railroad, they lived in company houses until they had earned money to buy homes.[2] New black arrivals during World War II were attracted to Watts, says McCone, partly because deed restrictions and other discriminatory practices often made it impossible for them to rent or buy homes in many other parts of the city and county. With blacks moving in, whites moved out, and the majority of the black population came to live in segregated areas.[3]

With Watts being a major rail junction, the Red Cars were a part of everyday life, and left a rich heritage of memories. Diane Watson, representing Watts in the California Senate, recalled, "I used to use that [the Red Cars] and it was just a tremendous way."

> Yeah, there are experiences, you see, as children. And I used to have a dressmaker that lived out in Watts. And she was very cheap, and I was very tall, so I used to have my clothes made, even when I was in school . . .
> I'll never forget them taking up the tracks to make way for more cars down Jefferson.

Grace Payne, Director of the Westminster Neighborhood Association, a major community service organization in Watts, also remembered the Red Cars as a positive element of Watts life: "I personally have had experience on the Red Car, and remember the joy of riding on the Red Car, so I think that would be a great experience for the younger generation, and very convenient for the older people."

The Watts Riots and After

When John McCone wrote his inquiry into the Watts riots in 1965, 88.6 percent of the black Los Angeles population lived in areas considered segregated, concentrated especially in the 46.5 square miles of Los Angeles placed under curfew when the riots erupted in August of that year.

The six days of riots left thirty-four people dead, 1,032 wounded, and $40 million of property damage, according to McCone. There were 3,952 arrests:

> The lawlessness in this one segment of the metropolitan area had terrified the entire county and its 6,000,000 citizens.
> In the ugliest interval, which lasted from Thursday through Saturday, per-

haps as many as 10,000 Negroes took to the streets in marauding bands. They looted stores, set fires, beat up white passerby whom they hauled from stopped cars, many of which were turned upside down and burned, exchanged shots with law enforcement officers, and stoned and shot at firemen. The rioters seemed to have been caught up in an insensate rage of destruction.[4]

The report identified core problems "sowing the wind" as a shortage of jobs, inadequate education, and resentment, "even hatred" of the police. In circumstances where law and order has only "tenuous hold, the conditions of life itself are often marginal; idleness leads to despair and finally, mass violence supplies a momentary relief from the malaise." Early in the report, McCone says: "Moreover, the fundamental problems which are the same here as in the cities which were racked by the [other] 1964 riots, are intensified by what may be the least adequate network of public transportation in any major city in America."[5]

The report called for improvements in law enforcement and police-community relations; in job programs, including efforts to end discrimination in employment; a reorganization and strengthening of programs for schools; control of business practices which put consumers at a disadvantage; improvements in health, welfare, and housing; and higher-quality transportation: "If the Los Angeles area as a whole and the Watts area in particular are to have better bus transportation service, it can only be provided through a public subsidy to accomplish three purposes: reduce fares, purchase or condemn the multiple uncoordinated bus system, and provide system-wide transfers." Additionally, McCone called for "immediate establishment of an adequate east-west cross town service as well as increasing the north-south service to permit efficient transportation to and from the area."[6]

McCone's report received heavy criticism. According to Los Angeles urban historian Robert Fogelson, for example: "The Negroes rioted because they could not passively accept conditions in the ghetto any longer and not because they were unprepared for urban life or because their leaders were contemptuous of law and order . . . The Watts vicinity is, by any physical or psychological criteria, a slum, in which Los Angeles' Negroes are rigorously and involuntarily segregated." McCone, Fogelson said, had erroneously led Los Angeles residents to the "conclusion that the rioting was meaningless or drawn the implication that the Negroes somehow lack the qualifications for responsible citizenship."[7] To Robert Blauner, "The spirit of the Watts rioters appears similar to that of anti-colonial crowds demonstrating against foreign masters." And, Blauner adds, "There was no attempt [by McCone] to view the outbreak from the point of view of the Negro poor."[8]

Rustin has a tale which nicely sums up the intensely political and revolutionary nature of the riots which McCone could not understand:

> At a street-corner meeting in Watts when the riots were over, an unemployed youth of about twenty said to me, "We won." I asked him: "How have you won? Homes have been destroyed, Negroes are lying dead in the streets, the stores from which you buy food and clothes are destroyed and people are bringing you relief." His reply was significant: "We won because we made the whole world pay attention to us. The police chief never came here before, the mayor always stayed uptown. We made them come."[9]

On McCone's specific proposals for transportation, Fogelson pointed out that while transit services were inadequate, a 1965 census showed that 65 percent of south-central families owned one or more cars, not the 14 percent reported by the McCone Commission. "The southcentral ghetto is indeed isolated, but not for reasons as simple and reassuring as dreadful bus service."[10]

McCone Revisited, a report to investigate the impacts of McCone two decades later found, however, "that the greatest progress since 1965 has been made in transportation."[11] RTD had received both state and federal funding, while—at the time of the report's writing—Proposition A was helping to keep local bus fares at a low fifty cents (the low fares ended after three years, however, as Proposition A money used to subsidize them was dedicated to the rail program after this period). Bus-to-bus transfer privileges, recommended by McCone, were in effect, while "RTD has acquired most of the smaller bus companies, which has standardized and simplified service and improved efficiency." A reorganization of RTD operations into a grid service had, meanwhile, "led to an immediate upsurge in ridership."

In contrast to the transportation improvements, hearing testimony showed that many problems were identical to those identified by McCone and that overall conditions were as bad or worse than nineteen years previously. According to one speaker, "A basic problem in South Central Los

Angeles in 1984, as it was in 1965, is poverty: grinding, unending, and debilitating for all whom it touches." The details were depressing. Police-community conflicts persisted, while "severe unemployment in 1984 continues unabated." Infant mortality in South Central Los Angeles remained over double the rate for white infants in Los Angeles: "One-third of the residents live in old, crime-ridden public housing projects. Of the 600 low cost housing units authorized, only 300 have been built during the 19 years since the riots, and no single family housing has been built in the area." Educational funding, meanwhile, was "inadequate compared to the need." Not only were there "several hundred classrooms in South Central Los Angeles . . . staffed by substitutes," but also a "chronic shortage of Math and Science teachers."[12]

Problems of Welfare and Social Services also remained "critical." Civil rights and community organization representatives characterized South Central Los Angeles as a low government priority. Government leaders lacked the commitment to take action, they said, and there was a general failure to initiate planning, strategy, or solutions to confront the problems cited by McCone.[13] Commented Larry Aubry, staff to the Los Angeles County Human Relations Commission:

> I don't think it's changed, because I don't think there's a public will to change . . . You have people there who are essentially disenfranchised . . . they're as disenfranchised in the greater South Central Los Angeles area now as they were in 1965 . . .
> I think that racism still obtains . . . The closer you are to White on this continuum, the better off you are. I'm sorry: that's the way it works. The closer you are to Black on the continuum, the worse off you are.

Felicia Bragg is a professional who grew up in Watts and joined the board of the community service Westminster Neighborhood Association. Only a small part of our meeting revolved around transportation questions, as Bragg pointed to the more pressing problems of the community, and how distant the majority white power structure was from understanding them. She told of being one of the brightest at Watts' Jordan High School and of the culture shock she faced upon entering the University of California at Santa Barbara:

> It was the first time I had really run into racism in its overt form, first time anybody called me nigger, first time I understood the absolute difference be-

tween poverty and non-poverty and what it could do for you, and it took a while to regain my confidence in my innate abilities, my own talents. It was also the first time a White teacher told me that I simply could not have produced a paper of that caliber because I was from Watts. You have those problems. But on that score I should tell you the same thing happened to me at USC a few years ago. I was in the MBA program, and it was a similar situation where they absolutely refused to believe that I produced something and had to call in the Dean to testify to my scores to kind of legitimize the whole thing.

And here I am, years later, I'm in business for myself, and I have to tell you frankly that we don't even hope to get White clients any more. I mean I did at first, but it just doesn't work. It isn't happening. When you think about racism, people think about the 60s and Martin Luther King and the days of the Civil Rights movement, but it's still very very much alive in some very insidious kinds of ways. And it isn't anything you can just put your hands on and say if we could only solve this then it would all be solved. It isn't all solved.

Some of the perceptions of the Watts community of which Aubry and Bragg complained emerged during certain other interviews. Walter King, Alternate Commissioner to Kenneth Hahn, for example, referred to transit riders as a "bunch of animals . . . We don't have discipline any more. You tell the guy to sit down and he won't do it." King's conception of the Blue Line was more as a defense mechanism against the people of Watts than as a service for them: "Because we're building in millions of dollars worth of things, it's going to make it safe." On the problems of Watts, he added: "But they're not poor people in Watts. They've got as much money, more than we do, and they've got better cars than I have." Norwalk Councilman and LACTC Alternate Commissioner Bob White, when asked about the effect of a fare increase on the poor, replied: "I get a kick out of that. My real opinion on that is that they'll have enough money always to ride the bus to get some place that they want to go because those same people sometimes are playing bingo every night, they're going to the race-track, they're doing everything they want. That's a cop-out. I think they really overdo that."

Gerald Leonard, former aide to Baxter Ward, drew a contrast between two mandates: "One is to carry the people we *need to carry*, the transit dependents, and the other is to carry the people we'd *like to carry*, the commuters." Long Beach Councilman Wallace Edgerton gave this reason as to why the car would remain important in South Central Los Angeles: "You take a young Black male, and he's got to have a car. The car's probably got to be yellow and big with light walls, horrendous ornament out there on the hood." Such comments do not represent the interview sample as a whole. They do provide evidence of how far the political establishment remains from understanding the problems of Watts.

Riots took place in Los Angeles once again in 1992, and civil disturbances were centered along the central and northern stretches of the Blue Line, and exacerbated the sense of decline in many Blue Line station neighborhoods.[14] Writing nine years after Blue Line opening, Loukaitou-Sideris and Banerjee found that eleven Blue Line stations were in areas with over 35 percent of households under the poverty line, and that the Blue Line route continued to encompass "some of the most depressed and neglected neighborhoods of Los Angeles County, which have suffered from poverty, abandonment and deterioration of their physical infrastructure."[15]

Perceived Benefits of Returning the Trolley to Watts

Claims were made that the Blue Line would bring benefits to Watts and Compton beyond mere transportation links. Light rail was seen as a path to employment, as a spur to economic development and civic improvement, and even as a catalyst to integrating the communities through which the service passes into greater Los Angeles as a whole.

Transportation and Employment

On April 1, 1966, four months after the McCone Commission report was issued, the California State Transportation Agency applied to the U.S. Department of Housing and Urban Development for a research and demonstration grant to study the relationship between a public transportation system, and job and other opportunities of low-income groups, and in particular to test what the McCone Commission "appeared to suggest . . . that the disadvantaged poor can be placed in jobs if a way is found to get them there, and that lack of adequate public transportation might be an important cause of unemployment."[16]

The report stated that the transportation-employment demonstration project, which monitored the effects of improved public transportation, was:

significant because it did not demonstrate that providing transportation for the jobless cures unemployment; but it did underline something that is at times forgotten—that adequate transportation is a necessary but not sufficient condition for people to have access to job opportunities. If jobs are not available, for whatever reason, no amount of transport will create them. Between the lines, one may also discern that discrimination, as well as lack of job skills, remain key barriers to job opportunity.[17]

The report additionally said that employers seeking low-skilled labor tended to prefer people living close to their plants and with their own cars to commute in. Furthermore:

> Many of the jobs that have been made available through the special job programs are low paying and at great distances from the project area. People willing to accept such employment cannot usually afford the high costs of travel by either bus or private automobile and do not generally stay on the jobs very long. Poor people who are placed in minimum wage jobs fifteen to twenty-five miles from their homes do not regard those jobs as permanent solutions to their employment problems.

Despite these findings, the McCone Commission's definition of isolation in terms of public transportation remained in good currency. In 1973, for example, a report examining the case for rapid transit for the RTD, declared that:

> Aside from having an easier, safer, quicker, and less costly trip to work, many members of the labor force will find better jobs, because rapid transit will make employment centers more accessible. More important, however, is the fact that rapid transit will enable some unemployed people to find jobs. Today, many of the unemployed do not have an auto. Although bus service is usually available, the service provided by the bus makes job hunting difficult. Rapid transit will take these people to major employment centers where they can find jobs.[18]

The October 16, 1983 *Los Angeles Times* recalled, furthermore, that:

> As far back as 1965, the McCone Commission investigating the Watts riots blamed bad public transit for isolating the area.
> Since then, bus transit has been improved, but county transportation commission officials said the light rail system will help even more.
> "We think it provides a real improvement to mobility in the inner areas because it provides top-quality service downtown (where thousands work in the garment district and in other jobs) and to Long Beach and to the other industrial and commercial development in the port area," said Rick Rich-

mond ... "a lot of the [bus] service in the area is east-west service," which does not take riders to jobs in Long Beach or downtown Los Angeles.

"The line is trying to significantly increase the transportation service for the people in South Central," said Dan Caufield, project manager for the commission ... "Today the buses that serve South-Central are not very frequent. Perhaps those buses are not any more inconvenient than buses anywhere else, but the fact is that in South-Central, people are more dependent on public transit, so public transit is more important to get to jobs."

There were several references to providing access to employment in interviews for this study. "It will give them an opportunity, perhaps, to hold jobs in Long Beach, and grab the light rail down here, and work at the Hyatt Regency, and be back home again," said Long Beach Mayor Ernie Kell. "More and more of the Black and Hispanic population will find employment affordable if they have affordable transportation," said citizen member of the LACTC Rail Construction Committee Allan Jonas. LACTC staff member Richard Stanger claimed that the light rail would open up "a very direct service" to "job opportunities that really are difficult now to get to."

Those interviewed for this study who were working in development programs, however, were unconvinced that the rail service would open job opportunities. While Richard Benbow of the Los Angeles Community Redevelopment Agency did acknowledge that transportation was a "very key factor" in making jobs accessible, he said that transportation improvements by themselves were "not going to do anything. It has to be a comprehensive approach ... Now I know where the jobs are and I can get there, but I still don't have the education and the qualifications to get the job, so what's the point of going out there?"

Others also drew attention to the deeper problems of the community, which transportation alone could not cure. These problems were receiving inadequate attention as transportation improvements went ahead, they said. Larry Aubry saw that people were out of work

> basically because they're ill-equipped. They're unskilled. They're unemployed ... They're not motivated, but they aren't motivated simply because they're not prepared, and that gets back to the education thing again ... You are talking about [people] here who, for the most part, are probably reading at the 6th grade level, albeit many of them have graduated from high school. That's a problem ...
>
> There's little doubt but that most of those students out there are essen-

tially dead by the time they're in third grade. It's over. Because they have very little expectations. You have this almost classic lack of expectations that goes on. That becomes a self-fulfilling prophesy. They don't feel that they can accomplish anything. Most of them have a low self-concept. They have very little experience for success. That's hardly—not only on their minds, it's not an agenda item.

And you've really got to be careful, because this has very little to do with intelligence. It has to do with degradation.

As Marilyn Lurie of the Community Development Department of the City of Los Angeles underlined, difficulties in getting people to jobs were far too complex to be solved by merely laying on transportation:

> They have a real hard problem with youth. Something like 40 percent of the job training funds in the city were supposed to go for youth training. The agency has a real hard time finding youth. One theory is the competing underground labor market. You know: drugs . . . Motivation is one [reason]. Seeing your way beyond the cycle. People who are on welfare, you've got the problem that unless the job at the end of the job training is going to be something more than a welfare payment, what incentive is there?

New rail service alone would not improve things. To look at evidence of simple transportation benefits, however, census data contradicts the assertions that the light rail service could provide connections to important centers of employment opportunity. As data prepared for a *Los Angeles Times* article show: "Less than 10% of the workers living in the so-called 'mid-corridor' between the two downtowns work in downtown Los Angeles, though many of those who do take public transit. By far, most residents in the mid-corridor either work within the area or travel to widely dispersed locations such as the Westside, South Bay and San Gabriel Valley"[19] (see fig. 11-1). Most mid-corridor bus ridership, the *Times* reported, was for local trips, and that was why then RTD General Manager John Dyer had explained that RTD did not offer express bus service from there to downtown Los Angeles. At the same time, the *Times* article quoted (without full citation) a 1982 Parsons Brinckerhoff study for the LACTC, which stated that "In the case of local trips, convenience in gaining access to service plays a more important role than the travel time or quality of service."

The claim that expenditures to develop the rail system had hurt minorities by damaging the bus system on which they were most dependent formed a major issue in a civil rights legal action against the MTA.[20] As part

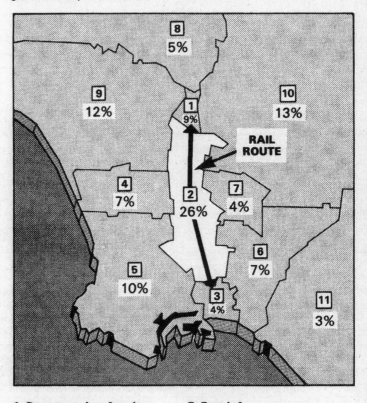

WHERE PEOPLE COMMUTE

Census data prepared for The Times by the Southern California Assn. of Governments shows where residents along the rail route travel to work. Highlighted are commuter patterns for those living along the middle of the route (Zone 2), where most boardings are expected. The percentages are Zone 2 workers employed in that area. Relatively few commuters travel between downtown Los Angeles and Long Beach. Critics say that the rail line will serve only a small percentage of commuters. Proponents say that those who do commute to downtown Los Angeles can make greater use of public transit.

8 — 5%

9 — 12%

1 — 9%

10 — 13%

RAIL ROUTE

4 — 7%

2 — 26%

7 — 4%

6 — 7%

5 — 10%

3 — 4%

11 — 3%

1. Downtown Los Angeles
2. Mid-Corridor Area
3. Downtown Long Beach
4. Century Freeway Corridor
5. Southwest L.A. County
6. Southeast L.A. County
7. East L.A.
8. San Fernando Valley
9. Westside and Ventura County
10. San Gabriel and North L.A.
11. Orange, Riverside and San Bernardino Counties

Source: Census and Southern California Association of Governments

Figure 11-1. Where people commute. (Copyright 1985, *Los Angeles Times*. Reprinted by Permission.)

of that case, evidence was given that, while minorities contributed 75.5 percent of Blue Line ridership,[21] minorities were nonetheless underrepresented on the Blue Line, given that 90 percent of the Blue Line corridor population was minority and that minorities accounted for 81 percent of overall MTA ridership over the 1991–1993 period,[22] while making up only 61 percent of county population.[23] This underlines the importance to minorities and those of low income in general of a dense network of buses for making the wide variety of dispersed local trips required by those who depend on transit, journeys which may require transfers if made by rail, or which may not be feasible via the rail network at all.

The statement by Caufield that buses were not very frequent sidesteps the question of why they could not be made more frequent, as against installing a capital-intensive rail project. The tenor of the interview statements in the 1983 *Los Angeles Times* article, however, rekindled the McCone concept criticized by Fogelson that community isolation is somehow a function of poor public transportation links, and helps explain the popular belief that community benefits would come from connecting low-income areas to light rail. It is poignant, given the enthusiasm for rail service prior to construction, to note rail's negative image in a lawsuit that complained about the deterioration in bus services which took place as the rail program was implemented, and about increases in fares and removal of pass programs, all of which had a disproportionately negative impact on minorities and those of low income.

Bringing Development to Watts

Just as blood flowing through arteries brings nutrients wherever it flows, the circulation metaphor tells us that rail will bring revitalization to communities through which it passes. "It does *go through* an area which has a potential for future development," said then Los Angeles Councilwoman Pat Russell. LACTC Commissioner Marcia Mednick agreed that "it *goes through* areas that are definitely transit-dependent; it may open up possibilities for different types of development in terms of growth in developing those areas." Then LACTC chair, Jacki Bacharach also stressed that it is "*going through* one of the most depressed areas in Los Angeles, a transit-dependent community." Debbie George of Supervisor Dana's office, meanwhile, was convinced that "it will just stimulate growth there, economic growth, by having a major light rail system *go through* there at grade, go through their city."

Los Angeles Councilwoman Joan Flores said that "part of the incentive to put the Watts shopping center in was the fact that there would be light rail. Once that starts happening and *people congregate around there*, that's where the development will be." And, according to Grace Payne of the Westminster Neighborhood Association: "If it's convenient for people to travel to an area, they would certainly do more shopping in that area, or seek more services for that area, so I'm sure that there would be a lot of improvements in the area of new economic growth."

Compton Councilman Floyd James "has argued that the trolley would allow commuters from throughout the region to see that Compton is rebuilding itself," while City Manager Laverta Montgomery backed the trolley system "as a way to 'showcase' a rebuilding Compton."[24] Mayor Tucker of Compton was among those who believed the city would derive such benefits: "we know that it would enhance our shopping centers for people to be able to come to Compton. We have a junior college, we have certain things." But Compton Councilman Maxcy Filer was skeptical:

> The City's saying, the City of Compton, that in essence we want it to travel at grade so that people will stop and shop in our shopping center. If I'm on my way to work in Los Angeles from Long Beach, I'm not going to stop in Compton just to shop. If I'm on my way home from Los Angeles to Long Beach from work, why should I stop in Compton and shop, as much as I love Compton, when the rapid transit as they call it—a light rail it really is—will actually stop in downtown Long Beach, right in the heart of the shopping center. So why would they shop in Compton? I don't see it. It doesn't logically follow.

LACTC staff were divided in their opinions over the likely development benefits of light rail. Although Long Beach light rail project manager, Dan Caufield, described the light rail as an "urban redevelopment project," Rick Richmond, Paul Taylor, and Richard Stanger were skeptical. According to Taylor:

> The development effect absent public intervention will probably be nil, apart from the ends. In the middle, where there isn't a market for development other than publicly catalyzed development, you're not going to find it. What it does is it provides an ingredient that makes it a little easier to catalyze de-

velopment through public intervention . . . You don't do it for development purposes.

Said Richard Benbow of the Community Redevelopment Agency, "I'm not at all optimistic about it [development]. I think if there were people, investors, developers who are looking to take advantage of this opportunity, we would know about them now, but I don't see them. They're not lining up outside my door, so I can't be sure that they're out there at all."

A report prepared by Sedway Cooke Associates for LACTC's environmental work declared that "In the Mid-Corridor, a small amount of retail development might be stimulated at each of the stations. In addition, the project could provide the support needed to expand the 103rd Street and Imperial Highway stations and to generate new development in the Florence Avenue Business District and in downtown Compton."[25] But, an evaluation of development potential for LACTC said, "market support for new industrial uses" in the northern part of the midcorridor, which contains several older industrial areas, "is very weak: lease rates are generally too low to support new space, and obsolescent buildings and vacant lots attest to the low level of reinvestment in the area." In addition, "new retail development in neighborhood centers of community centers generally has limited market support in the midcorridor" because of static or declining population; low income levels; a considerable inventory of already available low lease rate retail space; and sales competition from nearby areas. "The limitations listed above . . . are expected to severely restrict the future development of market-induced retail space in the midcorridor for the next several years."[26]

Loukaitou-Sideris and Banerjee conducted reviews of the effect of the Blue Line on development in neighborhoods near rail stations six and nine years after opening. At both stages of their work, they found that there has been almost no visible improvement or development in the neighborhoods around most Blue Line stations.[27] Station areas lacked significant concentrations of population and activities, with physical amenities such as parks, convenience stores, and restaurants "conspicuously absent"; station areas showed signs of abandonment and disinvestment; and high crime rates and negative perceptions of the areas only made these conditions worse.

Investors are unlikely to invest in redeveloping such decaying districts without incentives, the authors say, and these have been lacking. There has been a lack of strong commitment to restructuring development patterns at stations, with local initiatives lukewarm at best. The passage of the Blue Line through several local jurisdictions as well as unincorporated areas of Los Angeles County has, additionally, hampered the development of a co-ordinated response. As for the MTA, it has so far shown itself unwilling to invest in risky inner-city projects. The MTA's limited land holdings around stations and its reluctance or inability to undertake large-scale ventures, given fiscal constraints, make prospects for village-type developments dubious, the authors say.[28]

> The presumption of transit-induced development deeply rooted in many planners' visions of ideal community form and in the legacy of the streetcar suburbs does not seem to apply to inner-city neighborhoods . . . A transit system cannot by its mere presence catalyze miracles in the inner city. In that sense the notion of the modern transit village will remain a bourgeois utopia unless strong political and institutional commitments are made.[29]

The image of the train "going through" the community and bringing economic growth along with it is powerful, nonetheless, and was able to counteract any notions of the likely reality for those under its influence before the rail service had come into existence. As in the New Guinea cargo cult, where airstrips or docks are built without any reason to believe that cargo will arrive, the benefits to which light rail symbolically gives rise led to a desire to have light rail in place, despite the lack of evidence that such benefits would likely actually materialize.

Larger Understandings of Connection

The train is symbolically important in spirituals, representing the way to life, death, and redemption: a perfect example of casting the abstract in terms of the physical, and clearly tied to the train's role not only in nation building, but in taking people apart from each other, and bringing them to start out new lives. In "The Gospel Train is Coming":

> *The fare is cheap and all can go, The rich and poor are there,*
> *no second-class on board the train, No difference in the fare.*

In "The Railway to Heaven":

> *One grand first class is used for all, For Jew and Gentile, great and small;*
> *There's room for all the world inside, And kings with beggars there do ride.*

The egalitarian aspirations of the spiritual ring true in the comments of Watts community activist and longtime resident "Sweet" Alice Harris. "We need transportation," she said:

It's not only transportation. It's education. That's one way our children, even the small ones that have never been out of Watts, can get on the train and go from here to Long Beach. There's many educational sights that you can see between Watts and Long Beach, if it's no more that the people are the same as we are. Communities are the same as our communities. You have high class, low class.

The belief in the powers of trains as forces for social integration was, furthermore, promoted in a May 1, 1983 *Los Angeles Herald-Examiner* commentary article by David Israel:

A subway helps to develop a sense of community in a city. It establishes a common thread through diverse neighborhoods. It forces you to understand where and how different people live, it gives you a shared experience . . .

In the subway you aren't containerized separately, you're containerized together; you have to look the other fellow in the eye, and eventually you will probably wink or nod or maybe even engage in conversation.

This view of the socially unifying powers of transit has a long history, as discussed by Holt.[30] Holt cites an 1841 magazine article on the "leveling and democratic Omnibus" in which: "The statesman and politician . . . the greasy citizen who votes against him; and the zealots of different sectaries, dismounted of their several doxies, are compelled to ride, cheek-by-jowl, with one another."[31] While putting people in physical proximity may have metaphorically suggested social integration, however, Holt reports that the reality was otherwise. Gentlemen would refuse to give seats up to working-class women, while men with hand tools or muddy boots were consigned to ride on the platforms. Horsecars to bring workers to the suburbs on Sundays were forbidden in Philadelphia and St. Louis to avoid disturbing the prayers of those in churches along the lines. It took legal action in the former and as sit-in in the latter to get blacks the privilege to ride the cars at all. "The hope that public conveyances would somehow advance equality and democracy faded as the century aged."[32]

An understanding of connecting people to more than just places came through strongly in several interviews. Peter Ireland of Supervisor Deane Dana's office talked in terms of a metaphor of "connection" to opportunities: the physical transportation link symbolizes abstract connections to employment and the good things of life, even if there is no evidence such "connection" can actually be provided without change at a much deeper level.

> If you have an area of the city, that is, where [we] feel that we're trying to improve their lot in life and to do that they need to have access to jobs, have access to the *mainstream,* the other components of the city, then by a transportation system connecting them to the other parts of the city, indeed by virtue of having that access it certainly would lead to the conclusion that there would be some improvement there.

Providing access to the "mainstream" fits well with the image of restoring access to major blood arteries, and the concept is compatible with the circulation metaphor: the suggestion is that the transportation link will integrate area residents into mainstream society; their isolation and deprivation would then be removed.

Walter King suggests the same idea when he says "you'll find that those people wanted it because they wanted out of their ghetto." Again, the train is seen as doing more than simply providing for physical movement in and out of Watts: it symbolizes an escape from the deprivations of "ghetto" life. Baxter Ward also saw the train as a vehicle for upward social mobility. An article in *Reader* magazine cites a claim by Ward that: "A system of trains would 'raise the whole social tone' of the community." Ward recounts the tale told him by a vice president of a major corporation in town. "The man grew up in New York; he attributed his escape from the ghetto to the fact there were trains in New York that gave him ready access to the city's cultural, recreational, and commercial centers."[33]

Grace Payne of the Westminster Neighborhood Association, felt that the light rail is "going to enhance our community," while California Senator Diane Watson supported "the reinstitution of the old Red Car that used to run down Long Beach," because "I want my people, the people I represent, to have more opportunity to be *connected.*" But the most powerful symbolic imagery to extend the concept of "connection" from mere transportation to access to actual opportunities came from Mayor Tucker. "We know that poor people need it," he said:

> and we're trying to do everything we can to keep Compton alive and to substitute jobs in the place of dope . . .

We want to do things to make it so that our kids won't be set on drugs . . . I'm saying that if people don't have jobs, then that creates stagnation and causes problems.

Tucker went to college by Red Car: "If I hadn't had the Red Car, I don't know what—I probably wouldn't have been able to get in and out," he said.

This is one of the reasons why they had the Watts riots; it's because they didn't have transportation in and out to Watts to the hospitals and a lot of things . . . Everybody can't go to the beach because they don't have a car. Everybody can't go to the cultural events in L.A., you know, can't get to schools elsewhere . . . It would definitely serve a purpose, and the right-of-way is in place, it's just that they have to do justice with us. They want to spend a lot of money to put a Harbor City drive down in Long Beach instead of grade separation in Compton, you know, and it's unfair . . .

Whatever they do, we want to be included. But right now the problem is to bring the light trolley through. You understand. *We're trying to work with people, not against people.*

To Tucker, Compton is a locked pressure cooker, social problems expanding as the effects of increasing heat are exacerbated by an inability to get "in and out." Light rail is seen as an escape valve, even though there is little evidence that the range of desirable social changes with which it is associated would materialize. Tucker's main complaint about light rail is about a lack of planned grade separation for Compton. But he wants to work "with people," *within the system* to have that problem resolved.

Others also seem content to secure change *within* existing systems. Grace Payne was asked how she knew the light rail would provide the type of services Watts residents needed. She cited the work of LACTC planners in response. How did she know that work was reliable, she was asked. "I would not question that, because we try to put people in the places that we trust, have faith in, and are capable and knowledgeable to do the kind of thing that it's impossible for us to do." People had questioned the judgment of the commission, she was told. "Did you read the book that said the same thing about Jesus Christ? Did you remember that book? There are a lot of radicals, you know, that find fault with everything that people do. There are people who disagree with everything that they are not doing." Most significantly, Payne did say that "If I had access to that much money [being spent on the light rail], I might build another recreation center for the children, some more day care centers, if I had access to the money." But, given control of the money by someone else, Payne supports the light rail project.

The largest proportional vote for Proposition A of any of the five supervisorial districts was in Kenneth Hahn's district, where 68 percent were in favor, indicating broad community support. And Alice Harris worked to support the case for light rail there. She described how the LACTC had sent a representative to talk to the Watts community about plans for the project (LACTC had an extensive community outreach effort to convince residents of the benefits light rail would bring them). Following that visit, residents had called city council members to press for the light rail service, and teenagers were used to help spread the word to their parents and around the community: "Well, what I think you have to do is have teenagers going door-to-door with information and signatures. Would you like this? You live here. You buy your home here. You rent here. Would you like this light rail? Have you heard about the light rail transit that's coming through?"

Another longtime Watts resident and activist, Frieta Shaw-Johnson, played down the role of transportation in the riots: "I don't think transportation had anything to do with the riots. Nearly everybody owns some kind of car . . . It's true the Red Car doesn't run. But the bus'll take you downtown. I don't feel that transportation is it. I'm not a soft one to give excuses." Despite this, Shaw-Johnson nonetheless thought "the money is well spent putting the train in." Perhaps Pat Roche of the Community Development Agency summed up the community perception best in saying, "I think the community views anything that is built—a public facility—as a positive project . . . It also is a sign of confidence in the community, it is saying that we believe in this community, and that we're going to put this here."

Councilman Maxcy Filer of Compton, however, approached the problem from a different angle, unwilling to connect the symbolic benefits of light rail with prospects for real change. Filer, who personally filed suit to stop the trolley project, was not impressed by the argument that light rail would link Compton to downtown Los Angeles and Long Beach.[34] "It isn't taking you where you want to go," he said. "If I don't want to go to downtown Los Angeles, it's not taking me anywhere." People who might use the service need to go to a variety of destinations, he said. "Light rail is not going there. So it will not serve them. Those that are working do not again as best I know work in downtown Los Angeles . . . I know very few Comp-tonites that work in downtown Long Beach. I know many that work in scattered areas, out in the Veterans' Administration: would this get you to 11,000 Wilshire Boulevard? I don't think it would." Light rail will take you to a rail station, Filer said, but people aren't going to a station but to a destination which may be far from a station.

Filer did not see any development advantages, either. Refusing to accept that having light rail travel at-grade through Compton would induce people to get out and shop there, he also rejected the argument that light rail line would open up new job opportunities. If he had the money being spent on light rail, he "could build a plant where there are jobs in the community, then the people wouldn't have to travel so far . . . And I could build a better transit system as far as buses are concerned." Filer's community-oriented viewpoint extended to impacts the line might have on the city as a whole. He expressed concern that six-foot chain-link fences required on both sides of the tracks would divide the city into east and west, much as it was split two decades ago along racial lines.[35] Filer was offered a trip to San Diego to see how the light rail system there worked, but turned it down. "You might say, well, why does it work in New York and it doesn't work here? Why does it work in Chicago and it doesn't work here. Because actually, geographically things are different."

Mayor Tucker saw Compton as part of an *organic whole* that not only includes the set of opportunities available in the Los Angeles metropolitan area, but its system of governance. Compton is part of that larger community, and Compton is part of the government of the larger community. Tucker, seeing himself as a part of that larger system, wished to work *within it* to secure just treatment. He slots into place in a chain of command. The higher levels had decided to go ahead with the light rail project; his role was to secure the best deal for Compton, *given that the project was going ahead.* Filer, in contrast, took a view from *outside* the system. He saw Compton as a separate political entity, fighting against hostile forces from without. Filer's metaphor is one of self-determination.

"See, they're using us," he says:

When they took it [previous streetcar service] out, they say we're taking it out because it's obsolete and buses can get you there just as fast as cars, buses can get you to different places. You can't take the Red Car and then send it down Broadway or send it East-West when it's going North-South, but you can do that with buses. That's what they sold us on in order to take it off. Now, all at once, they're going to put it back, it's the same Red Car, just with a silver streak painted down it, that's all it is.

Dan Caufield, director of the Long Beach light rail project, "wouldn't answer my questions when I came before them, and not only that, they made sure in my opinion that they divided the Council, *our* Council," Filer said. Asked about Supervisor Kenneth Hahn's concern for the welfare of poor minority residents, Filer replied: "Tell him to take his paternalistic thinking and throw it in the river, as far as I'm concerned."

Filer felt that Compton was being duped in terms of job opportunities. "Now, if you're going to help the unemployment rate, I'll put them on the bus now," he says. "You find the jobs and I'll find the transportation." He talks about using the transit funds for other purposes, and of direct help for people in Compton, *within* the community-providing a plant in Compton, for example. Filer is a community activist, a "grassroots" man. If the imagery of the majority of politicians is of gleaming trains shooting across Los Angeles with the speed of bullets, and came from the experience of visiting vital cities seemingly transported to success by smooth, efficient rail transit, Filer's imagery came from the experience of close-up observation of the problems of his constituents. "I think what you've got to do is something that *they* haven't done," he says. "That's get on the buses. Then talk to people. And ride them for about 6 months or so."

Filer felt close to his constituents, poor, black, traditionally cheated by "the system," and that his duty was to fight the system, not work within it. The focus on the direct needs of constituents and distrust in the political system outside make him suspicious of symbols and provide a framework which allows him to move beyond the bounds of the circulation metaphor and escape the lure of the symbolic imagery of the train. Neither flashy trains nor free flow on a track are of value to him if people are not taken where they need to go; a physical artery through the community does not ensure the provision of nutrition if the problems, which go well beyond the realm of transportation, are left untouched.

Felicia Bragg, though not opposed to the light rail project per se, was one of the only others interviewed to see limited social change for Watts and Compton as a function of a lack of commitment by the political leadership, or of a lack of empowerment for local residents to force that change upon the leadership. "There's no real commitment in the United States to engineer the solution of human problems," she said.

> Part of it is we don't want to accept "ugly" in our lives. And we're living in a society where there is going to be ugliness. We train our children to go "ugh" when they see a bum on the street, instead of saying there is a human being who has not had the same breaks I've had. And if you carry it a step further, there's a human being that one day I may be replacing. We just don't really have any Christian ethic of brotherhood towards people.
>
> I'm a product of the riots. And I was one of those who was snatched up. And tell us how you feel and come and speak to us. I've never forgotten that I tasted my first ham as a result of the Watts riots, because a White family had me brought to their home. I was there basically to entertain them with stories of how it was to live in this war-torn city and grow up poor, and they felt good because they did that.
>
> So there was some of that rushing in and trying to do things. I can't tell you the number of programs set up after the riots. But people were people. And when they discovered a), that the natives were not suitably appreciative of their efforts and b), that it was damn hard work and was going to take a long, long time to turn even a tiny piece of that around, they kind of drifted away. Other things caught their attention.

Walter King, alternate commissioner and close friend to Kenneth Hahn, gave some background to the supervisor's interest in issues of poverty, stressing that the problems of the poor constantly caught Hahn's attention:

> Nobody that you'll ever meet was any poorer than Kenneth Hahn. His mother told me, well in the first place his father died when they moved from Canada down here, she raised 7 boys, she never had food to feed them many times, many times she didn't know where the next food was coming from. All his life he went to churches and places that had free birthday cakes for his birthday. He was a little fellow like this. He worked to bring in milk for the family by swamping, they called it, running the milk back and forth from the trucks. So, Kenny is the most conservative man financially, fiscally, you'll ever meet. But he'll see that the poor and the rich are taken care of—I'm sorry, poor and the elderly, makes him a liberal.

King pointed to Hahn's attention to basic things, to making sure that visibly the community looks good:

> One of the things is transportation. There's nothing closer to Kenny's heart than streets, paving. He started basically. You can tell when you're in and out of Kenny's district because the streets and alleys aren't paved. But every one

in his is. And he's got traffic improvements and signals; there's more—go check with the public works and see how much money is spent in his district—he works his district from a transportation angle.

Making the community look superficially smart may symbolically suggest that impacts at a deeper level have been achieved. But Bragg, while acknowledging that Hahn does give "a damn in a granddaddy kind of way," does not:

> think he's equipped to really talk about solutions that run deep through the layers of the problem . . .
>
> Hahn is sort of like a Great White Father to his district. You know. And he plays that role, too. It's too bad you can't go with him particularly when he goes out to Watts. He plays that role to the hilt. He's the Great White Father and he feels very paternalistic and speaks to them like that. You are my children kind of thing . . .
>
> He's sincere in that paternalism. He really feels that way. He really cares about Watts in the same way that a plantation owner cared about his darkies. He wanted them to be clean, and he would call the doctor when they got sick and all of that. Now he didn't care if they learned to read or write, he wouldn't want them to go get a job somewhere. And Hahn is the same way. You can almost see him patting people on the head and saying "be good, I'm going to take care of you . . ."
>
> The point is that he's taking care of them. He doesn't respond to what they want or in any way attempt to empower people to do for themselves. He has simply said I'll take care of you.

Bragg described the symbolic actions of the white political leadership, the decision "to completely tear down 103rd Street" (the center of the riots), for example. "They left nothing of the way it was before, and I think that that was a deliberate attempt to say it never happened . . . I'm simply saying that they could have left us something of our old community."

The light rail was also of symbolic importance, something which could be accomplished to show progress, Bragg said.

> It's easier to build a Metro Rail system with your billion dollars than it is to attack some other problems. If you have a billion dollars, for example, can you imagine the chaos, and you went into Watts and said ok, there's going to be a billion dollars spent here? Can you imagine the chaos that would ensue, the political blood letting that would happen by the time that decision how you're going to spend that billion dollars was made? And frankly, given the number of people and number of problems there, what would you really accomplish with a billion dollars? Maybe if you sunk it into education, but

there's 15 different views about that. Some people think Jordan High School should be shut down because it's the pits. Other people think it should be re-suscitated. The President of the school board is absolutely committed to bussing, others are absolutely committed that their children should go to school in their own neighborhood. There's nothing that could be easily decided.

The trolley, in short, is a response to complexity: it is a potent symbol of progress, is feasible, and a program that can be agreed upon. "Symbolically, it brings South Central Los Angeles into the twentieth century. It binds them closer with the rest of the world. . . . It's bright and beautiful and fast and modern and it's going to be clean for a while, it's going to be new."

Yet, despite her awareness, Bragg has memories of the Red Car in Watts. "I used to live a block from where the Red Car ran, and I remember our mother taking us on that very Red Car downtown Los Angeles to go to the movies or shopping, to Long Beach . . . sometimes just to go for a ride. We would visit my grandmother because she lived close to it." Trains were "cleaner and faster," Bragg said, and she did see some benefits from putting them in. "I guess I don't have any problem with them spending $1 billion on that." She did not have any confidence that if not spent on the trolley:

> they would spend it wisely, and if only a few people get a good ride, if my mom can get on that thing—she lives right near there—and ride down there in comfort and do her little shopping and stuff and then ride on back home, I'm happy. I'm happy. There are going to be some benefits, and I think the trick is not to expect so much. I'm not sure that I have a better idea anyway about what to do.

Ultimately, therefore, while Bragg would agree with Filer on the need for political change, she feels that "I'm not sure that he's thought through all of these issues." Bragg would take what the existing political system has to offer; only Filer, in the end, insists that it is the system that must change.

Transport of Delight

Two decades beyond the Watts riots, little had improved for that community—except for transportation. Yet, with endemic problems of unemployment, poor education, drugs and crime, the most significant project to come to town—the light rail service—was likely to offer the community few benefits. Light rail does not serve the principal employment destinations of the community, nor the needs of local trips in general. Transport is such a common metaphor for more abstract issues, however, that light rail appeared to offer the otherwise unattainable. When we talk of "social mobility" we are using a metaphor rooted in transport to understand what it means to change social status. The idea of "social integration" is also understood through the concept of movement: a "bringing together" of people. "Getting out of the ghetto" may mean far more than a physical displacement but the concept is—once more—understood in terms of transport.

It is not surprising in the light of such powerful metaphorical structuring that McCone's heavily criticized conception of isolation in terms of poor public transportation lived on in efforts to bring the Blue Line to Watts. It is easy to appreciate, given the metaphorical understanding, why bringing people together in a train might be seen as conducive to greater intercommunity understanding; and why providing a "connection" to jobs, schools, or concert halls might be seen as conducive to social improvements, especially when reinforced by memories of the "connections" the Red Cars did provide in days gone past. Unfortunately, the reality that poverty was just the same as always was only to emerge after light rail was implemented, bus services decimated, and the community finally made angry at the loss they had sustained through construction of the very rail system they had supported. Who would have thought at the time light rail was adopted that the shared community euphoria over light rail would be replaced by a disillusioned demonstration through the streets of downtown Los Angeles demanding "Billions for Buses . . . No More Rail Come Spring 1997?" Or that, as Loukaitou and Banerjee put it, the areas around light rail stations would remain "disinvested, forsaken and decaying"?[36]

The reality, of course, is that the attitudes of those in contrasting communities must change if they are to be brought together, and that provision of a physical "connection" to jobs and other opportunities is futile in providing *actual* "access" to such opportunities, barring changes in educational and other social factors which are the real barriers to such "connections" being made. Bringing light rail through Compton and Watts may, under the circulation metaphor, be akin, furthermore, to the provision of "new blood" to improve the health of the community, but without changes to the state of depression and lack of incentives for development to occur, there is no reason to suggest that any will materialize.

As in Bish Sanyal's Calcutta, so in Watts: While solutions to the deep underlying problems remained unavailable, a new light rail system lay within the scope of feasibility, and was symbolic of progress, even if it could bring few benefits compared to those which might come from other ways of spending the money. It was something people could rally around and, politicians and public alike, believe in. Perhaps most tragically, the trolley was a vehicle of massive self-delusion, not of bad intentions—those in power genuinely believed they were doing something for their constituents—the sort of delusion that allows existing political systems to stay in place and societies to operate with aspirations of hope in times of adversity.

The light rail project seemed like a good solution *within* existing systems of political power structure: systems where not only may there be a lack of commitment to deep social change, but a lack of a concept of where to start on such a massive endeavor. Mayor Tucker wanted to work *with* people, and play his part within the existing order. Grace Payne might have had

Figure 11-2. A spring 1997 protest demands "Billions for Buses" and "No More Rail." (Photo courtesy of author)

Light Rail and the Symbolic Promise of Community Renewal   371 higher priorities than light rail; but, given that other uses of the light rail money were not within her control, supported the project as something for the community. But while everyone else worked within the bubble of an existing political order, Maxcy Filer alone burst that bubble in applying his understanding of the problem as the need for self-determination. Perhaps before poverty can be displaced that bubble must be burst. In the interim, to communities that felt cut off from the world of opportunities that elusively seemed to wait just a few miles distant, the trolley's arrival was a sign of hope: in the midst of gloom, it was a transport of delight.

CHAPTER 12

Synthesizing the Political and the Mythical

Politics is the process by which the irrational bases of society are
brought out into the open . . . The rational and dialectical phases of politics
are subsidiary to the process of redefining an emotional consensus.
—Harold Lasswell

THE LOS ANGELES–LONG BEACH BLUE LINE light rail service is
not the result of a calculated, let alone reflective, effort to provide for the
transportation needs of Los Angeles. Ill-suited to the needs of Southern California's dispersed autopolis, rail service costs more but provides far fewer
benefits than alternative possible uses of the public expenditure. Studies conducted on behalf of the LACTC by SCAG concluded that light rail would not
provide relief to the congested freeways, or meaningfully reduce pollution or
energy consumption. Not only have capital costs for light rail been vastly
greater than for bus alternatives, but operating subsidy costs have proved to
be substantially higher than for innovative bus options. A comparison of
the performance of the Blue Line with that of the Rapid Bus program implemented in June 2000 is dramatic. With capital costs less than one-
seventeenth of the Blue Line, two Rapid Bus lines produced ridership increases in their respective corridors equivalent to half of Blue Line ridership
in just over a year of operation, while requiring less than half the subsidy per
ride of the Blue Line. Spending money on rail rather than on other more efficient options has resulted in a lower public transport ridership than could
otherwise be attained. The extraordinary level of public transport ridership
loss that took place when bus fare subsidies from Proposition A revenue were
diverted to funding rail provides the most alarming evidence of this.

Claims that light rail would significantly benefit depressed South Central Los Angeles and Compton are invalidated when it is seen that service
is ill-matched to the most pressing trip needs: far better service could be

373

provided for these communities with buses. There is no evidence, further-more, that the arrival of rail will stimulate development, given the lack of intrinsic attraction to developers to invest in these communities. Perhaps most fundamentally, the euphoria shown at the prospect of rail has turned into grief as the dream of rail has failed to materialize. Who could have foreseen that rail would have been cited as harmful in a civil rights lawsuit complaining about bus service deterioration and bus fare increase? Who would have predicted the sight of demonstrators in the street calling for an end to rail and a return to adequate funding for bus services?

Why was rail chosen? Complexity is unnatural to us—our minds reject it—and we fear the incurable, the intangible, the unknown. Rail offered a simple and reassuring solution to the unbearable problems of Los Angeles, one which appeared to have the potential to do much good. Members of the LACTC, growing desperate to *do* something about an overwhelming trans-portation problem, seized upon it as an answer: first as a part of the pack-aging of Proposition A of 1980, and then by bringing it to life in the form of the Long Beach light rail line. The public embraced rail, making Proposition A law, and thereby cementing rail further into the popular culture.

If the rail system is certainly the product of political action, the trolley's arrival in Los Angeles cannot be adequately explained from an account of either the pushings and shovings of interested parties or the actions of those elected to make public decisions: to appreciate the rebirth of rail, we need to see how its benefits are conceived and are understood by interested parties, decision makers, and the public at large. When we set rail in its mythical context, we see that it was no mere cynical political response to pressures, nor a deceptive attempt to fob off an illusory solution on an un-knowing public, but the product of a powerful myth, one lived by politi-cians and public alike. And, as Turbayne said, for those who conceive and transmit myth, "there is no make-belief, only belief."

Making Polis

It is characteristic of large numbers of people in our society that they see
and think in terms of stereotypes, personalization and over-simplifications,
that they cannot recognize or tolerate ambiguous and complex situations, and
that they accordingly respond chiefly to symbols that oversimplify and distort.
—Murray Edelman, *The Symbolic Uses of Power*

The decision to put Proposition A on the ballot and, subsequently, to select the light rail line to Long Beach as the first project for implementation, is a successful example of what Churchman calls "making polis," a term conceived after the activity of free Athenians meeting to engage in "polis" (literally "city-state"), "the actions of citizens in a nondictatorial society."[1] Making polis is the coming together of groups for action over a concern. As Churchman indicates, this process "has no self-consciousness about what the ideal planner calls 'overall progress,' or even about his 'measure of performance.'" Instead, it focuses on a specific issue, around which "polis makers" can gather. For polismaking to succeed, there must be something suitably solid and well-defined to anchor both thought and action. Success, Churchman says, "depends on the feeling of the polis; success is evidenced by expressions of joy, celebrations, parties, cheerful announcements, etc., even by monetary awards."

The case for developing a new rail system is a good cause around which polis might form: it is something quite tangible and solid. Yet, polis had not been successfully made around this issue right up until 1980. We can understand why if we trace both the actions of political systems and the development of popular conceptions of alternative policy options, which come from symbolic understandings—not systems of either analytical or reflective knowledge. We can then see that a window of political opportunity was coincident with the emergence of an available mythology. And, to judge from the ensuing "happiness," the polis making proved successful, whatever its merits—or lack thereof—in analytical terms.

Since 1925, Los Angeles central-city interests had lobbied for rail transit, but to no avail. To most other Angelenos, the thought of creating a city focused on a concentrated core—such as in the eastern cities from which they had fled—did not jibe with the image of the good life. As the streetcar declined and went out of existence, the road and car flourished in Los Angeles, becoming symbolic of the free-flowing, unfettered lifestyle of the Southland. The automobile had transcended the limitations of the metal wheel on fixed rails and become the transport of choice. Even so, rail transport's role in the creation of Los Angeles and its low-density suburban lifestyle could not be lost. If Henry Huntington's obituary talked of his "romantic and imaginative soul" having found an outlet in railroads, the romance of the rails—which opened up Los Angeles as well as the West as a whole—lingered on, leaving behind a repository of symbolic associations to reawaken and fuel the myth-driven dream later on.

The 1925 Kelker, De Leuw proposal, with its imagery of speed and technological prowess overcoming barriers of space and time, also generated mythical understandings of the power of rail even if, for now, they lurked in the shadows. The 1948 RTAG proposal, which brought further claims of easy technological answers to complex social problems, emphasized the alleged luxury, speed, and sex appeal of rail service, projecting trains yet more powerfully as mythical saviors. But the automobile dominated the public imagination by now.

A report issued in 1950—at a time when *free*ways remained symbolic of *free*dom, and none of the negative connotations had hit home—stated that while 40 percent of those surveyed in a study favored one of three fixed-rail options, 47 percent called for buses on freeways. Highway construction meanwhile continued at a rapid pace, uncontroversially, and with broad public support. The time for forming polis around rail had not yet arrived.

Donald Schön defined "ideas in good currency" as

> ideas powerful for the formation of public policy . . . They change over time; they obey a law of limited numbers, and they lag behind changing events, sometimes in dramatic ways . . .Characteristically, what precipitates a change in that system of powerful ideas is a disruptive event or sequence of events, which sets up a demand for new ideas in good currency. At that point, ideas already present in free or marginal areas of the society begin to surface in the mainstream, mediated by certain crucial rules . . . When the ideas are taken up by people already powerful in society this gives them a kind of legitimacy and completes their power to change public policy. After this, the ideas become an integral part of the conceptual dimension of the social system and appear, in retrospect, obvious.[2]

Highway building was an "idea in good currency" of the 1950s but, as the 1960s and 1970s brought increasing highway congestion and pollution, the image of the freeways was transformed from one of liberty into one of entrapment and disease. With concern, also, over issues such as the environment, poverty, and unmet social service needs, highway building went out of vogue, and an opportunity was created for a new "idea" to emerge. Federal attention became increasingly focused on transit programs, and the concept of transit as a commodity to be supplied by the private sector according to market forces was transformed into one of transit as a social service to be publicly supplied and subsidized.

A series of Los Angeles center-city backed propositions, starting in 1968, nonetheless failed. The 1968 proposition restimulated the previous debates over core domination versus dispersed urban forms. The same happened in 1974. Yet, if one thing is clear, it is that opponents as well as supporters felt rail would have an impact on patterns of accessibility and hence locational utility and property values, an impact which, when compared with the accessibility provided by existing transportation links, was at best questionable: there was something about rail which made it seem powerful, yet there were enough people who saw this power as contrary to their interests to vote it down.

Nineteen seventy-four was, also, the year of the Snell Report and its highly emotional and equally unsubstantiated conspiracy theory. The Los Angeles of the Red Cars, Snell said, became an "ecological wasteland" largely because "the noisy, foul-smelling buses turned earlier patrons of the high-speed rail system away from public transport and, in effect, sold millions of private automobiles." If the Snell Report ignored the reality of rail's obsolescence in the new autopian Los Angeles and rail's decline due to its failure to meet public needs, it nonetheless captured the public imagination, and its broad-brush impressions of good and evil, of heroic trains being trampled under by iniquitous automobiles, and filthy buses, were to emerge again later on. If we ask why the report found a sympathetic reception, we need look no further than our daily experiences: the freeways are congested, many local bus services are less than pleasant to use, and rail systems we see in cities elsewhere appear impressive. Here was a theory with simple, visceral appeal.

Baxter Ward's 1976 "Sunset Coast Line" proposal encapsulated the virtues Snell attributed to rail, adding futuristic imagery: here was the system that would get you where you were going in no time at all, with elegance as well as speed, without pollution or discomfort. While Ward's proposal played a significant role in myth building, it was an expensive one and not one which the public would yet buy. By then, however, a subtle shift had occurred. With Los Angeles Mayor Tom Bradley giving his support for the Wilshire subway in August 1974, and Long Beach Mayor Clark criticizing him for endorsing "a system that will be totally within Los Angeles," the debate moved from whether there should be any rail at all to arguments over whether rail should serve mainly the core, or reach out to serve the more dispersed communities. Despite the local infighting over this issue—which prevented a consensus from emerging—rail was now firmly accepted and on the agenda.

As we build up to the dramatic events of 1980, we see the LACTC, in place since 1977, under pressure to show accomplishment. Its then chairman, Supervisor Kenneth Hahn, in a statement of goals in January 1978, reflected a feeling that the public "want their leaders to act . . . It is time that we act decisively together to get on with the needed transportation improvements."

By 1980, rail was a regular favorite on television and in news articles, not only exposing its symbolic appeal to the public, but rekindling the appeal that already lay locked in memories. Pictures of speeding Red Cars suggested quick and easy answers to the snarled-up roads; and the understanding that the Red Cars had suffered a "painful murder" at the hands of "rubber interests"—rather than because they no longer served the needs of a dispersed Southern California—was disseminated along with the images. This gives the impression that if only the Red Cars had been kept, Angelenos would not have been turned into "gas addicts." People in San Diego were said to have realized that buses could not carry as many people as trains, and the triumphant "unveiling" of a light rail train in San Diego gave a vivid impression of progress in that city, compared to paralysis in Los Angeles.

A survey conducted by LACTC in July 1980, via advertisements in the *Los Angeles Times* and *Los Angeles Herald-Examiner,* elicited the response that, while 20 percent called for a rail-based system, 60 percent wanted a mixed rail/bus system. Given this expression of interest, Kenneth Hahn's packaging of Proposition A to include both rail and bus elements was entirely logical. The myth of rail had taken hold, and the popular images flashing across the TV screen needed to be turned into a renewed reality.

The initial formulation of Proposition A did not specify that rail was to be put in place: the money, should the proposition pass, was to be available for any sort of transit improvements, and bus-on-freeway systems were particularly favored by staff. Rail was still, however, very much at the forefront of the thoughts of elected officials. At a critical LACTC meeting on

August 20, 1980, Supervisor Hahn finally forged a consensus. To secure Baxter Ward's vote, he had the proposition's language changed to *mandate* that rail be built should the proposition pass. The heady influence of the myth of rail now made this readily acceptable. Hahn, furthermore, included a 25 percent "local return" element to secure the support of those from dispersed communities. A reduction in bus fares for three years, and a map designed to show service provision to as many corners of Los Angeles County as possible, contributed to making the package sellable to the public; yet rail was only sellable in the first place because a mythical belief in its powers was by now firmly in place. Also highly significant was that, unlike earlier proposals, the system did not appear to cater primarily to the interests of downtown. Rail now seemed to reach out to the dispersed lifestyle as well. In this form, the proposition was approved by the commission, put on the ballot, passed by the voters, and became law.

The popular theory (put forward by Whitt, for example), that rail is a product of a central-city conspiracy and geared primarily to increasing land values in the urban core, holds no water when we appreciate that when these interests were the prime pushers for rail, the public refused to endorse their proposals, and that, in the case of Proposition A, the central-city lobby was hardly involved at all: it was the broad appeal of the measure, based on both the three-year bus fare reduction it incorporated and the belief in the powers of rail to alleviate the transportation problems of Los Angeles, that carried the day.[3]

Despite the claims that Hahn engaged in "bribery" in putting the Proposition A package together, the process could arguably be said to be politics working as it should: providing something the voters liked enough to democratically approve. Yet, to simply assert that the proposition passed because of the successful operation of a political process, and because it appeared to meet a need, begs the question—one typically ignored in political analyses—why did the elected officials and the public feel it met a need. The answer lies in myth.

At a key August 1981 meeting of the California Assembly Transportation Committee in Long Beach—which paved the way for the Long Beach light rail project to become reality—we see the mythology of rail in full operation. We hear of the "untimely death" of the Red Car system, which until the 1940s had provided a "good balanced transportation system"; that light rail provides for comfort and convenience without polluting the air and at low cost, while buses are "dirty, smelly, and definitely not rapid"; and that light rail was successful in San Diego, and could similarly succeed in Long Beach, the city which had "kept faith" with the Pacific Electric the longest. And there were the childhood memories of growing up with the Pacific Electric, which speakers wanted revived today.

At the crucial LACTC meeting at which the decision was made to proceed with the Long Beach line, we heard that the light rail "connects two major population centers," that it "will provide a key transportation link to recreation and employment centers," and that "we need look no further than our clogged freeway system any working day to see the need for transportation alternatives."

The Coherence of Myth

How do we know such statements are not simply isolated rhetorical devices? They are rhetorical, of course, for we have heard them from people with a political case to make. But what makes such statements vastly more interesting and indicative of deep underlying beliefs is the extraordinary set of coherences represented within them: the logic of the symbolic system which enables it to all make sense.

A study of the elements comprising the myth of rail shows why the idea of rail systems developed a great symbolic appeal in Los Angeles, one little-related to the benefits rail might actually bring. These elements paint bold pictures, drawing clear-cut answers from out of a web of otherwise intolerable complexity: they fulfill the human need for simplicity. The associations mesh together coherently not with the logic of analytical reason, but according to a symbolic logic which draws on our experiences and emotions to create its own far more powerful picture of apparent reality. As de Neufville suggests, images act as a proxy for unstated assumptions, and serve to anchor inferential logic. The logic acts subconsciously and synthetically, putting together impressions, rather than taking apart facts. History and experience paint evocative imagery of the potential of rail to provide benefits, while the metaphorical ways in which understanding takes place provide interpretations of such images which conclude that rail is the best way ahead.

These conceptions are the work of experience: experience is powerful because it presents the evidence of our bodily senses. We see traffic locked in congestion; we thrill to the acceleration of a fast train; we smell the fumes of a bus. Because we get close to these technologies, they become objects of emotional attachment—and hatred—and are desired or spurned with the logic of a mythical world with its special set of rules.

Experience of freeways and buses in Los Angeles, of trains elsewhere, and images of the trains which might come to L.A., tell us that trains are fast, comfortable, safe, clean, classy, and efficient, while buses are slow, uncomfortable, crime-ridden, dirty, low-class, and inefficient. The fact that trains are seen as fast, and buses slow, makes trains win out in our metonymic understanding of a trip, which focuses on the time spent on the main vehicle, and ignores the time getting to and from rail stations. Because the conclusions we are so apt to erroneously draw come from the obvious evidence of our everyday world and common sense assumptions about how it runs, such inferences are as compelling as they are natural.

Gaining this experience is a form of natural experimentation in which the merits of the different transportation systems are "tested" by trying them out. As we have seen in Baxter Ward's test for the case between rail and bus—taking the judge for a ride to evaluate the experience—and in the accounts of visits to other cities—this type of experimentation seems to have widespread acceptance. The superficial impressions formed by the immediately experienced technology may be misleading when extrapolated to wider social and economic domains, but they not only contribute to the formation of evocative imagery, but provide for the mass of associations that the technology-as-symbol then acts to recall.

These images and symbols take part in a process which provides for the making of inferences based on experience. Because other cities appear to have successful rail transit systems and are seen to be successful cities, it is inferred that a rail transit system will also make Los Angeles successful. As Walter King said, "all I'm saying is good sound logic": central to the functioning of the myth is the belief of those making such inferences that they are being perfectly logical.

The symbolic act of crossing contexts translates experience from one setting to another. Monorail, experienced at Disneyland, becomes a symbol of speed, efficiency, and flair: the evident success of monorail at Disneyland is thereby extrapolated to likely success elsewhere, without a testing of the assumptions upon which that success would depend. Images of trains "working" in other cities similarly evoke images of successful rail in a rejuvenated Los Angeles.

A principal act of symbolism and imagery is the displaying of partial impressions as if they made up the whole picture. The parts, we have seen, get chosen according to their visibility and familiarity. Just as we take "the face for the person," we take the trolley car ride for the transit trip. Complete journeys are metonymically wholly understood in terms of the time spent on the principal vehicle, thereby giving the speed of the principal vehicle undue weight: and the train's high-speed imagery easily wins out here over the slow coach reputation of the bus. Where people actually wish to travel is not investigated, and problems of a total trip—in particular getting to and from the main transit vehicle—are overlooked because the problem is structured simply in terms of getting the "right" line-haul vehicles. Everything else is expected to fall into place, just as a person's stomach is expected to be below their face. The metonymy explains why the vehicle that is to provide the "free flow" that the circulation metaphor demands must be the train.

The choice of parts which come to symbolize the whole reflects the tendency to see abstract concepts in physical terms. There are assumptions about economic and social conditions implicit in particular technological choices, but they are not examined directly: instead, they are seen as a function of the physical performance characteristics so much more readily perceived by the eye. Buses observed spewing fumes are seen as necessarily bad. The gleaming surface imagery of Washington's METRO or San Francisco's BART suggests that the rail concept as a whole is good, and that concept-seen-as-good is symbolically transported to Los Angeles. Because trains are observed to have only one driver who transports several hundred people, it is assumed that they are more efficient than buses; this idea is further reinforced by the perception that there would be lower maintenance costs for electric (read clean) versus diesel (read dirty) equipment. Trains are seen to be more comfortable as well as more glamorous than buses; buses, in addition to being perceived as slow and causers of congestion and pollution, are seen to be uncomfortable and associated with crime.

There are several ways we have seen symbols acting as gateways to larger patterns of associations: the bus has its myriad associations of poverty and vice even if a bus service need not in reality be any less pleasant than a train; in contrast, trains symbolize middle-class values and convey associations of cleanliness and security. These understandings suggest not only that trains are safer to ride than buses, but that they might also bring social reforms because people apparently won't vandalize nice vehicles. A complex social problem is thereby seen as soluble merely by changing the physical infrastructure.

Because trains have been seen in European and concentrated core-based city contexts, the concept of a European-style city founded on the central place is symbolized by the train. Through memories of Red Cars conveying Watts and Compton residents to college, employment, and to the beach, the train comes to symbolize connection in more than a transportation sense: it is a link to a world of opportunities outside, a way to escape unemployment, drugs, and other urban problems. In this way, the train acts as a symbol which serves to remember and recall, promising that the supposedly good things of the past can be here again today. We also see here the power of the physical symbol to convey abstract ideas: providing "connections" to opportunities in fact requires far more than the laying of a steel rail, yet the steel rail is seen as the road to attainment of these more complex connections.

We have seen how symbols can act in the manner of the "imperious and infantile" and can link the idiosyncratic and the personal with the general and the universal. The train given to children in train sets is a friendly thing. And because kids like playing with toy trains, it is assumed that when grown up they will ride real ones. The childish thrill of hearing a train in a distant city go "*whoosh*" makes it seem that trains which go "*whoosh*" are universally good. The fact that San Diego has a bright, shiny red toy, in front of which people like to have their photos taken, and that San Diego appears to be a bustling, successful city, suggests that if only Los Angeles had a similar transport, it would result in delight there, too.

Personal experiences of modern railways and of the "grief and hassle" of overcrowded freeways give the technologies symbolic power, a power which acts with a refusal to take difficulties into account. "I wanted a train," said Baxter Ward and, when the route he initially wanted was unavailable, he went ahead and took any route available simply to have the train, whatever the merits of the train for that particular route.

If symbols represent emotions, there is no shortage of these displayed in the interview transcripts. We have seen a love of trains, and trains carrying romantic connotations. The technology becomes an object of emotional attachment; it becomes a god for religious worship when the virtuosity of its performance captures our imagination. Roy Donley's "high-powered locomotion" has more than transportation value: the power has allure of its own, and the train as symbolic of power satisfies a central human desire. The sexual connotations of the train also exert influence. San Diego "whips it out" first, while cuckolded Los Angeles must stand by. All these associations are unconnected with any transportation advantages rail service might have.

All of this symbolism and imagery not only meshes coherently with, but serves as the inputs for, a set of metaphors which shape fundamental understandings of transportation systems and of how—when in disrepair—they should be fixed. Most fundamental is the circulation metaphor with its entailments of an interconnected system of arteries focused on the heart of the city. The circulation metaphor explains the desirability of employing major arteries to link major cities even though there is little demand for travel between them. The metaphor, furthermore, dictates the need for "free flow." With freeways seen as choked, or the city strangled, the metaphor prescribes a new system, isolated from contamination from the old system, through which free flow can take place. The image of the bus— slow and stuck in traffic—and of the train—dashing at high speed on its own right-of-way—tells us why, under the circulation metaphor the way to go is by train and not by bus.

Like blood, traffic is seen as a homogeneous substance which is to be kept flowing round the system, giving life wherever it passes: the actual origins or destinations of particular units of traffic are not seen as being of great importance. No mention is made of real equity issues—"going through" poor areas is seen as enough. While the metaphor is guiding thought in the interviews, no consideration is given to the possibility that transportation might be improved by *regulating flow*—charging tolls, encouraging the use of high-occupancy vehicles, or implementing other forms of road management—because the goal is simple "free flow." It is this picture—which highlights certain features compatible with the metaphor and leaves out others, and which makes solutions consistent with the metaphor seem obvious—which suggests that the metaphor is operating at a deep level.

Two metaphors, of "balance" and of an "escape valve," provide further ways of seeing why a new rail system is desirable. Common to both metaphors is the notion of providing a separate system in which free flow can be maintained: both metaphors align with the free-flow ideal of the circulation metaphor. In both cases, the way the technologies are understood in being "processed" by the metaphors, is determined by the symbols and imagery associated with the technology in question. The balance metaphor tells us that predefined transportation technologies need to be assembled to create a state of equilibrium in which all function according to their place in the "natural order." The apparent oddity of interviewees who malign buses, but approve of their use as feeders to rail transit, is explained when it can be seen that the feeder function puts buses in their correct place in the envisaged "natural order": it is not that no buses are required, but that they cause imbalance when they try to perform the line-haul function of rail.

Under the balance metaphor it is believed that an equilibrium is attainable, with free flow on both road and rail systems. Under the escape valve metaphor, it is accepted that this is not within reach. Under the "balance" metaphor, rail is part of a natural ordering of transportation systems, and is desired as part of an ideal balance. Under the "escape valve" metaphor, however, rail is an alternative needed only because of the inadequacies of the road system: it is not natural, but an artificial device installed to take excess pressure with which the "natural" road system cannot cope.

The two metaphors were not generally used together by one interview subject, indicating an important difference between the ways in which each frames the problem, possibly a difference between an idealism from those promoting "balance" and a resignation on the part of those subject to the "escape valve" metaphor. But of greater importance than the differences between the assumptions of the metaphors is the same conclusion they both indicate: that a rail system is required, whether to return free flow to the entire system or to provide it only on the new rail part of it. In providing for at least some new free flow, both metaphors fit coherently with the circulation metaphor; both fit coherently with and use as inputs the symbolism and imagery of freeways, buses, and trains. The image of the bus caught in congestion, to give one example, helps put it on the "roads" side of the balance scale.

The "addiction metaphor" helps explain in personal terms how the transportation system became "out of balance." It tells us that *we* have become addicted to gasoline, and that our reliance on the substance has unbalanced the transportation system as well as ourselves. As with the other metaphors, images of road and rail provide input, with the Red Cars in a heroic role, and the snarled-up roads representing the very incarnation of evil.

A further metaphor, one of evolution, (which charts the development of transportation technology on a fixed path of progress) also fits with the above: if the evolutionary step up to rail has not occurred in Los Angeles, then of course the system is out of balance; of course there's no surprise that the freeways are clogged. If there are no trains which go "*whoosh*" it's no surprise that L.A. grinds to a halt. The image of the train as a vehicle of both high capacity and modernity gives the impression that it is on a genetically more advanced level according to the tenets of the "evolution" metaphor. This complex system of metaphors, symbols, and images, then, comes together to coherently structure the myth of rail.

It is important to note the compelling rather than compulsory nature of symbolism, and to see that those not predisposed to be influenced by a particular symbol or aspect of symbolism can escape its influence. Kenneth Hahn, though subject to most aspects of the myth of rail, was a supporter of the bus: his daily contact with constituents who depend on buses makes him aware of the buses' key role in transportation.

Los Angeles Councilman Ernani Bernardi was immune to the myth as a whole because he started with an image of a dispersed Los Angeles, which is antithetical to the core-based circulation metaphor, and because he was unwilling to believe that just because the Proposition A map shows rail reaching out throughout Los Angeles County, actual interaction patterns would be served. Because his framing of the problem—which started with the assertion that there was too much development taking place at downtown—is entirely at odds with a rail system and its associations of the concentrated city, the myth had no meaning to him and was seen as myth, not reality.

Roy Donley and Eunice Sato, on the other hand, while apparently opposed to light rail, were not immune to the myth: their problem with the light rail was that it would not go fast enough. Both believed in high-speed rail systems: while Donley might not "worship" at the "altar" of light rail, he was taken with the virtuosity of true high-speed rail. While Sato was unimpressed by the slowness of the "streetcar," she—like the light rail supporters—concentrated her attention on the physical link between major centers and on the vehicle to provide that link, but saw the speedier monorail as superior to light rail.

Christine Reed and a number of others, meanwhile, had critical comments about the Long Beach light rail by itself, but became supportive when the line became seen as part of a total "system," in other words, as part of the network of interconnected lines the circulation metaphor would demand. These respondents—under the metaphor's spell—assumed the believed benefits of interconnection would come to fruition, and did not verify whether the system would actually likely be attractive for the sorts of longer trips for which connections would be necessary.

But doubtless the most interesting outsider was Maxcy Filer. He was entirely outside the mythical world of rail—indeed while the rail supporters saw their perceptions as truth, Filer understood them as myth. Filer did not

Figure 12-1. Blue Line station. (Photo courtesy of author)

operate under the influence of any of the metaphors examined above, nor was he swayed by the symbolic imagery of rail cars. Nor, despite his long-term connections with his community, was he influenced by nostalgic memories of the "good old days." Because his framing of the problem was at odds with a framing supportive of the rail myth, he was not open to persuasion by any of the arguments supporters put forward, which build up the myth.

The key to Filer's different orientation lies in his political conception—working for constituents whom he viewed as an island seeking self-determination. His view is a "grassroots" community view. He saw the problems people face daily, and wanted jobs immediately and in Compton. Riding the buses, he saw where people go, and sought to directly serve their needs. While others, such as Mayor Tucker, worked *within* "the system," Filer functioned outside it. His metaphor (for it is not that Filer operates *without* metaphor, but that his metaphors are different) is of self-determination, of Compton's need to free itself from a colonial periphery-center relationship with Los Angeles; and it allowed him to view the assumptions of the myth of light rail from the outside, and thereby see the myth for what it was.

The mythology we have seen is linked with the politics. The set of available political actions is driven by the mythology in good currency. Elements of the mythology—in this case the vast storehouse of symbolic associations of rail—may stay dormant or only partially active and, in turn, be spurred to action when the politics demands a vehicle for forming polis. As the old savior—the freeway—faded from favor and good currency, rail emerged in response to the need for a new mythical hero, a new symbol around which to form polis. The train: concrete, sexy—transport of intimate memories and powerful ideas—provided a firm basis for polis formation. Technologies with negative symbolic connotations cannot do that, although, we saw, at the tail end of this story, that the Los Angeles politicians, forced in desperation to find an alternative to the now financially impossible rail system of their dreams, finally rallied around the Rapid Bus, when made to appear to them like a train, like a "subway on wheels" as Supervisor Zev Yaroslavsky put it. Complex, abstract ideas that would reformulate the way transportation systems as a whole are organized, provide no basis for arriving at polis, either.

As Proposition A came to reality and the decision to build light rail to Long Beach ensued, rail had an almost universal appeal, as did the highways in the 1950s. And, just as the highways were built on the dream of a better way, so the railway also came to bear us on a fantasy to a renewed life built on our experience of life in the past and our imagination of how it might be better in the future. As rail building became a stable element in the policy realm, it took on an ascendancy that put it above other ideas, placing it in a bubble of its own, impenetrable by other ideas, and unable to see itself as the product of imprisonment in a limited set of erroneous concepts.

CHAPTER 13 ## Concluding Implications— On the Need to Burst Bubbles

[I]t may be safe to say that at the present time inquiry has become part of the unconscious life of most people: they are unaware of the ways in which they function as inquiring systems. nor is there a strong inclination for them to give expression to this function so that its nature appears at the conscious level.

—C. West Churchman

FROM THE THEME OF THE PRESOCRATIC philosopher Anaxago-ras—"in everything there is everything"—Churchman finds an important lesson: "In systems planning, in any problem are to be found all other problems."[1] In world malnutrition, for example, "one will eventually be forced to recognize the relevance of militarism and energy and all other problems." Churchman's "systems approach" demands the ever-broadening of inquiry to reach beyond the superficial and into the profound to identify and tackle the roots of universal problems, rather than merely masking their symptoms. Yet, if a question of transportation is to be broadened into questions of urban form, employment, education, equity, poverty, and race—as it must to be meaningfully answered—it becomes increasingly complex. And we find complexity unnatural. We yearn for simplicity. And in doing so we leave the core problems facing us untouched.

Metaphor renders the abstract concrete, the complex simple and more amenable to comprehension and action. At the same time, it acts to mask the real, complex, nature of problems and, by producing distorted accounts of possible solutions, misleads those under its power into thinking they have answers when they have none. Metaphor helps us simplify, and in doing so, it can mislead us. The myriad symbols and images which impress us in our daily lives also tell partial, if compelling stories, which can lead us to make bad decisions. Our symbolic world provides our primary, most elemental way of understanding. Living as we do within it, it is hard to escape its boundaries and view it from the outside.

Central to the definition of the transportation problem in Los Angeles and the prescription for its cure, is a focusing on technologies: the "preselected" possible solutions to problems become the center of attention at the expense of discussion of the problems such "solutions" are supposed to solve. The questions of technology act as proxies for more abstract and difficult social and economic questions to which there are no easy answers. Technologies provide a sharply defined focus of attention, one of simplicity and seeming certainty. Technologies provide a ready source of imagery: They are easy to imagine, and leave concrete—and lasting—impressions. These impressions, operating within the realm of understandings available in the symbolic world where we live, depend upon the assumptions we build for ourselves out of our experiences and history within a particular culture, and lead to solid commonsense conceptions of what action should be taken to cure the transportation malaise.

While political actors preoccupy themselves with the most superficial and misconceived questions of technology choice, planners in government agencies succumb to an equally myopic narrowness in the approaches they take. Their analysis begins with given technologies to be "evaluated," rather than social problems to be addressed. It then proceeds to operate through the lens of given techniques, rather than by arriving at research approaches through reflection on the framing of the problems at hand. Such standard quantitative techniques generate results with an aura of certainty, even though they depend on assumptions which must be subjectively chosen, the applicability of which for forecasting future conditions is at best uncertain.

In the case of the Long Beach light rail forecasting exercise, the results obtained were simply meaningless, not only because of the use of "unreasonable" assumptions, but because the model could not represent the system it purported to simulate. By "correctly" following standard procedures, however, planners avoid responsibility for the inadequacy of their work, just as the Vietnam bombardier avoided responsibility for killing people by simply "correctly" following externally specified instructions on how and where to drop bombs. The wider issues just do not appear to be relevant in either case: the planner or bombardier acts just as a cog in a larger machine. But, perhaps even more disturbingly, the planning process does not exist—as popularly believed—for the making of choices, but for the legitimization of choices already made on other grounds.

Decision makers nonetheless justify their choices on rational grounds, unaware that they have been tacitly guided by erroneous symbolic understandings. Technical analysis is believed to provide solid substantiation, even by many of those at least partly aware of its problems: the lure of data with the appearance of certainty is a beguiling invitation to self-delusion. When some evidence does emerge which is not supportive of the favored technology—as did occur in the Long Beach line evaluation—it is discounted, disbelieved, or ignored.

The reduction of complex problems to simple ones, whether by politicians or planners, is a natural function of the mind: not only does it appear to clear away ambiguity, but also to create "solutions" which are attainable. The problems of freeway congestion cannot be eliminated overnight; but a rail system, symbolic of free flow, can indeed be installed. Rail was also something which could be promised and delivered within a predictable time frame. The reformation of life in Watts could not materialize so fast. Tragically, the rail project becomes a symbol for the solution of deeper problems and one around which polis could be successfully formed, but leaves them untouched. Its impact enters the political realm in a subtle but insidious way: as noted from reports on the lack of overall social and economic progress in Watts, there is no real political commitment for dramatic action to better the community's lot.

Installing the rail system gives the impression of action, not only to constituents but to politicians making the move: they delude themselves into believing that a solution is on the way, and it is a delusion which allows them to continue in office without any real commitment to social change. As Edelman says, "emotional commitment to a symbol is associated with contentment and quiescence regarding problems that would otherwise arouse concern," but the effect is not on the "victims" of the symbol alone: it targets and influences politicians and constituents alike.[2]

We note, significantly, that Compton Councilman Maxcy Filer, who did show real commitment to social change, was immune from the symbolic powers of rail. It is because he was aware of the deep problems affecting his community, working on the basis of the need for self-determination for his constituents, that he rejected the plans which came from outside. His political view kept him outside the mythical world of rail, just as both those prepared to offer no real change and those content to work "within the system" and accept whatever is given are imprisoned within it.

The process we have observed is deeply conservative, and on a number of levels. Despite some of the futuristic images of rail solutions to urban problems, it is far from innovative, in that it reflects technological and other conceptions from the past: rail is a technology of the nineteenth century, being implanted in a non-centralized city of the twenty-first century. Yet symbols are generally created out of past memories, experiences, and identities, and so ideas of the past—with past associations of good and bad—become inappropriately transplanted to the future.

Conceptions of existing systems depend on experience of them today, not on how they might be rearranged in the future. This failure to explore possible innovations to these systems is intrinsic to the "weighing" that takes place as part of the "balance" metaphor: when objects are weighed they are already in their final form, and the act of weighing does not include consideration of how they might be changed. With powerful impressions of buses causing congestion and providing uncomfortable and unsafe rides, there was little to draw the imagination to the possibility of buses operated better so that they could provide a more attractive service.

Los Angeles' newly inaugurated Rapid Bus program may seem to provide a hopeful sign for the future, but its origins leave reasons for concern. As in the case of Seattle, which adopted a bus tunnel project when lack of federal funding prevented the immediate implementation of rail, the financial impossibility of continuing with the Los Angeles rail program created a vacuum as the twentieth century came to a close, and a trip to Curitiba, Brazil, organized by a local architect provided exposure to a sexy bus technology of a similar type of allure to the now unavailable rail. Members of the MTA staff not only focused on comet-like designs to connote speed and make the project seem exciting to board members, but the Rapid Bus program was presented to the board as an "interim step until we organized the rail program."

While the Rapid Bus does promise to make a valuable and cost-effective contribution to improving transit services in Los Angeles, the MTA continues to resist fulfilling its obligations to augment overall bus service, as required by a consent decree resulting from a legal action protesting the destruction of bus services occasioned by the diversion of funds to rail. The Rapids provide a sexy symbolic solution, which more readily seizes the imagination of board members than adding more local buses to serve the many local areas of poverty where board members are unlikely to ever tread.

The political conservatism implied not only by the primacy of technological imagery in influencing transportation decisions, but by the focus on providing symbolic transportation benefits as against directly addressing social problems, is deep-rooted. By acting only on symptoms of social problems (and not even eliminating those), rather than going to their root causes, the problematic status quo is preserved: the city remains polluted, the freeways congested, the poor uneducated and unemployed, despite any slight extra mobility which might be provided to reach opportunities from which they cannot benefit; and political power remains concentrated among those who have created symbolic solutions which, to all everyday appearances, represent progress.

Those at the bottom of the social ladder are made to feel that they have been given something, while what they really need is a voice. While discontent over the loss of bus service and increasing fares was to later set in, the opening of rail service made just about everyone happy with the symbol of success, despite the reality that nothing basic had changed. It is that ability to convey impressions of salvation while leaving root problems untouched that lies at the heart of the symbolic power of the train. And to all caught in the mythical world of rail, the illusion of progress is very real.

In the end, if we follow Churchman's "systems approach" and direct ourselves to broaden our scope to universal problems, we may come to appreciate that, compared to the other, more pressing difficulties of Los Angeles, transportation is hardly the most urgent problem at all: the resources expended on rail will not only produce virtually no benefits—other than symbolic ones—but would be more effectively spent elsewhere, such as on education or employment development. Yet, our political, budgetary, and mental processes put these problems all in separate bubbles, protecting us from the complexity of considering them together, and making us all the poorer as a result.

Voices for Change

In the first chapter, "analytical rationality" was differentiated from "reflective rationality." Analytical rationality implies a taking apart and examination of a set of parts; it operates within a bubble of given variables and assumptions.

Much of the technical analysis we have seen has been bad analysis—biased and insubstantial even according to the standards of the trade. Other analysis has faltered simply because it either cannot answer the questions it purports to address or because, in its narrowness, it asks the wrong questions. Analytical rationality cannot, furthermore, itself provide help with the framing of evaluative questions. Too often issues of framing are ignored, and the frame becomes a tacit consequence of, rather than a basis for, the analytical technique in use. If, however, framing is explicitly considered up-front, and the use of a particular technique follows from that framing, there are many questions which analytical rationality can usefully address. While the fragility of assumptions necessarily makes large-scale forecasting futile, we can analyze the nature of urban transport patterns to deduce the type of transportation most conducive to serving them; we can compare the costs of different transportation options based on past experience; we can evaluate claims that particular types of transportation will reduce congestion based on past performance. In all these instances, analytical rationality, honestly applied, has its place.

Yet, to ask which questions should be asked and to inquire into how to ask them, we must shift to a reflective mode where the task is to burst bubbles rather than to merely live inside them. Breaking free from a warm and familiar environment is rarely a comfortable proposition, but may be our only path to innovation and wisdom. Bubble bursting may tell us to steer away from our obsession with questions of technology to instead ask ourselves about needs to be met and about the values we would have steer those needs. Bubble bursting may tell us that our questions of transportation must become larger questions of urbanity and society, and that ultimately the larger questions must be addressed before the transportation ones will fall into place. In the end, it may seem that we can never escape from some sort of bubble. We must always interpret the world through the artifacts of our culture and lifetime's experience. But we do have the ability to burst at least the smaller bubbles which constrain our effectiveness. If we may draw comfort, it is through the knowledge that while, in our narrowness, we spurn reflection, reflection is a uniquely human quality. By definition it is in antithesis to the preprogrammed quality of the machine.

The political decision makers whose minds we have observed at play follow the paths of neither analytical nor reflective rationality: they tacitly follow the tenets of a myth-creating system of symbols, images, and metaphors that works with a "logic" of its own. It is not logical to give the name of "reason," for it operates subconsciously: while it may wear the guise of thought, it is not thoughtful.

One clue to possible progress is the mismatch between claims about rail transit and how benefits of transit modes are understood. While inferences may be based on mythical understandings, claims decision makers make are generally couched in rational economic terms. A claim may be made that trains cost less to operate than buses. It may be based on the observation that high-capacity trains require only one driver. That observation may provide "evidence," but verification—according to the structure of the claim—requires a statement in terms of dollars. It can be shown that the cost of a given unit of rail service is not lower than the equivalent cost of using buses.

Claims on ridership may be tacitly based on high-tech imagery, and claims to congestion relief on a metaphor of balance restoring free flow, but each claim—as stated—depends for verification on a measurement that is the preserve of the analytical world. There is a tension, which may be used to advantage, to show that the decision makers' claims are quite at odds with the reality of actual performance. The tension is dissipated, however, by the power of myth to lead those under its power to reject "facts" which run counter to the mythology. While, following the stated nature of the claim, verification could be expected in terms of dollars or riders, it is actually subconsciously supplied by the tacit network of symbols, images and metaphors that comes together to constitute myth.

While the academic community has long questioned rail solutions for dispersed western cities in terms of economic efficiency, such criticism has been almost entirely ignored by those making decisions. This study has shown why: decision makers do not act according to a logic of either conscious analytical or reflective reason, but subconsciously according to their experience in the symbolic world in which they live. Economic analysis—abstract, academic, distant—has only a very limited role to play in such a world, compared to vivid images, meaningful symbols, and the powerful tacit metaphors which guide everyday life. Showing politicians that their images do not match the "reality" therefore proves to be insufficient: the problem is that the images constitute their reality. If we are to exploit the tension between claims made in economic terms and inferences drawn in mythical terms, we therefore need more than analyses and facts. And the reflection we must engender must involve not only reflection on how the problems to be solved are to be framed, but on how they are framed now: we must surface the subconscious.

If the greatest need is for reflection on our assumptions, such thought must recognize the reality of the symbolic forces which give our perceptions and actions meaning. One approach involves using alternative symbols to have decision makers change their views. This is what happened when the surface imagery of Curitiba's virtuoso buses was used to sell a sexy vision for bus development to visiting Los Angeles politicians. Such an approach, taken in the shadow of the financial demise of the rail program, while helpful in exposing politicians to alternative technological approaches, does little to encourage a more mature approach to overall problem conception and decision-making.

To engage the thoughtful attention of decision makers, we have to do more than tell them they are wrong: we have to find a way to enable them to understand how they have formed the conceptions which hold their attention, and to ask themselves whether, with that knowledge, those conceptions are still desirable.

We learn the importance of this if we contrast the thinking of Mayor Tucker with that of Maxcy Filer. Tucker was not aware of how his thought was molded by a myth, which deceitfully led him to believe rail would bring benefits to his community. Although the benefits were symbolic, to him they appeared real. While Filer did not believe in rail, he was not exempt from metaphor. His thinking-he talked in terms of bringing jobs to Compton and casting it as a center rather than a periphery, for example—was deeply metaphorical. Filer was aware, however, of the way he framed his analysis, and his metaphors were in league with his objectives. Mayor Tucker's were not.

The most refreshing interview of this study was conducted with a strong rail advocate. After most of the questioning was over, I started talking with him about my theories. As he was interested, not offended, I pointed out the nature of the symbolic and metaphorical impressions behind many of the statements he had given me. We had already been talking for two hours, but we continued for over two hours more, the subject fascinated by the symbolism, if somewhat disturbed that it meshed so well with the claims he had made. As we progressed, economic facts were presented which discounted the alleged benefits of rail. With a growing awareness of the symbolic nature of the understandings previously held, the subject seemed to increasingly feel the tension between his previous views and what the economic data suggested. While at interview's end he was not wholly "converted," the subject (whose honesty I will not betray by naming them) had shifted to a substantially more critical view of rail.

Most importantly, then, it follows that we need to instill a process of psychotherapy to bring to the surface the assumptions which constrain our creativity, for it is only through being made to realize that we are in a prison that we can be persuaded to try to escape. It is to be hoped that the account provided here might make for a modest start to such a psychotherapeutic process. As Will Glass-Husain pointed out, however, at the defense of the MIT dissertation that provided the basis for this book, another word for psychotherapy, in the sense used here, is "education," and perhaps, to burst the bubble of the story which has been unraveled here, the biggest constraint to our creativity is an educational system that focuses on delivering skills at solving bounded problems, rather than instilling an ability to criticize, go beyond boundaries, and think for ourselves with the power of reflective rationality. If we are to equip ourselves to burst bubbles, then, we may have to reform our very systems of learning.

But, not only must we learn to reflect on the sources of the stories in which we daily dwell; we must find ways to convey alternative and richer stories. If Maxcy Filer lives in a realm different from the mythical world of rail, we need to find ways to illuminate his experience and understandings for others. And that means telling stories not only of economically wasteful rail services but of disenfranchisement and the need for political change; stories those in power will be reluctant to hear. Perhaps, ultimately, it is only through grassroots organization and the development of a voice for the deepest of urban problems that different stories can be heard, the myth of rail displaced, and real change effected.

E-1. Blue Line Train Number 1 approaches Pico Station in downtown Los Angeles on July 14, 1990. (Photo courtesy of author)

Epilogue

Yet by June 1990 the taxpayers of Los Angeles will have been paying sales tax for 9 years and for 9 years they will have been reading about the promised beginning of the end of their gridlock nightmares. For over 4 years they will have been negotiating construction disruption in downtown LA and elsewhere. For as long as they can remember, they have been fed promises of this great new set of trains that will whisk them to their jobs conveniently. If the day arrives and the trains are not there, the taxpayers will vent their wrath in the press and at the polls. Our duty is not to let the public down when the big day arrives. That big day is what we are referring to as "Show Time."

—Norman Jester, LACTC staff (1989)

In real terms today, children's lungs will be healthier.
Their immune systems will be stronger . . .
[I]t's going to reduce those traffic and smog
problems in a very significant way.

—California Lieutenant Governor Leo McCarthy

"AND THROUGHOUT THIS COUNTY 150 miles will be started here today," announced LACTC Chairman and County Supervisor Ed Edelman as he addressed his fellow commissioners and other dignitaries sitting wearing engineer's caps on the platform of Pico Station in downtown Los Angeles, as well as the crowds who had gathered to watch from stands sited across the tracks. "Now, it's hard to believe it was just 1960 that we had the last of the Red Car line running on this virtually almost identical route, but here we are today knowing now how important providing an *alternative* to the automobile is," he said. "Back in the 1950s and 1960s, I think we were not as aware of the congestion problems that would be. But today we know what the congestion problems are every day in our lives. So we're here today to celebrate a very important event in the transportation history of Los Angeles County." [italics are my emphasis throughout this chapter]

Edelman acknowledged Supervisor Kenneth Hahn for his leadership role, to great applause from the crowds, and then praised the voters for what they had done: "For this line the people of Los Angeles County had the wisdom, had the wisdom back in 1980 to vote a half-cent sales tax and that money *is you can see today* very well spent." Edelman now introduced Los Angeles Mayor Tom Bradley, who in turn thanked those responsible for bringing "those *beautiful* Blue Cars" to Los Angeles: "It's been a long journey, but this is the beginning of a new and happy journey."

Long Beach Congressman Glenn Anderson was on next to announce that:

> The opening of the Blue Line, the first of the Metro lines to open, marks the beginning of a new era, which will bring benefits of *fast, clean and efficient* rail transportation to millions of residents of Los Angeles County, and never have we needed it more than we do right now . . .
>
> And I know that all of you share my *pride* in this *modern, world-class* transportation system, which will serve a *world-class* community.

Anderson had brought remarks from President George H. W. Bush from Washington, and read these next:

> I am pleased to extend my warm greetings and congratulations to everyone gathered for the inaugural run of the Metro Blue Line *connecting* Long Beach to Los Angeles. The residents of Los Angeles County can take great *pride* in the opening of this line, the area's first new rail line in 30 years, and by approving a local sales tax to fund this project, you not only demonstrated an exemplary sense of civic responsibility, but also *helped to alleviate highway congestion and improve air quality.* I applaud the outstanding spirit of civic cooperation that helped to make this project a reality. Your efforts will benefit Southern Californians for years to come and will long stand as a shining example of the kind of local commitment that we need as we work to build a better America.
>
> Barbara joins me in sending our best wishes for many years of safe and successful operation. God Bless You. Signed by George Bush.

Nicholas Patsaouras, president of the RTD, was next on the podium.

> Kenny, you remember, you held my coat when *I danced at the groundbreaking ceremonies for Metro Rail, and I feel like dancing again here today.* Last time I was President of the Board, there were a lot of skeptics out there. Bring

rail transit to L.A.? Come on, it will never happen, they said. Well, I take great pleasure to say here with you today triumphantly, the skeptics were wrong.

The *beautiful trains* . . . are the culmination of long planning, hard work, and dedication of thousands of people, many of you in the audience, who *had a vision for L.A.*, and they knew it could be realized. The dream is a reality. The rhetoric is over. The trains are back. But this is the beginning of a *rail renaissance* for L.A. as we look into the twenty-first century.

Commissioners Christine Reed and Jacki Bacharach, representing the cities of Los Angeles County, were introduced next, and Reed stated that: "We're as excited as everyone else to be here today. *This M* that you see on these hats and on the signs *represents movement all over Los Angeles County that we're introducing today . . . We are ready to ride.*"

California Lieutenant Governor Leo McCarthy began his speech by praising Kenneth Hahn, then lamented the "very severe health problems because of smog and traffic. This today helps change that. In *real terms today, children's lungs will be healthier.* Their immune systems will be stronger."

"You bet," chimed in Kenneth Hahn, as McCarthy continued:

Fewer than the 300 people who die of cancer each year will suffer because of that. What's happened here today because of good leadership is an extraordinary thing that should be imitated throughout the State of California. When 12,000 passengers ride this Blue Line next year, *it's going to reduce those traffic and smog problems in a very significant way,* and many more in years beyond that, so today I say thank you Ken and thank you Ed, thank you Jacki and Chris and Tom and all the others that have made this possible, but particularly to the people of Los Angeles County. *You made cleaner skies, clearer roads and healthier children a reality.* Thank you. From the state to all of you, to the people of Los Angeles: congratulations!

In rain dances and victory dances men achieve symbolically something they collectively need or want by reaffirming their common interest, denying their doubts, and acting out the result they seek. The motor activity, performed together with others, reassures everyone that there are no dissenters and brings pride and satisfaction in a collective enterprise. A simplified model or semblance of reality is created, and facts that do not fit are screened out of it. Conformity and satisfaction with the basic order are the keynotes; and the

Figure E-2. RTD President, Nick Patsouras, Supervisor Kenneth Hahn, and Supervisor Ed Edelman ring the bell to open the Blue Line on July 14, 1990 at the Pico station in downtown Los Angeles. LATC member Jackie Bacharach is to the far left, former Commissioner Christine Reed is next to her. Mayor Tom Bradley is behind the bell. (Photo courtesy of author)

> *acting out of what is to be believed is a psychologically effective mode of instilling conviction and fixing patterns of future behavior.*
> —Edelman (1964)

Supervisor Edelman once more acknowledged that "We're here today really because of Kenny Hahn's efforts, and the other people who supported Prop. A," then he waxed nostalgic:

> I'm going to in a moment here declare the first Metro line going by ringing this historic bell. This was a bell that was used on the Red Car line from Long Beach to Los Angeles. I as a young boy lived in Long Beach during the war, and *I remember riding that Red Car* to visit my relatives in Eagle Rock and Hollywood. It was a *wonderful line.* We're here today, some 30 years later, du-

plicating that line but bringing it up to the current technology, making it *swift*, making it *efficient*, and so we're back in a sense *in an old era with new technology*, and hopefully the people of L.A. will ride this Metro Rail . . . It's a pretty good investment for a ride free of congestion, a ride free of discomfort. So I'm going to ask . . . Kenny Hahn, the other speakers, if they would join me as I kick off the opening of the Metro line of Los Angeles, today May 14th [*sic*], we Los Angeles County Commission, elected officials, dedicate this light rail line. Alright, Kenny, why don't you ring it.

As the bell rings, the crowd cheers, then spacey, "futuristic" sounds are broadcast over loudspeakers and voices are heard, as if direct from *Star Trek*:

Metro Control, this is Pico Station. Requesting the status of the Metro Blue Line.

Pico Station, this is Metro Control. The Blue Line system is fully operational. *All systems are go* for the inaugural ride of America's newest rail transit line.

Roger, Metro Control, and thank you. This is Pico Station. Blue Line Train Number 1, this is Pico Station. Metro Control informs us the system is ready. You are cleared for the inaugural ride.

Roger, Pico Station, this is Blue Line train Number 1 standing by.

There are now assorted further strange sounds, and then the light rail train is sighted (the moment is referred to in an LACTC release as the "reveal" of the train) and, in a cloud of blue smoke and with much hooting, breaks through the banner stretched across the track and glides into the station. The dignitaries excitedly board to consummate Los Angeles' new marriage, and the train sets off southward, escorted out of downtown L.A. by a platoon of motorcycle-borne sheriff's deputies.

There are crowds awaiting the train's arrival in Watts, the line for free rides on the train stretching two hours into the distance. The dignitaries— and accompanying press entourage—disembark for a ceremony at a site close to the center of the Watts riots. Supervisor Edelman begins by telling Watts residents that:

This is the beginning of 150 miles of light rail line that will link this community to the other communities of Pasadena, the San Fernando Valley and Norwalk and the L.A. International Airport . . .

One person in particular I want to salute and that is Supervisor Kenny Hahn . . . [cheers]. Kenny Hahn made it possible to get the money to build this line. He had the wisdom and courage and the tenacity.

Watts' representative on the Los Angeles City Council, Joan Milke Flores, now declares that "There is probably no part of Los Angeles that has *missed* the old Red Car more than this community," expressing her pleasure at the

> return of that transportation to our community. Today that *dream* is realized . . . For the first time in almost thirty years, people who live and work in Los Angeles, especially in the Watts community, will have an *alternate mode of transportation, one which will help reduce gridlock and pollution.* The Blue Line is a welcome addition to our community. This line will *connect* downtown Los Angeles and downtown Long Beach and provide a tremendous opportunity for immediate access to either civic center and cities along the line. And the Blue Line will provide . . . *access to transportation and to shopping centers and employment.*

Referring to the project as an "investment in this community," she said it "gives us a . . . solid sign of progress."

Tom Bradley now appeared to talk of the beginning of "this new era with the Metro Blue Line," and Kenneth Hahn was then introduced, amidst many cheers: "The Blue Line is here . . . Thank you Tom Bradley. Now, here is my speech. It is the shortest political speech I have ever made: Here's the Blue Line: use it!"

The train continued to Compton, where another massive crowd was waiting to try out the Blue Line. "I see the citizens of Compton want to ride the train," observed Ed Edelman.

> People have been standing out waiting to ride the train for two hours already. *That shows you the need that the people will be able to ride quickly wherever they want to go,* and I wanted to simply tell you that this would not have been possible without Kenny Hahn . . . It was Kenny Hahn who moved this Proposition A to the ballot, he got the funds to get it approved, and thanks to him, we have the beginning of a light rail line that will bring *Los Angeles County to the world-class cities that have adequate ways to move people other than automobiles,* so we know the people of Compton are going to use it.

Kenneth Hahn spoke again: "Well, this is a great day for everybody. It was a hard fight, but we won, and I'm going to make the second shortest speech of my life, time me: We won the fight, the Blue Line is here, now use it!"

"Let the Blue Line roll!" shouted Tom Bradley.

The next stop was Carson, and Mayor De Witt said that:

> It's a bit of irony that today's events in celebrating the opening of the Blue Line that the Red Car was retired some 30 years ago as people took to their

auto and to the freeways, and now it's ironic that we're *relying on the new Blue Line to get people out of their cars and off our freeways.* It too will become, the Blue Line will become, a part of our history, the part that has yet to be ridden [sic] and we're hopeful it will be a chapter that speaks of success and solutions to mass transit challenges, solutions that will make our lives better all around. I have with me here a proclamation that I'd like to present to Supervisor Edelman. It's a proclamation proclaiming "Blue Line Days" here in the City of Carson, and I'd also like to congratulate the Los Angeles County Transportation Commission on the Blue Line's opening and wish great success to the track of the future. We'd like to present you with this proclamation.

Ed Edelman thanked her and asked the crowd:

How many of you would like to ride the new trains? Raise your hands. I can see that you're going to enjoy it. I've ridden the first train coming from Los Angeles. We stopped off in Compton, we stopped off at Willowbrook, we stopped off at 103rd St. and people along this way are excited. They are excited because this is a new day. It's a new day in terms of *linking our communities together in a fast and efficient manner.* How many of you don't like the heat today? Raise your hands. Alright, *even if you don't have to travel anywhere, get on one of these cars and cool off!* It's like 65 degrees on those cars. It's wonderful! So, we're celebrating here today a new beginning, that's 150 miles are going to be built after this 19 miles, going to Pasadena, going to the San Fernando Valley, going to the Airport.

I remember as a young person—I grew up in Long Beach—not very far from here, and I remember in 1945, 47, riding the Red Car—that one right there, from Long Beach to downtown, and then I went up to Hollywood to visit some relatives. Well, you're all going to be able to do the same thing . . .

Kenny Hahn was the real force behind Proposition A . . . The people of L.A. County, believe it or not, voted a tax increase. You know, today we're not supposed to use the T-word, though back in 1980, thanks to Kenny Hahn, the people realized that the T-word was a pretty good word. It wasn't so bad, because *look* what we've got. We've got this new light rail line. We're going to get the rest of it built, too, thanks to the leadership of Kenny Hahn.

Kenneth Hahn repeated his "short" speech. Next stop Long Beach, where a marching band and majorettes meet the train. Ed Edelman started the ceremonies there:

We've returned to Long Beach. The old Red Line is here, but it's now the Blue Line, it's now the Blue Line, and it's returned to the second largest city of this county, the city that has a proud history and a proud heritage. One of those heritage was that it was the base for the Red Line that ran from Long Beach to Los Angeles back in the 30s, 40s, 50s, till someone had the idea that we ought to do away with it. *Well, that wasn't a very good idea then, and it's a worse idea now, because we have more traffic, more congestion, more need to move people, to link communities than ever before . . .*

People are waiting two and three hours, four hours, to get a chance to ride this new light rail system, and I can tell you riding it, *it's the coolest place to be on a hot day, it's one of the quietest places to be on a hot day, the ride is smooth, the ride is fast, it's wonderful . . .* So I'm just pleased to be here today to see a *dream* come true.

Over to Kenneth Hahn: "We had a hard fight. We won that fight and Proposition A, we have the Blue Line, now, the people of Long Beach: you use it!" Long Beach Mayor Ernie Kell declared that, "We're proud and honored that this is going to be the first *link* of a 150 mile system." Deane Dana spoke next and was followed by Glenn Anderson, who talked about

the benefits of fast, clean and efficient rail transportation to millions of residents of Los Angeles County. And never have we needed it more than we do now. . . .

The residents of Long Beach will be able to travel comfortably throughout Los Angeles County, from L.A. to the Universal City, the Airport, all around . . . And I know that all of you share my pride in this *modern, world-class* transportation system which will *connect* two *world-class* communities. Of course, we all know that the living is going to be a little better at the end of the line, and once again congratulations to everyone involved.

The message from President Bush is read again. More speeches follow, and the dignitaries adjourn to a celebration lunch. The Blue Line opens, the crowds pack on to try out the newest, most exciting ride in the park—and with no admission fee! Everyone is happy.

After lunch, the dignitaries board specially scheduled RTD buses to return them express to downtown in half the time it would have taken by Blue Line light rail.

Closing Thoughts

The ceremony is heavy in political posturing: a combination of tribute paying and self-congratulation. But it is the symbolic world we have observed above that gives it meaning and gives a sense of real achievement to the participants.

Rail is seen as an "alternative" to the roads (according to the "escape valve" metaphor): the fact that trains are "fast, clean, and efficient" is taken to mean that they can provide benefits, without having to consider whether they provide service to where people want to go. In the implicit metonymic reduction, the trip is seen in terms of time on board the train, without the problems of getting to and from the train registering on the imagination.

The train is said to be capable of reducing "those traffic and smog problems in a very significant way"—and thereby improving the condition of children's lungs—even though consulting work done for the commission said it would make no significant difference. The politicians do not generally read the details of consulting reports and, even when they do, they tend to reject conclusions which go against the logic of the everyday symbolism and imagery to which they are exposed and the common sense through which they understand it. The 'M' in the engineer's hats "*represents* movement all over Los Angeles County," and that symbolic understanding is very real.

The train is seen as providing "access" to employment for people in Watts, even though no real "access" can be provided if Watts residents lack the skills to get the jobs. The train is seen as "linking" communities in a physical sense, even if there is no economic or social basis upon which such links might be secured. It is a "sign of progress" even without any evidence that "progress" will result from providing rail service.

There is a nostalgic boyhood remembering of the Red Cars, now defunct and "missed" as if they were people. There is a recalling of what a bad decision it was to scrap them, as if congestion could have been prevented were they to have been retained. The powers of a speeding train appear strong; any abstract academic understanding of the changing patterns of transportation interaction in Los Angeles, which made it impossible for rail to keep fulfilling a significant role, is absent.

The old bell is rung: a reliving of something with good associations from the past which, through the transport of memories, is taken to symbolize good even today. An "old era" is re-evoked with supposedly "new technology," even though the basis for the technology and its mismatch with modern needs is—except on the most superficial level—as old as the era from which it originates.

Proof is ever in the eyes: The "T-word" wasn't so bad because "*look* what we've got. We've got this new light rail line." "*As you can see today* the sales tax money voted with Proposition A is "very well spent." How can this be seen? By looking at and riding on the trains, not from conducting an economic evaluation. And, to the eye, the trains look very good. The rail cars are "beautiful" as well as "quiet," "smooth," and "fast," and they have a sensual allure. The train is something out of Hollywood (viz. the *Star Trek* presentation), an exciting extraterrestrial experience, a Disneyland ride. It is something to experience "even if you don't have to travel anywhere" because the cars have wonderful air-conditioning. The train is something to be "enjoyed," rather than a necessity. The fact that people are standing in line waiting to ride "shows you the need" even though those people are waiting to partake of a one-time fantasy experience, rather than to go anywhere in particular.

Having such a "world-class transportation system" of which everyone can be "proud" is symbolically linked to having a "world-class community." The train is the result of a vision, not of analytical reason: it stands for the kind of Los Angeles that is sought, a Los Angeles flourishing in a "rail renaissance," and whether or not it can actually bring that Los Angeles into existence is irrelevant, given the power of the dream to appear real. The opening is an occasion for dancing, a denial of doubt, as in rain or victory dancing. It is a display of pride and an establishing of conformity.

With "expressions of joy, celebrations, parties, cheerful announcements," polis had been successfully formed.[1] Most disturbingly, the political success of the rail project has little or nothing to do with whether it is of any real value. The symbolic success is for all intents and purposes real. And so what if the trolley cannot be useful as a transport of necessity? It is a transport of delight.

Are all the speeches "just" rhetoric? No, the logic connecting the threads of understanding is too palpable; the history, the memories, the experiences, and the visual imagery all offer evidence to support what is being said, and they offer evidence according to the way most of us think most of the time. We think naturally according to our human experiences; unawares, we create realities for ourselves that can lead us deeply astray. But in our rain dances we sometimes transcend a reality we would rather escape. And, for the moment, with the happy crowds of disenfranchised and poor people cheering the train that is sailing through Watts, that is all that matters. Even if the cruel elements of an unchanged reality will nonetheless return once the train has disappeared into the distance.

APPENDIX A   # List of Interview Subjects

List of participants in interviews conducted over 1985/86 Titles/Positions as of 1985/86.

Board of Supervisors, County of Los Angeles

Supervisors

Antonovich, Michael, Supervisor, County of Los Angeles
Hahn, Kenneth, Supervisor, County of Los Angeles

Staff to Supervisors

Roche, Burke, aide to Supervisor Kenneth Hahn
Mull, Howard, aide to Supervisor Mike Antonovich
George, Deborah, aide to Supervisor Deane Dana
Ireland, Peter, aide to Supervisor Deane Dana
Lewis, Mike, Deputy to Supervisor Pete Schabarum, former Chairman of the Board, RTD, Los Angeles

Former Supervisor and Staff

Ward, Baxter, Former Supervisor, County of Los Angeles
Leonard, Gerald, former aide to Baxter Ward

LACTC

Commissioners and Alternate Commissioners

Bacharach, Jacki, Chair
Donley, Roy, Alternate Commissioner to Supervisor Michael Antonovich (1986)
King, Walter, Alternate Commissioner to Supervisor Kenneth Hahn
Mednick, Marcia, Commissioner
Pierce, Ted, Alternate Commissioner to Supervisor Michael Antonovich (1985)
Reed, Christine, Mayor, City of Santa Monica, Commissioner, LACTC
Sanborn, Blake, Alternate Commissioner to Supervisor Peter Schabarum
Szabo, Barna, Alternate Commissioner to Supervisor Deane Dana
White, Bob, Councilman, City of Norwalk, Alternate Commissioner to Christine Reed
One interview not for attribution and off the record: not cited in this study.

Citizen Members, LACTC Rail Construction Committee

Jonas, Allan, Citizen Member, Rail Construction Committee
Perez, Manuel, Citizen Member, Rail Construction Committee

Staff

Caufield, Daniel, Project Manager, Long Beach light rail (1985)
McSpedon, Ed, Project Manager, Long Beach light rail (1986)
Richmond, Rick, Executive Director
Sims, Jim
Stanger, Richard
Taylor, Paul, Deputy Executive Director

Former Staff

Premo, Jerry, former Executive Director

Department of Public Works, County of Los Angeles

Staff

Burger, Roger, LA County, Department of Public Works
Kaufman, John, LA County Division of Public Works
Willis, Alan, LA County Division of Public Works

RTD

Board Members

Estrada, Carmen
(Hall, Jan, Councilwoman, City of Long Beach, Long Beach listing)
Holen, Marvin

Staff

PLANNING

Bevan, Leo
Killough, Keith
Lee, Byron
Odell, Anne
Parry, Steve
Spivack, Garry
Woodhull, Joel

MODELING

Stopher, Peter, Demand Modeler, Schimpeler Corradino, under contract to and working at RTD, Los Angeles.

COMMUNITY RELATIONS

Flagg, Wanda

LOBBYIST, WASHINGTON, D.C.

Slagle, Roger, Lobbyist for RTD, Washington, D.C.

SCAG

Staff (All interviews not for attribution)

DEMAND MODELING

Three staff

OTHER TRANSPORTATION PLANNING
Three staff

OTHERS
Two staff

Commuter Computer

Staff
Widby, Tad

City of Los Angeles

Members of the Council
Bernardi, Ernani
Ferraro, John
Flores, Joan
Russell, Pat, Chair, City Transportation Committee

Staff to the Mayor
Lawson, Craig, Transportation aide to Mayor Bradley, City of LA

Staff to the Council
Goldthwaite, Princess, aide to Councilwoman Pat Russell
Hedrick, Bob, Assistant to Councilman John Ferraro
Lovejoy, Tracy, office of Councilman Joel Wachs
McCarberry, Dennis, office of Councilwoman Flores
Stewart, Mike, office of Councilwoman Flores, Watts office
Melnick, Jay, aide to Councilman Ernani Bernardi

Department of Transportation
Rifkin, Allyn
Lepis, Alice

City of Long Beach

Members of the Council
Kell, Ernie, Mayor
Wallace Edgerton, Councilman
Hall, Jan, Councilwoman
Sato, Eunice, Councilwoman

Staff
Paternoster, Bob, Director of Planning

Long Beach Merchants Interests
Caso, Bob and Lee, Robert

City of Compton

Members of the Council
Tucker, Walter, Mayor, City of Compton

Filer, Maxcy, Councilman, City of Compton

Staff

Johnson, John (joint interview with Ed Satello)
Satello, Ed

Westminster Neighborhood Association, Watts

Payne, Grace, Director
Corbett, Ed
Harris, "Sweet" Alice

Other Watts-related interviews

Shaw-Johnson, Freita
Saucedo, Robert, Young People of Watts
(See also Mike Stewart interview under City of L.A., council staff)

Other Black Community

Bragg, Felicia

Automobile Club of Southern California

Barrett, Dick
Gilbert, Keith
Ortner, Jim

Other Interests

Central City

Steven Gavin, Chairman, Gavin Associates, member, Central City Association, Los Angeles
Lamb, Ron, Los Angeles Chamber of Commerce
Williams, Pam, Central City Association, Los Angeles
(Caso, Bob and Lee, Robert, Long Beach, included under Long Beach)

Rail Advocates

Falick, Abraham
Hart, Stanley (Sierra Club)
(Roger Slagle, RTD lobbyist included with RTD)
Multiple interviews of membership during a Los Angeles Railway Historical Society meeting

Others

Shay, Paul, L.A. Tax
(Yaksik, Nick, APTA included with UMTA)

Community Development

Community Development, City of Los Angeles

Lurie, Marilyn

Community Redevelopment Agency, Los Angeles

Benbow, Richard
Roche, Pat

Human Relations Commission, Los Angeles County
Aubry, Larry

Other
Dukett, Steve, County of Los Angeles

Air Quality
Wuebben, Paul, Southern California Air Quality Management District

Other L.A.-related interviews
Brodsly, David, City of Los Angeles
Comapana, Lori
One interview not for attribution

Bus trip
Interviews on LA-Orange County express bus

California State Legislature

Members

Democrats

SENATORS
Alquist, Alfred, San Jose (Santa Clara)
Deddeh, Wadie, Bonita (San Diego)
Foran, John F., San Francisco, Chairman, Senate Transportation Committee
Greene, Bill, Los Angeles
Torres, Art, Los Angeles, Senate Transportation Committee
Watson, Diane, Los Angeles

MEMBERS OF THE ASSEMBLY
Areias, Rusty, Los Banos (Santa Clara), Assembly Transportation Committee
Clute, Steve, Riverside, Assembly Transportation Committee
Elder, Dave, Long Beach
Harris, Elihu, Oakland, Assembly Transportation Committee
Katz, Richard, Sepulveda (Los Angeles), Assembly Transportation Committee
Killea, Lucy, San Diego

Republicans

SENATORS
Beverly, Robert, Manhattan Beach (Los Angeles), Senate Transportation Committee
Ellis, Jim, Senator, El Cajon (San Diego), Senate Transportation Committee (ranking minority)

MEMBERS OF THE ASSEMBLY
Brown, Dennis, Long Beach, Chairman, Subcommittee on Transportation, Assembly Committee on Ways & Means
Frizzelle, Nolan, Huntington Beach, Assembly Transportation Committee, (ranking minority)
Lancaster, William, Covina (Los Angeles), Assembly Transportation Committee
Mojonnier, Sunny, Encinitas (San Diego)

Legislators' Staff

Democrats

Brokaw, Barry, Chief of staff, office of Senator Daniel E. Boatwright, Concord
Higginbotham, Keith, assistant to Senator Art Torres, Los Angeles
Medina, Leslie, assistant to Assemblyman Rusty Areias, Los Banos (Santa Clara)

Republicans

Jerome, Gary, assistant to Senator Marian Bergeson, Newport Beach (Orange County)

COMMITTEE STAFF

Schifferle, Patricia, Consultant, Assembly Transportation Committee, Richard Katz, D-Sepulveda, (Los Angeles), Chair
Lange, L. Eric, Consultant, Assembly Transportation Committee, Richard Katz, D-Sepulveda, (Los Angeles), Chair
Lind, Alan, Counsel, Subcommittee on Transportation, Assembly Committee on Ways & Means, Dennis Brown, R-Long Beach, Chair
Watts, Mark, Principal Consultant, Assembly Committee on Ways & Means

California Transportation Commission

Commissioners

Duffel, Joe
Hulitt, Stanley
Leonard, Bill
(Nestande, Bruce, Supervisor, Orange County, listed under Orange County)
Romero, Richard

Staff

Nielson, Robert, Executive Director
Bohlinger, Linda

Caltrans

Sacramento

Cross, Don
Deter, Lee
Hendrix, Alan
Lynch, Ray
Miller, Steve
Schaeffer, Bill, Deputy Director
Shirley, Earl
Simpson, Phil
Tolmach, Rich
Vostrez, John
Weber, Warren
Wieman, Larry

District 7, Los Angeles

Dove, Don
Heckeroth, Heinz, District Director (1985), Commissioner, LACTC, ex-officio
Watson, Donald, District Director (1986), Commissioner, LACTC, ex-officio

Michalak, Steve, with Daniel Butler and Dale Ratzlaff
Roper, David

U.S. Congress

Members of Congress

Democrat

Bates, Jim, Representative, San Diego, former member MTDB

Republican

Fiedler, Bobbi, Representative, Los Angeles (San Fernando Valley)

Staff

Democrats

Dodgson, Jerry, Office of Rep. Henry A. Waxman, Los Angeles
Hershman, Mark I., Legislative Assistant, Office of Rep. Robert T. Matsui, Sacramento
Rideau, Rodney M., Legislative Assistant, Office of Rep. Julian C. Dixon, Los Angeles
Roberts, Glenn E., Legislative Director, Office of Rep. Norman Y. Mineta, San Jose
Sampson, Phil, staff, Senate Urban Affairs & Banking Committee, Majority
Schlesinger, Paul L., Professional Staff Member, Subcommittee on Surface Transportation, (Committee on Public Works and Transportation)

Republicans

Staff member to Senator Proxmire
Bobeck, Jeff, Legislative Assistant, Office of Rep. David Dreier, Los Angeles (La Verne, Pomona)
Ives, Lori, Legislative Assistant, Office of Rep. Daniel E. Lungren, Long Beach, Washington, DC
Kraft, Ken, Office of Rep. Coughlin, Subcommittee on Transportation
Malakoff, Bob, staff, Senate Urban Affairs & Banking, Minority
Rehr, David K., Executive Assistant, office of Rep. Vin Webber, Minnesota

US Department of Transportation

UMTA

Davis, Aubry, Seattle
Emerson, Don, Washington, D.C.
Harrant, Don, Washington, D.C.
Stout, Bob, Washington, D.C.
Thomas, Ed, Washington, D.C.
Zimmerman, Sam, Director of Planning, Washington, D.C.

Federal Highways Administration

Clinton, Glen

American Public Transit Association

Yaksik, Nick, Washington, D.C.

Board of Supervisors, Orange County

Supervisors

Clark, Ralph
Nestande, Bruce, Member California Transportation Commission

Orange County Transportation Commission

Staff

Greene, Sharon

Mortazemi, Kia

Orange County Transit District

Staff

Ordway, Jeff

Board of Supervisors, County of San Diego

Supervisors

Bilbray, Brian

City of San Diego

Member of the Council

Struiksma, Ed, Councilman, City of San Diego

San Diego Metropolitan Transportation District

Board Members

Mills, Jim

Staff

Larwin, Tom, Executive Director

Liebermann, William

Robenheimer, Bob

SANDAG

Staff

Franck, George

Hultgren, Lee

Santa Clara Transportation Commission

Commissioner

Fadness, David, Santa Clara, Chairman

Santa Clara Transit District

Staff

Miller, Les and Gordon Smith (joint interview)

Pierson, Jim

Sacramento

Regional Transit

Wiley, Mike

Other

Bauer, Arthur

Metropolitan Transportation Commission (San Francisco Bay Area)

Staff

Dahms, Larry, Executive Director

San Francisco Municipal Railway

Staff

Bernhard, Bruce, Assistant to General Manager
Natvig, Carl

Tri-Met, Portland

Staff

Mr. Higbee, Director, Banfield light rail
Arrington, G. B.

METRO, Portland

Staff

Katugno, Andy

City of Portland

Staff

Dotterer, Steve
Siegel, Steve, Portland Development Commission

Oregon Department of Transportation

Spence, Ted

METRO, Seattle

Board Member

Neir, Bob, Chair, Transit Committee

Staff

Graczyk, Dan, Downtown Tunnel Manager
Kalberer, David
Postuma, Ron

Puget Sound Council of Governments

Staff

Fryztacki, Wes

City of Seattle

Member of the Council

Williams, Jeannette, Seattle

Staff
Stafford, Bill, Director, Government Relations Office

Vancouver

London Transport International
Calver, Dave

Academics
Churchman, C. West, University of California, Berkeley
Edner, Sheldon, Portland State University
Gordon, Peter, University of Southern California
Jones, David
Webber, Mel, University of California, Berkeley

Questionnaire

Four different questionnaire plans were used in the conduct of this research during 1985 and 1986. Each one provided a guideline for proceeding with interviews. Interviews were generally allowed to proceed free-form, however. In many instances, only a part of the material on the guideline was covered, and in some cases several topics not on the guidelines were included.

The most commonly used questionnaire—for decision-makers in Los Angeles County—is reprinted below.

Do you support the plans for the development of a light rail system between Los Angeles and Long Beach (in the Los Angeles area)? Why? (prompt) Congestion? Pollution? Conservation of energy? Providing mobility to the poor/minorities/elderly?

What particular areas would be helped?

Any advantages for Compton/Watts?

What sources of information were helpful to you in reaching this view? How do you know that they were reliable?

What is the role of forecasting in such a decision? Suppose that forecasts suggested that few people would use the Long Beach trolley. Would you have still supported its construction?

What has been the relationship of technical analysis to political decision-making? What do you think it should be?

How important was public opinion in reaching a decision? How was it canvassed? How could one be sure that it was reliably canvassed? Do you consider it to be important?

What influence should be accorded to vocal special interests? For example, downtown interests?

(Explain move back to more general questions).

What is a transportation problem? How should we go about making choices of transportation facilities? On what criteria? How do we choose the criteria? How do we select transportation alternatives for consideration?

Suppose you are told to design a new transportation system for an imaginary new city. What sort of questions would you want to ask? How would you go about finding answers to them (prompts: agency research, forecasts, public participation, interests?)

(Return to Long Beach)

Why is it important for people to use the train rather than to travel by some other means?

The document prepared by SCAG to project patronage for the Long Beach line says that there will only be 1,600 extra work trips made by transit in the Los Angeles-Long Beach corridor (2,900 in Los Angeles County), only 2 percent of auto traffic. It also says: "From a county-level or even a corridor level, the LB/LA project has only a very minor positive impact on traffic." The other passengers on the trolley would otherwise have gone by bus. Why is it better for them to go by train than by bus? (Explain problems of access/distribution in Southern California. Why "telephone" network the best . . .)

Were alternatives of express buses seriously considered? Possibilities for increasing carpooling?

Blue collar worker job opportunities are mostly in an East-West direction, so the trolley can't be of great use in helping to serve them. The trip time from LB-LA will be as long as the current express bus takes . . . Express buses retained in Northern California. Might happen here. Lower ridership?

(Proposition A)

Why did Proposition A specify that a rail network be built? Didn't that limit options rather? Was there a case for doing this?

Evidence suggests that people voted for Prop. A for the lower bus fares, rather than for rail transit. The increased bus fares will probably lose far more transit riders than the Long Beach trolley will gain. The fare increase will also cause hardship to those of low income. In this context, why is the trolley more important than lower bus fares?

We rarely look at transportation in a larger context. We say that we should build this facility, rather than that one. What is the case for examining transportation in the context of larger social needs? There may, for example, be programs of community revitalization of greater importance to Compton/Watts. How should they be weighed against transportation expenditures?

Resolution of
Proposition A

RESOLUTION CALLING SPECIAL ELECTION
PROPOSING A RETAIL TRANSACTIONS AND
USE TAX FOR PUBLIC TRANSIT PURPOSES BE
SUBMITTED TO THE VOTERS OF THE COUNTY AT
THE SPECIAL ELECTION AND ORDERING THE
CONSOLIDATION OF THE SPECIAL ELECTION
WITH THE NOVEMBER GENERAL ELECTION

BE IT RESOLVED by the Los Angeles County Transportation
Commission, that a special election is hereby ordered and
called to be held on Tuesday, November 4, 1980, and that the
following Proposition be submitted to the electors of the
County of Los Angeles at the special election.

BE IT FURTHER RESOLVED that the Los Angeles County
Transportation Commission requests that the Board of Supervisors
of the County of Los Angeles, State of California, order the
consolidation of the special election with the November General
Election and orders that the Proposition be placed upon the
same ballot as shall be provided for the General Election to
be held on the 4th day of November, 1980, and that the same
precincts, polling places, and precinct board members as used
for the General Election shall be used for the Special Election
pursuant to Elections Code Section 23300 et seq.

The exact form of the Proposition as it is to appear on
the ballot and a complete text of the proposed ordinance is as
follows:

BALLOT PROPOSITION. (See attached Exhibit A)

ORDINANCE. (See attached Exhibit B)

Proclamation. Pursuant to Section 2653 of the Elections
Code the Los Angeles County Transportation Commission hereby
PROCLAIMS that a special County-wide election shall be held on

Tuesday, November 4, 1980, to vote upon the Proposition set forth in this resolution. The polls shall be open for said election from 7:00 a.m. to 8:00 p.m. The Registrar-Recorder shall cause this proclamation to be published in a daily newspaper of general circulation, printed, published, and circulated in Los Angeles County, for at least one (1) time not less than fifty (50) days before the 4th day of November, 1980, pursuant to Elections Code Section 2554.

The Acting Executive Director of the Los Angeles County Transportation Commission is ordered to file a copy of this resolution with the Registrar-Recorder at least seventy-four (74) days prior to the date of the election.

ANALYSIS OF ORDINANCE. The County Counsel of the County of Los Angeles is hereby requested to prepare an analysis of said ordinance pursuant to Section 3781 of the Elections Code.

I certify that the foregoing Resolution was adopted by a majority vote of all members of the Los Angeles County Transportation Commission, at its meeting held on the 20th day of August, 1980.

RICK RICHMOND
Executive Director
Los Angeles County
Transportation Commission

LOS ANGELES COUNTY TRANSPORTATION COMMISSION – PUBLIC TRANSIT: To improve and expand existing public transit Countywide, reduce fares, construct and operate a rail rapid transit system serving at least:	
San Fernando Valley West Los Angeles South Central Los Angeles/Long Beach South Bay/Harbor Century Freeway Corridor Santa Ana Freeway Corridor San Gabriel Valley and more effectively use State and Federal funds, benefit assessments, and fares for those purposes, shall the Commission approve an ordinance authorizing a Countywide ½ percent sales tax?	YES
Revenues will be allocated: 25 percent to local jurisdictions for local transit; a specified reduced fare structure for SCRTD for 3 years; and specified allocations for rail rapid transit and to the Commission for public transit purposes.	NO

EXHIBIT A

424 Appendix C

ORDINANCE NO. 16

AN ORDINANCE ESTABLISHING A RETAIL TRANSACTIONS
AND USE TAX IN THE COUNTY OF LOS ANGELES
FOR PUBLIC TRANSIT PURPOSES

The Los Angeles County Transportation Commission do ordain
as follows:

SECTION I

A retail Transactions and Use Tax is hereby imposed in the
County of Los Angeles as follows:

SECTION 1. DEFINITIONS. The following words, whenever used in
this Ordinance, shall have the meanings set forth below:

    (a) "Commission" means the Los Angeles County Transportation
        Commission.

    (b) "County" means the incorporated and unincorporated territory
        of the County of Los Angeles.

    (c) "Transaction" or "Transactions" have the same meaning,
        respectively, as the words "Sale" or "Sales"; and the word
        "Transactor" has the same meaning as "Seller", as "Sale"
        or "Sales" and "Seller" are used in Part 1 (commencing
        with Section 6001) of Division 2 of the Revenue and Taxa-
        tion Code.

SECTION 2. IMPOSITION OF RETAIL TRANSACTIONS TAX. There is hereby
imposed a tax for the privilege of selling tangible personal property
at retail upon every retailer in the County at a rate of one-half
of 1% of the gross receipts of the retailer from the sale of all
tangible personal property sold by him at retail in the County.

SECTION 3. IMPOSITION OF USE TAX. There is hereby imposed a
complementary tax upon the storage, use or other consumption in the
County of tangible personal property purchased from any retailer for
storage, use or other consumption in the County. Such tax shall be
at a rate of one-half of 1% of the sales price of the property whose
storage, use or other consumption is subject to the tax.

SECTION 4. APPLICATION OF SALES AND USE TAX PROVISIONS OF REVENUE
AND TAXATION CODE. The provisions contained in Part 1 of Division 2
of the Revenue and Taxation Code (Sales and Use Taxes, commencing
with Section 6001), insofar as they relate to sales or use taxes and
are not inconsistent with Part 1.6 of Division 2 of the Revenue and
Taxation Code (Transactions and Use Taxes, commencing with Section
7251), shall apply and be part of this Ordinance, being incorporated
by reference herein, except that:

    (a) The Commission, as the taxing agency, shall be substituted
        for that of the State;

    (b) An additional transactor's permit shall not be required
        if a seller's permit has been or is issued to the
        transactor under Section 6067 of the Revenue and Taxation
        Code; and

    (c) The word "County" shall be substituted for the word "State"
        in the phrase, "Retailer engaged in business in this State"

in Section 6203 of the Revenue and Taxation Code and
in the definition of that phrase.

A retailer engaged in business in the County shall not be
required to collect use tax from the purchase of tangible personal
property unless the retailer ships or delivers the property into
the County or participates within the County in making the sale of
the property, including, but not limited to soliciting or receiving
the order, either directly or indirectly, at a place of business
of the retailer in the County or through any representative, agent,
canvasser, solicitor, or subsidiary or person in the County under
authority of the retailer.

All amendments subsequent to January 1, 1970, to the above
cited Sales and Use Taxes provisions relating to sales or use taxes
and not consistent with this Ordinance shall automatically become
a part of this Ordinance; provided, however, that no such amendment
shall operate as to affect the rate of tax imposed by the Commission.

SECTION 5. USE OF REVENUES RECEIVED FROM IMPOSITION OF THE TRANSACTIONS
AND USE TAX. The revenues received by the Commission from the
imposition of the transactions and use tax shall be used for public
transit purposes, as follows:

    (a) Definitions:

        1.   "System" or "Rail rapid transit system" means all
land and other improvements and equipment necessary
to provide an operable, exclusive right-of-way, or
guideway, for rail transit.

        2.   "Local transit" means eligible transit, paratransit,
and Transportation Systems Management improvements
which benefit one jurisdiction.

    (b) Purpose of Tax.

This tax is being imposed to improve and expand existing
public transit Countywide, including reduction of transit
fares, to construct and operate a rail rapid transit
system hereinafter described, and to more effectively
use State and Federal funds, benefit assessments, and fares.

    (c) Use of Revenues.

Revenues will be allocated as follows:

        1.   For the first three (3) years from the operative
date of this Ordinance:

            a.   Twenty-five (25) percent, calculated on an
annual basis, to local jurisdictions for local
transit, based on their relative percentage
share of the population of the County of Los
Angeles.

            b.   To the Southern California Rapid Transit Dis-
trict (District), or any other existing or
successor entity in the District receiving
funds under the Mills-Alquist-Deddeh Act, such
sums as are necessary to accomplish the following purposes:

                (1) Establishment of a basic cash fare of
fifty (50) cents.

                (2) Establishment of an unlimited use transfer
charge of ten (10) cents.

(3) Establishment of a charge for a basic
     monthly transit pass of $20.00.

(4) Establishment of a charge for a monthly
     transit pass for the elderly, handicapped
     and students of $4.00.

(5) Establishment of a basic cash fare for
     the elderly, handicapped and students of
     twenty (20) cents.

(6) Establishment of a comparable fare
     structure for express or premium bus
     service.

c. The remainder to the Commission for construction
    and operation of the System.

2. Thereafter:

a. Twenty-five (25) percent, calculated on an
    annual basis, to local jurisdictions for local
    transit, based on their relative percentage
    share of the population of the County of Los
    Angeles.

b. Thirty-five (35) percent, calculated on an
    annual basis, to the Commission for construction
    and operation of the System.

c. The remainder shall be allocated to the Com-
    mission for public transit purposes.

3. Scope of Use.
    Revenues can be used for capital or operating
    expenses.

(d) Commission Policy.

1. Relative to the Local Transit Component:

a. Allocation of funds to local jurisdictions
    shall be subject to the following conditions:

    1. Submission to the Commission of a descrip-
        tion of intended use of the funds, in order
        to establish legal eligibility. Such use
        shall not duplicate or compete with exist-
        ing transit service.

    2. The Commission may impose regulations to
        insure the timely use of local transit
        funds.

    3. Recipients shall account annually to the
        Commission on the use of such funds.

b. Local jurisdictions are encouraged to use
    available funds for improved transit service.

2. Relative to the System Component:

a. The Commission will determine the System to be
    constructed and operated.

b. The System will be constructed as expeditiously
    as possible. In carrying out this policy, the
    Commission shall use the following guidelines:

    1. Emphasis shall be placed on the use of
        funds for construction of the System.

    2. Use of existing rights-of-way will be
        emphasized.

c.   The System will be constructed and operated in
     substantial conformity with the map attached
     hereto as Exhibit "A". The areas proposed to
     be served are, at least, the following:

          San Fernando Valley
          West Los Angeles
          South Central Los Angeles/Long Beach
          South Bay/Harbor
          Century Freeway Corridor
          Santa Ana Freeway Corridor
          San Gabriel Valley

SECTION 6. EXCLUSION OF TAX IMPOSED UNDER BRADLEY-BURNS UNIFORM
LOCAL SALES AND USE TAX LAW. The amount subject to tax under this
Ordinance shall not include the amount of any sales tax or use tax
imposed by the State of California or by any city, city and county,
or county, pursuant to the Bradley-Burns Uniform Local Sales and Use
Tax Law, or the amount of any State-administered transactions or
use tax.

SECTION 7. EXEMPTIONS FROM RETAIL TRANSACTIONS TAX.

(a) There are exempted from the tax imposed by this Ordinance
     the gross receipts from the sale of tangible personal
     property to operators of waterborne vessels to be used
     or consumed principally outside the County in which the
     sale is made and directly and exclusively in the carriage
     or persons or property in such vessels for commercial
     purposes.

(b) There are exempted from the tax imposed under this
     Ordinance the gross receipts from the sale of tangible
     personal property to the operators of aircraft to be
     used or consumed principally outside the County in which
     the sale is made, and directly and exclusively in the
     use of such aircraft as common carriers of persons or
     property under the authority of the laws of this State,
     the United States, or any foreign government.

(c) Sales of property to be used outside the County which
     are shipped to a point outside the County pursuant to the
     contract of sale, by delivery to such point by the retailer
     or his agent, or by delivery by the retailer to a carrier
     for shipment to a consignee at such point, are exempt
     from the tax imposed under this Ordinance.

          For purposes of this Section, "delivery" of vehicles
     subject to registration pursuant to Chapter 1 (commencing
     with Section 4000) of Division 3 of the Vehicle Code,
     the aircraft license in compliance with Section 21411 of
     the Public Utilities Code and undocumented vessels regis-
     tered under Article 2 (commencing with Section 680) of
     Chapter 5 of Division 3 of the Harbors and Navigation
     Code shall be satisfied by registration to an out-of-
     County address and by a declaration under penalty of
     perjury, signed by the buyer, stating that such address
     is, in fact, his principal place of residence.

          "Delivery" of commercial vehicle shall be satisfied
     by registration to a place of business out of County,
     and a declaration under penalty of perjury signed by the

428   Appendix C

buyer that the vehicle will be operated from that
address.

(d) The sale of tangible personal property is exempt from
tax, if the seller is obligated to furnish the property
for a fixed price pursuant to a contract entered into
prior to the operative date of this Ordinance. A lease
of tangible personal property which is a continuing sale
of such property is exempt from tax for any period of
time for which the lessor is obligated to lease the
property for an amount fixed by the lease prior to the
operative date of this Ordinance. For purposes of this
Section, the sale or lease of tangible personal property
shall be deemed not to be obligated pursuant to a contract or lease for any period of time for which any
party to the contract or lease has the unconditional
right to terminate the contract or lease upon notice,
whether or not such right is exercised.

SECTION 8.   EXEMPTIONS FROM USE TAX.

(a) The storage, use or other consumption of tangible
personal property, the gross receipts from the sale
of which have been subject to a transaction tax under
any State administered transactions and use taxes ordi-
nances, shall be exempt from the tax imposed under this
Ordinance.

(b) The storage, use or other consumption of tangible
personal property purchased by operators of waterborne
vessels and used or consumed by such operators directly
and exclusively in the carriage of persons or property
in such vessels for commercial taxes is exempt from the
use tax.

(c) In addition to the exemption provided in Sections 6366
and 6366.1 of the Revenue and Taxation Code, the storage,
use, or other consumption of tangible personal property
purchased by operators of aircraft and used or consumed
by such operators directly and exclusively in the use of
such aircraft as common carriers of persons or property
for hire or compensation under a certificate of public
convenience and necessity issued pursuant to the laws of
this State, United States, or any foreign government, is
exempt from the use tax.

(d) The storage, use, or other consumption in the County of
tangible personal property is exempt from the use tax
imposed under this Ordinance if purchaser is obligated
to purchase the property for a fixed price pursuant to
a contract entered into prior to the operative date of
the Ordinance. The possession of, or the exercise of
any right or power over, tangible personal property under
a lease which is a continuing purchase of such property
is exempt from tax for any period of time for which a
lessee is obligated to lease the property for an amount
fixed by a lease prior to the operative date of this
Ordinance. For the purposes of this Section, storage,
use or other consumption, or possession, or exercise of

any right or power over, tangible personal property shall be deemed not to be obligated pursuant to a contract or lease for any period of time for which any party to the contract or lease has the unconditional right to terminate the contract or lease upon notice, whether or not such right is exercised.

SECTION 9. PLACE OF CONSUMMATION OF RETAIL TRANSACTION. For the purpose of a retail transaction tax imposed by this Ordinance, all retail transactions are consummated at the place of business of the retailer, unless the tangible personal property sold is delivered by the retailer or his agent to an out-of-State destination or to a common carrier for delivery to an out-of-State destination. The gross receipts from such sales shall include delivery charges, when such charges are subject to the State sales and use tax, regardless of the place to which delivery is made. In the event a retailer has no permanent place of business in the State, or has more than one place of business, the place or places at which the retail sales are consummated for the purpose of the transactions tax imposed by this Ordinance shall be determined under rules and regulations to be prescribed and adopted by the State Board of Equalization.

SECTION 10. DEDUCTION OF LOCAL TRANSACTIONS TAXES ON SALES OF MOTOR VEHICLE FUEL. The Controller shall deduct local transactions taxes on sales of motor vehicle fuel which are subject to tax and refund pursuant to Part 2 (commencing with Section 7301) of this division, unless the claimant establishes to the satisfaction of the Controller that the claimant has paid local sales tax reimbursement for a use tax measured by the sale price of the fuel to him.

If the claimant establishes to the satisfaction of the Controller that he has paid transactions tax reimbursement or Commission use tax measured by the sale price of the fuel to him, including the amount of the tax imposed by said Part 2, the Controller shall repay to the claimant the amount of transactions tax reimbursement or use tax paid with respect to the amount of the motor vehicle license tax refunded. If the buyer receives a refund under this Section, no refund shall be made to the seller.

SECTION 11. ADOPTION AND ENACTMENT OF ORDINANCE. This Ordinance is hereby adopted by the Commission and shall be enacted upon authorization of the electors voting in favor thereof at the special election called for November 4, 1980, to vote on the measure.

SECTION 12. OPERATIVE DATE. This Retail Transactions and Use Tax Ordinance shall be operative the first day of the first calendar quarter commencing not less than 180 days after the adoption of said Ordinance.

SECTION 13. EFFECTIVE DATE. The effective date of this Ordinance shall be August 20, 1980.

PASSED AND ADOPTED by the Los Angeles County Transportation
Commission this 20th day of August, 1980, by the following vote:

AYES: Geoghegan, Hahn, Rubley, Russ, Szabo, Ward, Zimmerman.

NOES: Cox, Remy, Russell , Schabarum.

ABSENT: None

                          The Los Angeles County
                          Transportation Commission

                          By ⟨signature⟩ Peter F. Schabarum
                             ─────────────────────────────
                                       Chairman

ATTEST:

Executive Director
of the Los Angeles County
Transportation Commission

⟨signature⟩
─────────────────────
   RICK RICHMOND

I hereby certify that at its meeting of August 20, 1980, the
foregoing Ordinance was adopted by the Los Angeles County
Transportation Commission.

                          Executive Director
                          of the Los Angeles County
                          Transportation Commission

                          ⟨signature⟩
                          ─────────────────────
                             RICK RICHMOND

APPROVED AS TO FORM:
JOHN H. LARSON
County Counsel

By ⟨signature⟩
   ─────────────────────
   RONALD L. SCHNEIDER
   Principal Deputy County Counsel

Analysis of Proposition A

ANALYSIS OF PROPOSITION A

By John H. Larson, County Counsel

Present law provides that the Los Angeles County Transportation Commission may, by ordinance, and subject to voter approval, impose a retail transactions and use tax (commonly called a "sales tax") in the incorporated and unincorporated area of the County of Los Angeles for public transit purposes.

The Commission has adopted such an ordinance imposing a one–half cent sales tax, the revenues of which would be used to a) improve and expand existing public transit Countywide, including reduction of transit fares, b) construct and operate a rail rapid transit system, and c) more effectively use State and Federal funds, benefit assessments, and fares. The revenues would be allocated as follows:

1. For the first three years from the operative date of the ordinance:
   a) Twenty–five percent, calculated on an annual basis, would be allocated to local jurisdictions for local transit, based on their relative percentage share of the population of the County of Los Angeles.
   b) From the remaining seventy–five percent, sufficient funds would be allocated to the Southern California Rapid Transit District or any other existing or successor entity in the District receiving funds under the Mills–Alquist/Deddeh Act to accomplish the following purposes:
      (1) Establishment of a basic cash fare of fifty cents.
      (2) Establishment of an unlimited use transfer charge of ten cents.
      (3) Establishment of a charge for a basic monthly transit pass of $20.00.
      (4) Establishment of a charge for a monthly pass for the elderly, handicapped and students of $4.00.
      (5) Establishment of a basic cash fare for the elderly, handicapped and students of twenty cents.
      (6) Establishment of a comparable fare structure for express or premium bus service.
   c) The remainder would be allocated to the Commission for construction and operation of a rail transit system.

2. Thereafter:
   a) Twenty–five percent, calculated on an annual basis, would be allocated to local jurisdictions for local transit, based on their relative percentage share of the population of the County of Los Angeles.
   b) Thirty–five percent, calculated on an annual basis, would be allocated to the Commission for construction and operation of the rail transit system.
   c) The remainder would be allocated to the Commission for public transit purposes.

Revenues would be used for capital or operating expenses.

The rail transit system would be constructed and operated in substantial conformity with a map attached to the ordinance and areas proposed to be served, at the least, are described.

The ordinance also indicates Commission policies relative to local and rail transit.

Provisions detailing the imposition and application of the tax as well as certain tax exclusions, exemptions and a deduction are indicated in the ordinance.

The ordinance would be operative, and the tax would be imposed beginning July 1, 1981.

(For full text of Proposition, see Sample Ballot; the full text of the proposed sales tax ordinance is available at the Los Angeles County Transportation Commission, 311 South Spring Street, Suite 1206, Los Angeles, California 90013.)

ARGUMENT IN FAVOR OF PROPOSITION A

It's time for some plain talk about transportation in Los Angeles County. It's a sick patient and needs some strong medicine to heal itself.

Let's face facts:

– Gasoline doubled in price last year; many experts feel that it will soon exceed $2.00 per gallon;

– Mideast instability could result in more gas lines or gas rationing;

– Because we don't have an adequate public transit system, most people must continue to drive on increasingly clogged streets and highways: that costs them time and money and wastes energy.

Proposition A is a realistic, sensible approach to these problems. It will:

– Improve and expand existing public transit countywide by providing substantial funding directly to each city for transit improvements;

– Guarantee a 50¢ countywide bus fare, 10¢ transfer, and $4 monthly pass for seniors, handicapped and students for at least 3 years;

Construct and operate a rail rapid transit system serving the areas shown on the attached map. This will create productive jobs to boost our economy;

Help Los Angeles County get back its fair share of Federal and State tax dollars for transportation. Currently, Atlanta receives over $728 per capita from the Federal government for transit, Los Angeles receives $54.

Transportation is like the nervous system in the body of our community. If that system is sick, we all suffer the consequences. We pay the price in increased energy costs, more smog and congestion, fewer jobs, and a stagnating economy.

Because of previous inaction, Los Angeles County is the only urban area of its size that does not have a rapid transit system. Approval of this 1/2% sales tax will bring immediate action through lower bus fares and development of rapid transit countywide.

Proposition A will end the delay and get Los Angeles moving in the right direction.

Vote YES on Proposition A.

KENNETH HAHN
Los Angeles County Supervisor
Past Chairman and current member of the
Los Angeles County Transportation Commission

BURKE ROCHE
Executive Vice President
Los Angeles Taxpayers Association

LEAGUE OF WOMEN VOTERS OF LOS ANGELES
COUNTY

WILLIAM R. ROBERTSON
Executive Secretary–Treasurer
Los Angeles County Federation of Labor, AFL–CIO

REBUTTAL TO ARGUMENT IN FAVOR OF PROPOSITION A

Proposition A is a bad tax! It would finance a transportation proposal that will cost too much — and serve too few!

Proposition A may sound good and look good; but when you read the fine print — this proposal is nothing but air. It has no specific program to permanently improve public transportation and there are no assurances or protections for the taxpaying public.

Where does it guarantee that the rail lines on the map will actually be built???

Where does it say that there will be no more bus strikes if this tax passes???

Where is the assurance that your neighborhood will really get more frequent bus service???

What will keep the politicians from squandering all the dollars on their own pet projects???

Finally, will this new tax really make it any easier for you and your family to travel to their desired destinations???

Proposition A is the bureaucrats usual answer to solving a complex problem "throw more money at it and may be it will go away."

Don't be fooled by their smooth sales pitch. Save your tax dollars for a better proposal — VOTE NO ON PROPOSITION A.

> PETE SCHABARUM
> L. A. County Supervisor and
> Chairman, L. A. County Transportation
> Commission
>
> MICHAEL W. LEWIS
> Director, RTD
>
> J. EDWARD MARTIN
> President, Southern California
> Transportation Action Committee

ARGUMENT AGAINST PROPOSITION A

Proposition A will raise your taxes every time you make a purchase. In return, if you live in the suburbs, it offers you nothing.

Proposition A will make a lot of politicians very happy because they'll have the chance to spend this new tax money almost as they please.

To call this a transportation improvement tax is very misleading.

1. It offers lower fares, but not for long distance commuters.

2. It offers money to local cities, but only if they don't "duplicate or compete" with existing transit service. That could create up to 81 more bus operators in this County.

3. If there is any money left over, the politicians will "expeditiously" construct and operate the fixed rail system as it appears on the map.

This new tax money for transit won't necessarily put more bus service on the street and it won't put an end to transit strikes.

By the Transportation Commission's own admission, it will take 75 years to construct the system envisioned on their map. That's a long time to wait for an alternative to our next energy crisis.

The public is already being taxed heavily for transportation. We now pay over 17¢ per gallon in gas tax. Unfortunately, much of it goes into the State's General Fund rather than the transportation fund. We should be seeking the return of our tax money, not asking the taxpayers to provide more tax money.

The taxpayers wisely rejected these multi–billion dollar transit proposals in 1968, 1972 and 1976. Yet some how we have managed to purchase 1200 new buses, acquire funding for a downtown people mover, a Wilshire subway, and a Century Freeway with exclusive bus lanes.

Tell the politicians they'll have to make do with the millions they've already got – vote NO on Proposition A.

> PETER F. SCHABARUM
> Supervisor, 1st District
> County of Los Angeles
>
> MICHAEL W. LEWIS
> Director, RTD
>
> J. EDWARD MARTIN
> President, Southern California
> Transportation Action Committee

REBUTTAL TO ARGUMENT AGAINST PROPOSITION A

You can tell a lot about a proposition by who supports it and who opposes it.

Proposition A is supported by economy–minded taxpayers, the League of Women Voters, labor and civic leaders.

The opponents of better transportation consist of a politician, (Peter Schabarum), his deputy, and the highway lobby. These obstructionists are trying to block the citizens' last hope for effective transportation.

The obstructionists offer continued crowded freeways, wasteful use of energy, long gas lines, and the prospect of our entire local economy grinding to a halt in the event of Middle East oil shut–offs.

These prophets of doom make several false statements.

The truth is, this balanced transportation plan:
1. WILL reduce bus fares EVERYWHERE in Los Angeles County.
2. WILL NOT create 81 more bus companies.
3. WILL begin rapid transit construction immediately.

There are always some politicians who hesitate and do not act courageously. They never offer positive alternatives, but only excuses. These are the obstructionists who defeated transit proposals in 1968, 1974 and 1976 (if only 5% more voters had supported transit in 1968, we would have rapid transit today). Let's not be fooled again.

"Where there is no vision, the people perish." Do what is best for your family, do what is best for Los Angeles County, and do what is best for America.

Fight inflation. Conserve energy. Create jobs. Solve our transportation mess.

Now, in 1980, your vote must count.

Vote for Proposition A. It will put us back on the right track.

> KENNETH HAHN
> Los Angeles County Supervisor
> Past Chairman and Current Member of the
> Los Angeles County Transportation Commission
>
> BURKE ROCHE
> Executive Vice President
> Los Angeles Taxpayers Association
>
> LEAGUE OF WOMEN VOTERS OF LOS ANGELES
> COUNTY
>
> WILLIAM R. ROBERTSON
> Executive Secretary–Treasurer
> Los Angeles County Federation of Labor, AFL–CIO

List of Abbreviations

AC: Alameda-Contra Costa
ARCO: Atlantic Richfield Corporation
ATA: American Transit Association
BART: San Francisco Bay Area Rapid Transit District
CALTRANS: California Department of Transportation
CBD: Central Business District
CST: Cable Suspended Transit
DEIS: Draft Environmental Impact Statement
DMJM: Daniel, Mann, Johnson & Mendenhall
DOT: City of Los Angeles Department of Transportation
EBA: East Busway, Pittsburgh
EIR: Environmental Impact Report
FHWA: Federal Highways Administration, U.S. Department of Transportation
FTA: Federal Transit Administration, U.S. Department of Transportation
HOV: High Occupancy Vehicle
ICM: Idealized Cognitive Model
IPA: Institute of Public Administration
ISTEA: Intermodal Surface Transportation Efficiency Act
LACTC: Los Angeles County Transportation Commission
LARTS: Los Angeles Regional Transportation Study
LATAX: Los Angeles Taxpayers Association
LAX: Los Angeles International Airport
LRT: Light Rail Transit
MAX: Metropolitan Area Express, Portland, Oregon, light rail system
MBTA: Massachusetts Bay Transportation Authority
METRO: Metropolitan Service district, Portland, Oregon
MTA: Los Angeles County Metropolitan Transportation Authority
MTBD: San Diego Metropolitan Transit Development Board
NTD: National Transit Database
PAT: Port Authority of Allegheny County Pennsylvania
PBHM: Parsons, Brinkerhoff, Hall & McDonald
PBTB: Parsons Brickerhoff Tudor Bechtel
PCC: Presidents Conference Committee cars,
P.E.: Pacific Electric
RTAC: Rapid Transit Advisory Committee, formed by Southern California Rapid Transit District
RTAG: Rapid Transit Action Group
RTD: Southern California Rapid Transit District
RTP: Regional Transportation Plan

SACOG: Sacramento Area Council of Governments
SANDAG: San Diego Associations of Governments
SCAG: Southern California Association of Governments
SPTC: Southern Pacific Transportation Company
STIP: Statewide Transportation Improvement Program, State of California
TRI-MET: Tri-County Metropolitan Transportation District, Portland, Oregon
TSM: Transportation Systems Management
UMTA: Urban Mass Transportation Administration, U.S. Department of Transportation
USC: University of Southern California
UTU: United Transportation Union

Notes

Chapter 1

1. Melanchthon, *Initia Doctrinae Physicae* quoted in Andrew D. White, *A History of the Warfare of Science with Theology in Christendom* (New York: Appleton, 1896), 1, 126; Thomas S. Kuhn, *The Copernican Revolution* (Cambridge: Harvard University Press, 1957), 191.

2. Kenneth E. Boulding, "The Ethics of Rational Decision," *Management Science* 12, no. 6 (1966): B-161–B-169; Harold J. Laski, "Limitations of the Expert," *Chemtech* (April 1974): 99; Russell L. Ackoff, *The Art of Problem Solving* (New York: Wiley, 1978), 53.

3. Jonathan E. D. Richmond, "Introducing Philosophical Theories to Urban Transportation Planning or Why All Planners Should Practice Bursting Bubbles," *Systems Research* 7, no. 1 (1990): 47–56, Jonathan E. D. Richmond "Introducing Philosophical Theories to Urban Transportation Planning," in *Planning Ethics: A Reader in Planning Theory, Practice, and Education*, ed. Sue Hendler (New Brunswick, N.J.: Center for Urban Policy Research, 1995), 301–20.

4. Alan A. Altshuler, *The Urban Transportation System: Politics and Policy Innovation* (Cambridge: MIT Press, 1979), ix.

5. Extending themes explored in Jonathan E. D. Richmond, "Perpetuum Mobile-AMTRAK: The Original Sin" (master's thesis, Massachusetts Institute of Technology, 1981); and Jonathan E. D. Richmond, "Mental Barriers to Learning and Creativity in Transportation Planning," *Carolina Planning* 13, no. 1 (fall 1987): 42–53.

6. Donald A. Schön, *Beyond the Stable State* (New York: W. W. Norton, 1971), 123.

7. Summarized in Jonathan E. D. Richmond, "The Mythical Conception of Rail Transit in Los Angeles," *Journal of Architectural and Planning Research* 15, no. 4. (winter 1998): 294–320.

8. For examples see Rogers Brubaker, *The Limits of Rationality: An Essay on the Social and Moral Thought of Max Weber* (London: GeorgeAllen & Unwin, 1984); Yehezkel Dror, *Public Policymaking Reexamined* (Scranton, Pa.: Chandler Publishing Company, 1968).

9. John Friedmann and Barclay Hudson, "Knowledge and Action: A Guide to Planning Theory," *AIP Journal* (January 1974): 3; Donald A. Schön, *The Reflective Practitioner: How Professionals Think in Action* (New York: Basic Books, 1983), 21.

10. C. West Churchman, *The Systems Approach and Its Enemies* (New York: Basic Books, 1979), 71.

11. C. West Churchman, *Thought and Wisdom* (Seaside, Calif.: Intersystems Publications, 1982), 132, 129.

12. Ibid., 7, 8, and 117.

Chapter 2

1. "Henry E. Huntington Dead," *Electric Railway Journal* (May 28, 1927): 981.

2. Theodore E. Treutlein, "Los Angeles, California: The Question of the City's Original Spanish Name," *Southern California Quarterly* 55 (spring 1973): 1–8.

3. W. W. Robinson, *Los Angeles: From the Days of the Pueblo* (San Francisco: Chronicle Books, 1981), 29.

4. Glen E. Holt, "The Changing Perception of Urban Pathology: An Essay on the Development of Mass Transit in the United States," in *Cities in American History*, ed. Kenneth J. Jackson and Stanley K. Schultz (New York: Knopf, 1972), 234.

5. Robinson, *Los Angeles*, 88.

6. Kirkman-Harriman, *The Kirkman-Harriman Pictorial and Historical Map of Los Angeles County, 1860 A.D.* (Los Angeles: Kirkman-Harriman, 1937).

7. Howard J. Nelson, "The Spread of an Artificial Landscape over Southern California," in *Man, Time, and Space in Southern California*, ed. William L. Thomas Jr., special supplement to *Annals of the Association of American Geographers* 49 (September 1959): 85.

8. David Brodsly, *L.A. Freeway: An Appreciative Essay* (Berkeley: University of California Press, 1981), 63.

9. Martin Wachs, "Autos, Transit, and the Sprawl of Los Angeles: The 1920s," *Journal of the American Planning Association* 50, no. 4 (summer 1984): 298.

10. James E. Vance Jr., "California and the Search for the Ideal," *Annals of the Association of American Geographers* 62 (1972).

11. Robert M. Fogelson, *The Fragmented Metropolis: Los Angeles 1850–1930* (Cambridge: Harvard University Press, 1967), 144–45.

12. Rockwell D. Hunt and William S. Amert, *Oxcart to Airplane* (Los Angeles: Powell Publishing Co., 1929), 190.

13. Randolph Karr, "Rail Passenger Service History of Pacific Electric Railway Company," available in Los Angeles County Metropolitan Transportation Authority Library (1973): 1.

14. Laurence R. Veysey, "The Pacific Electric Railway Company, 1910–1953, A Study in the Operations of Economic, Social and Political Forces Upon American Local Transportation," available in Los Angeles County Metropolitan Transportation Authority Library (1953): 3.

15. Wachs, "Autos, Transit," 300; Scott L. Bottles, *Los Angeles and the Automobile. The Making of the Modern City* (Berkeley: University of California Press, 1987), 22–51; Mark S. Foster, "The Decentralization of Los Angeles During the 1920s" (Ph.D. diss., University of Southern California, 1971).

16. *Los Angeles Examiner*, December 12, 1904.

17. Spencer Crump, *Henry Huntington and the Pacific Electric* (Corona del Mar, Calif.: Trans-Anglo Books, 1978), 10.

18. Wachs, "Autos, Transit," 300.

19. "Henry E. Huntington Dead," 981.

20. Crump, *Henry Huntington*, 13.

21. Spencer Crump, *Ride the Big Red Cars: How Trolleys Helped Build Southern California* (Corona del Mar, Calif: Trans-Anglo Books, 1970), 76–79.

22. Fogelson, *The Fragmented Metropolis*, 104.

23. Crump, *Henry Huntington*, 12.

24. Pat Adler, "Watts: A Legacy of Lines," *Westways* 58 (Aug. 1966): 22–24, quoted in Brodsly, *L.A. Freeway*, 68.

25. Charles N. Glaab and A. Theodore Brown, *The History of Urban America* (London: Macmillan, 1967), 282, quoted in Brodsly, *L.A. Freeway*, 71.

26. Carey McWilliams, *Southern California: An Island on the Land.* (Santa Barbara: Peregrine Smith, 1973), 169.

27. Crump, *Henry Huntington*, 21.

28. Veysey, "The Pacific Electric Railway Company," 4–5.

29. Ibid., 25.

30. Brodsly, *L.A. Freeway*, 69.

31. Karr, "Rail Passenger Service History."

32. Bradford C. Snell, "American Ground Transport, A Proposal for Restructuring the Auto-

mobile, Truck, Bus, and Rail Industries," Part 4A, Appendix to Part 4, *Hearings* before the Subcommittee on Antitrust and Monopoly of the Committee on the Judiciary, United States Senate (1974).

33. Ibid., A-1.

34. Ibid.

35. Ibid.

36. Ibid., A-28.

37. Ibid., A-28, A-29.

38. Ibid., A-3.

39. Ibid., A-31.

40. Ibid., A-32.

41. General Motors, "The Truth About 'American Ground Transport'—A Reply by General Motors," Part 4A, Appendix to Part 4, *Hearings* before the Subcommittee on Antitrust and Monopoly of the Committee on the Judiciary, United States Senate (1974), A-107; Ibid., A-112; Ibid; Ibid., A-113.

42. Ibid., A-114.

43. Ibid., A-116.

44. George W. Hilton and John F. Due, *The Electric Interurban Railway in America* (Stanford: Stanford University Press, 1960), 4.

45. David W. Jones Jr., *Urban Transit Policy: An Economic and Political History* (Englewood Cliffs, N.J.: Prentice-Hall, 1985), 15.

46. Veysey, "The Pacific Electric Railway Company," 9; Ibid., 98.

47. Ibid., 158–59.

48. Brodsly, *L.A. Freeway*, 82.

49. Veysey, "The Pacific Electric Railway Company," 103–4.

50. "Street Cars Coming Back," *Electric Traction* (Oct. 1925): 522; Joseph P. Hallihan, "Relation of Rapid Transit to Community Development," *Electric Railway Journal* 73 (Sept. 14, 1929): 912.

51. Eli Bail, *From Railway to Freeway: Pacific Electric and the Motor Coach* (Glendale, Calif.: Interurban Press, 1984), 12.

52. Fogelson, *The Fragmented Metropolis*, 92.

53. Mark S. Foster, "The Model-T, The Hard Sell, and Los Angeles' Urban Growth: The Decentralization of Los Angeles during the 1920s," *Pacific Historical Review* 44 (Nov. 1975): 476.

54. Brodsly, *L.A. Freeway*, 82–83.

55. George W. Hilton, "Rail Transport and the Pattern of Modern Cities: The California Case," *Traffic Quarterly* 3 (July 1967): 379–80.

56. Fogelson, *The Fragmented Metropolis*, 179–80.

57. Jones, *Urban Transit Policy*, 44.

58. Brodsly, *L.A. Freeway*, 92.

59. Bottles, *Los Angeles and the Automobile*, 15.

60. Fogelson, *The Fragmented Metropolis*, 179.

61. Veysey, "The Pacific Electric Railway Company," 146.

62. Crump, *Ride the Big Red Cars*, 203.

63. Veysey, "The Pacific Electric Railway Company," 168.

64. Ibid.

65. Ibid., 226.

66. Bail, *From Railway to Freeway*, 23.

67. Veysey, "The Pacific Electric Railway Company," 29.

68. *Electric Railway Journal* 67, (Jan.–June 1916): 715

69. Ibid., 71.

70. Bail, *From Railway to Freeway*, 23.

71. Ibid., 95.

72. Veysey, "The Pacific Electric Railway Company," 10.

73. J. R. Stauffer, "Opportunities for Profits in De Luxe Bus Operation," *Electric Railway Journal* (Feb. 1930): 91–96.

74. Carl W. Stocks, "Improving the Bus to Increase Its Usefulness," *Electric Railway Journal* (June 14, 1930): 387–88.

75. A. T. Warner, "Development of the Bus for Mass Transportation," *Electric Railway Journal* (Dec. 1930): 745–46.

76. Ibid., 746.

77. "How the New Gas Buses Have Improved Earnings." *Transit Journal* (Sept. 15, 1936): 337.

78. General Motors, "The Truth About 'American Ground Transport,'" A-114.

79. Veysey, "The Pacific Electric Railway Company," 131.

80. Ibid., 134.

81. Ibid., 183, 185.

82. Ibid., 208, quoting *Transit Journal* (1937): 11.

83. General Motors, "The Truth About 'American Ground Transport,'" A-115.

84. Bail, *From Railway to Freeway*, 113.

85. California Railroad Commission, Decision No. 33088 (May 14, 1940), 631–32.

86. California Railroad Commission, Decision No. 33984 (Mar. 11, 1941), 379.

87. Veysey, "The Pacific Electric Railway Company," 219.

88. Jones, *Urban Transit Policy*, 74–75.

89. Veysey, "The Pacific Electric Railway Company," 276.

90. Ibid., 283.

91. Karr, "Rail Passenger Service History," 16.

92. Bail, From Railway to Freeway, 190.

93. Veysey, "The Pacific Electric Railway Company," 297, quoting *Los Angeles Daily News*, Aug. 22, 1948.

94. Veysey, "The Pacific Electric Railway Company," 297.

95. California Railroad Commission, Decision No. 44161 (May 9, 1950), 661–62.

96. Veysey, "The Pacific Electric Railway Company," 304.

97. Karr, "Rail Passenger Service History," 17.

98. Crump, *Henry Huntington*, 27.

99. Ibid., 28.

100. Jones, *Urban Transit Policy*, 63.

101. Ibid., 63–64.

102. Sy Adler, "The Transformation of the Pacific Electric Railway: Bradford Snell, Roger Rabbit, and the Politics of Transit in Los Angeles," *Urban Affairs Quarterly* 27, no. 1 (Sept. 1991): 83.

103. Fogelson, *The Fragmented Metropolis*, 92.

104. Bail, *From Railway to Freeway*, 190.

Chapter 3

1. Editorial, *Los Angeles Times*, Jan. 21, 1991.

2. LACTC, *The Long Beach-Los Angeles Rail Transit Project, Final Environmental Impact Report* (Mar. 1985), III-86.

3. Norman J. Jester, "Preparations for 'Show Time:' The Los Angeles Story," in *Light Rail Transit, New System Successes at Affordable Prices* (Washington, D.C.: Transportation Research Board, Special Report 221, 1989), 337.

4. "MTA Adopts Austere FY 04 Budget That Still Pushes Forward with Major Transit Projects." *MTA press release*, May 22, 2003.

5. SCAG, *Transcript of SCAG Executive Committee Regional Transportation Plan Workshop* (Sept. 1, 1983), 12–16.

6. Gordon and Eckert, eds., *Transportation Alternatives for Southern California*, 2.

7. Jonathan E. D. Richmond, *New Rail Transit Investments—A Review*, Taubman Center for State and Local Government, John F. Kennedy School of Government, Harvard University (June

1998). Summarized in Jonathan E. D. Richmond, "A Whole-System Approach to Evaluating Urban Transit Investments," *Transport Reviews* 21, no. 2 (Apr.–June 2001): 141–179. The current chapter makes substantial use of the results of this work.

8. Don H. Pickrell, *Urban Rail Transit Projects: Forecast Versus Actual Ridership and Costs. Final Report,* Prepared for Office of Grants Management, Urban Mass Transportation Administration, U.S. Department of Transportation, Report DOT-T-91-04, Oct. 1990, x. Summarized in Don H. Pickrell, "A Desire Names Streetcar: Fantasy and Fact in Rail Transit Planning," *Journal of the American Planning Association* 58, no. 2 (spring 1992).

9. National Transit Database, except San Diego, which is MTDB. See Richmond, "A Whole-System Approach," 152–54.

10. Email from Dana Woodbury, Deputy Executive Officer, Operations Planning & Scheduling, MTA, July 21, 2003. "MTA Says Subway Riders Over-Counted," *Los Angeles Times,* June 18, 2003. Note also that FY 2003 bus ridership is based on only eleven months of data for the small amount of contract service purchased by the MTA. This is unlikely to make more than a negligible difference to accuracy.

11. UMTA/RTD, *Los Angeles Rail Rapid Transit Project Metro Rail, Final Supplemental Environmental Impact Statement/Subsequent Environmental Impact Report* (July 1989), S-4-2.

12. Using National Transit Database data.

13. Using MTA monthly data.

14. MTA.

15. RTD.

16. SANDAG (San Diego Association of Governments), 1990 *San Diego Regional Transit Survey,* vol. 2 (Oct. 1991), 41.

17. Baltimore Metropolitan Council, *Central Light Rail Survey, Volume 1: Methodology and Results,* Cornelius Nuworsoo, project manager, *Task Report* 94-2 (1994), S-3.

18. MTA, "The Metro Green Line Turns One," news release (Aug. 12, 1996).

19. Howell Research Group, *Regional Transportation District Light Rail Passenger Survey* (May 1995).

20. PBHM, Regional Rapid Transit, *Report to the San Francisco Bay Area Rapid Transit Commission* (Jan. 5, 1956), 38.

21. Board of Public Utilities, *Annual Report* (1926), 23.

22. SCAG, *Transcript,* 8, 10.

23. John W. Dyckman, "Riding the Sunset Coast Line," in *Transportation Alternatives for Southern California,* ed. Peter Gordon and Ross D. Eckert, Conference Proceedings of a Symposium, The Institute for Public Policy Research, Center for Public Affairs, University of Southern California (Apr. 12, 1976), 31, 34.

24. David A. Quarmby, "Choice of Travel Mode for the Journey to Work," *Journal of Transport Economics and Policy* 1 (1967): 273–314; Lisco cited in Martin Wachs, "Consumer Attitudes Toward Transit Service: An Interpretive Review," *Journal of the American Institute of Planners* (Jan. 1976): 98; C. Henderson and J. Billheimer, *Manhattan Passenger Distribution Project: Effectiveness of Midtown Manhattan System Alternatives* (Menlo Park, Calif.: Stanford Research Institute, 1972); Daniel McFadden, "The Measurement of Urban Travel Demand," working paper no. 227 (Berkeley: Institute of Urban and Regional Development, University of California, Berkeley, 1974).

25. Allan N. Nash and Stanley J. Hille, "Public Attitudes Toward Transport Modes: A Summary of Two Pilot Studies," *Highway Research Record* 233 (Washington, D.C.: Highway Research Board, 1968); *National Analysis, A Survey of Commuter Attitudes Toward Rapid Transit Systems,* vol. 1 (Washington, D.C.: National Capital Transportation Agency, 1963); *Los Angeles Times,* June 8, 1973.

26. S. Algers, S. Hansen, and G. Tegner, "Role of Waiting Time, Comfort, and Convenience in Modal Choice for Work Trip," *Transportation Research Record* No. 534 (Washington, D.C.: Transportation Research Board, 1975); Wachs, "Consumer Attitudes," 99.

27. Melvin M. Webber, *The BART Experience—What Have We Learned?* monograph no. 26

(Berkeley: Institute of Urban and Regional Development, University of California, Berkeley, Oct., 1976), 34.

28. Martin Wachs, "The Case for Bus Rapid Transit in Los Angeles." in *Transportation Alternatives for Southern California*, ed. Peter Gordon and Ross D. Eckert, Conference Proceedings of a Symposium, The Institute for Public Policy Research, Center for Public Affairs, University of Southern California (Apr. 12, 1976), 71.

29. Neil Peterson to LACTC Planning and Mobility Improvement Committee, memorandum "Interim Downtown METRO Blue Line Shuttle Update," prepared by Alan E. Patashnick (Los Angeles: Los Angeles County Transportation Commission, Sept. 14, 1990).

30. Edward McSpedon, "Building Light Rail Transit in Existing Rail Corridors—Panacea or Nightmare? The Los Angeles Experience," in *Light Rail Transit, New System Successes at Affordable Prices*, Washington, D.C.: Transportation Research Board, Special Report 221 (1989), 427.

31. Anastasia Loukaitou-Sideris and Tridib Banerjee, "The Blue Line Blues: Why the Vision of Transit Village May Not Materialize Despite Impressive Growth in Transit Ridership," *Journal of Urban Design* 5, no. 2 (2000): 119.

32. MTA data supplied to author.

33. Webber, *The BART Experience*, 30.

34. McFadden, "The Measurement of Urban Travel Demand."

35. Wachs, "Consumer Attitudes," and "Policy Implications of Recent Behavioral Research in Transportation Demand Management," *Journal of Planning Literature* 5, no. 4 (May 1991): 333–41.; Gerald K. Miller and Keith M. Goodman, *The Shirley Highway Express-Bus-on-Freeway Demonstration Project: First Year Results* (Washington, D.C.: Technical Analysis Division, National Bureau of Standards, 1972).

36. Wachs, "The Case for Bus Rapid Transit," 101.

37. Maritz Marketing Research, 1989.

38. J. R. Stauffer, "Opportunities for Profits in De Luxe Bus Operation," *Electric Railway Journal* (Feb. 1930): 91–92.

39. Ibid., 93, 91.

40. Seth Mydans, *New York Times*, Nov. 21, 1990.

41. Miller and Goodman, *The Shirley Highway Express*; Wachs, "Consumer Attitudes," 102; Wachs, "The Case for Bus Rapid Transit," 72.

42. Texas Transportation Institute, *Houston High-Occupancy Vehicle Lane Operations Summary* (Dec. 1996).

43. METRO management, interviewed by author, Houston, Tex.

44. John Crain and Associates, *First-year Report: San Bernardino Freeway Express Busway Evaluation* (Menlo Park, Calif.: John Crain and Associates, 1974).

45. MTA, response to data request from Supervisor Antonovich's office (Sept. 13, 1996).

46. John Bonsall, "A Bus For All Seasons" (paper presented at seminar on *The Canadian Experience: Making Transit Work in the Golden Gate Corridor*, co-sponsored by the Golden Gate Bridge, Highway and Transportation District and the Canadian Consulate General, San Rafael, Calif., Oct. 3, 1985), 9.

47. OC Transpo, *Operating Statistics*, OC Transpo Research & Development, Ottawa (1996); Richmond, "A Whole-System Approach," 153.

48. Crain & Associates, Inc. *The Martin Luther King East Busway in Pittsburgh, PA, Final Report*, UMTA/TSC Evaluation Series, U.S. Department of Transportation, Urban Mass Transportation Administration, UMTA Technical Assistance Program (October 1987), xvii.

49. PAT, *Light Rail and Busway Statistical Overview* (Based on Fiscal Year 1995 Operations) (May 1996); Crain, Martin Luther King East Busway, xiv; data supplied to author by Port Authority Transit.

50. John F. Kain and Zhi Liu, *Secrets of Success: How Houston and San Diego Transit Providers Achieved Large Increases in Transit Ridership*, prepared for Federal Transit Administration (May 1995).

51. Tri-Met, *Tri-Met Five Year Plan FY 93–97, Phase I. Foundation Document No. 1. Comprehensive Service Analysis* (Sept. 1991), 3-3; From table supplied to author by Tri-Met entitled: "Fixed Route Service and Ridership"; Tri-Met, letter with comments on draft of Richmond (June 12, 1998).

52. MTA, Draft 2001 Long Range Transportation Plan for Los Angeles County (Feb. 2001), 2–32.

53. MTA.

54. MTA, *Draft 2001 Long Range Transportation Plan*, 2–37; Ibid., 2–34; Ibid., 2–38; Ibid., 2–35.

55. Ibid., 2–38.

56. Webber, *The BART Experience*.

57. Ibid., 8, 12.

58. Altshuler, *The Urban Transportation System*, 434.

59. Wohl, "The Case for Rapid Transit," 57.

60. George W. Hilton, "What Did We Give Up With The Big Red Cars?" In *Transportation Alternatives for Southern California*, ed. Peter Gordon and Ross D. Eckert, Conference Proceedings of a Symposium, The Institute for Public Policy Research, Center for Public Affairs, University of Southern California (Apr. 12, 1976), 95.

61. Hilton, *Transportation Alternatives*, 95.

62. Charles A. Lave, "The Mass Transit Panacea and Other Fallacies About Energy," *The Atlantic* 244, no. 4 (Oct. 1979): 40.

63. SCAG, *Los Angeles–Long Beach Light Rail Transit Project, Patronage Estimation and Impacts* (Mar. 1984).

64. Ibid., table 2, 7.1–6.

65. Ibid., 7.1–9.10.

66. Russell H. Henk, Daniel E. Morris, and Dennis L. Christiansen, *An Evaluation of High-Occupancy Vehicle Lanes in Texas*, 1994 (College Station: Texas Transportation Institute Rept. FHWA/TX-96/1353-3, 1995), xxxiii.

67. Transportation Research Board, *Highway Capacity Manual*, Special Report 209 (1985), 12-10.

68. Ibid., 12-4.

69. Ibid., 12-10.

70. *New York Times*, Nov. 21, 1990.

71. Lave, "The Mass Transit Panacea," 39.

72. Ibid., 40.

73. SCAG, *Los Angeles–Long Beach*, table 10, 8.2-14-8.2-15.

74. Ibid., 8.1–18, 8.1–27.

75. G. Warren Heenan, "The Economic Effect of Rapid Transit on Real Estate Development in Toronto," *The Appraisal Journal* (Apr. 1968); Robert L. Knight and Lisa L. Trygg, "Evidence of Land Use Impacts of Rapid Transit Systems," *Transportation* 6 (1977): 233; Carol Kovach, "On Conducting an 'Impact' Study of a Rapid Transit Facility—The Case of Toronto," meeting preprint, presented to Joint Transportation Engineering Meeting (Montreal: July 15–19, 1974).

76. Heenan, "The Economic Effect of Rapid Transit," Marcou, O'Leary & Associates, *Transit Impact on Land Development* (Washington, D.C.: 1971); Kovach, "On Conducting an 'Impact' Study,"

77. Knight and Trygg, "Evidence of Land Use Impacts," 235.

78. City of Toronto Planning Board, *Air Rights and Land Use Adjacent to the Bloor–Danforth Subway* (June 1963), and City of Toronto Planning Board, *Report in a Proposal to Rezone Lands Along the East/West Subway Route to Permit High Rise Apartments* (Oct. 12, 1971).

79. Knight and Trygg, "Evidence of Land Use Impacts," 235.

80. Webber, *The BART Experience*, 5.

81. Webber, *The BART Experience*, 13, 14.

82. Douglass B. Lee Jr. and D. F. Wiech, "Market Street Study," *BART Impact Studies* (Berkeley: Institute of Urban and Regional Development, University of California, 1972).

83. Martin C. Libicki, "Land Use Impacts of Major Transit Improvements," *Urban Analysis Program, Office of the Secretary of Transportation* (Washington, D.C.: Mar. 1975).

84. Gruen Associates, Inc., "Indirect Environmental Impacts," TM 244-77; BART Impact Program (Berkeley: Metropolitan Transportation Commission, 1977), Douglass B. Lee Jr., "Impacts of BART on Prices of Single-Family Residences and Commercial Property," *BART Impact Studies* (Berkeley: Institute of Urban and Regional Development, University of California, 1973); William R. Wells, *Rapid Transit Impact on Suburban Planning and Development, Perspective and Case Study* (Department of Industrial Engineering, Stanford University, 1973); Knight and Trygg, "Evidence of Land Use Impacts," 236.

85. Webber, *The BART Experience*, 15; Wells, *Rapid Transit Impact*; John Landis and Robert Cervero, "Middle Age Sprawl: BART and Urban Development," *Access* no. 14 (summer 1999): 4, 6.

86. Landis and Cervero, "Middle Age Sprawl," 5, 9, 10.

87. Knight and Trygg, "Evidence of Land Use Impacts," 240; William L. Garrison, "Urban Transportation and Land Use," in *Public Transportation: Planning, Operations and Management*, ed. George E. Gray and Lester E. Hoel (Englewood Cliffs, N.J.: Prentice-Hall, 1979), 525.

88. Webber, *The BART Experience*, 16.

89. SCAG, Transcript (Sept. 1, 1983), 9.

90. G. B. Arrington, "Beyond the Field of Dreams: Light Rail and Growth Management in Portland," *Seventh National Conference on Light Rail Transit*, vol. 1 (Washington, D.C.: Transportation Research Board, 1995).

91. Steve Buckstein, *Top Ten Light Rail Myths*, Cascade Policy Institute, (January 30, 1996) http://www.cascadepolicy.org/ transit/topten.htm; John A. Charles, "The mythical world of Transit Oriented Development," Cascade Policy Institute, *Policy Perspective No. 1019* (Portland, Ore.: 2001); and Michael J. Cunneen, *MAX, Lies and Videotapes: How "Oregonians for Roads and Rails" is Lying to Oregonians to Sell the Measure 32 Gravy Train* (Portland, Ore, n.d.).

92. Peter Gordon and Harry W. Richardson, "Compactness or Sprawl: America's Future vs. The Present," paper presented at conference of the Association of Collegiate Schools of Planning, Atlanta, Nov. 2000, 17; Hartman, "The Round," available at http://www.hartmanco.com/about/awards/round.htm.

93. *The Oregonian*, Apr 28, 1999.

94. City of Beaverton, Mayor's Report (June/July 2001), http://www.ci.beaverton.or.us/mayor/welcome_june_july01.html

95. Robert Cervero, "Light Rail Transit and Urban Development," *Journal of the American Planning Association* (spring 1984): 146.

96. David E. Wohlwill, *Development Along a Busway. A Case Study of Development Along the Martin Luther King, Jr. East Busway in Pittsburgh*, PAT (Pittsburgh, Pa: June 1996); MTA, *Draft 2001 Long Range Transportation*, 2–36.

97. Jonathan E. D. Richmond, "Simplicity and Complexity in Design for Transportation Systems and Urban Forms," *Journal of Planning Education and Research* 17, no. 3 (spring 1998): 220–30.

98. Altshuler, *The Urban Transportation System*, 435.

99. Thomas F. Larwin, "1988 Transportation Achievement Award (Operations). San Diego's Light Rail System: A Success Story," *ITE Journal* (Jan. 1989): 19.

100. MTDB, *East Urban Corridor Alternatives Analysis/Environmental Impact Statement. Capital Cost Estimates. Technical Report*, submitted to UMTA, (Sept. 24, 1984), 7; MTDB, *The San Diego Trolley. South Line and East Line Summary* (Sept. 1995). Updated with letter/phone call with MTDB, June 1998.

101. NFTA, *Light Rail Rapid Transit System*, memorandum (1987); Pickrell, *Urban Rail Transit*, 33; Bonsall, "A Bus For All Seasons."

102. Tri-Met, *Portland's Metropolitan Area Express: Forecast Versus Actual Ridership and Costs. Public Services Division* (June 1990) unnumbered cover and three text sheets; Bob Post, interoffice memorandum to R. Higbee, D. Porter, and G. Arrington on the subject of "U.S. DOT Banfield Visit," Tri-Met (Sept. 16, 1988), 1.

103. Tri-Met, *Staff Recommendation to the Tri-Met Board of Directors on the Banfield Transitway Project* (Aug. 1978).

104. Tri-Met, *Portland's Metropolitan Area Express*; Pickrell, *Urban Rail Transit Projects*, 33.

105. See *Los Angeles Times*, Nov. 11, 1984, for example.

106. MTA, *Facts At A Glance* (Feb. 1996).

107. Caltrans, *Long Beach to Los Angeles Light Rail Transit Feasibility Study* (Los Angeles: Caltrans District 7-Public Transportation Branch, Oct. 1981); PBQD, *Summary Report, The Los Angeles–Long Beach Light Rail Project and Evaluation of Other Rapid Transit Opportunities* (Santa Ana, Calif.: 1982), 10, 18; LACTC, transcript of commission meeting (Mar. 24, 1982); *Long Beach Press-Telegram*, Apr. 15, 1982.

108. Rich Connell, *Los Angeles Times*, October 29, 1985; *Los Angeles Times*, October 16, 1983; LACTC, *The Long Beach–Los Angeles Rail Transit Project, Draft Environmental Impact Report* (May 1984), I–80.

109. McSpedon, "Building Light Rail Transit in Existing Rail Corridors," 428–30.

110. Ibid., 430–31.

111. Ibid., 432.

112. Ibid., 437–39.

113. RTD; MTA, *Facts At A Glance*.

114. McSpedon, "Building Light Rail Transit in Existing Rail Corridors," 441.

115. MTA, *Executive Summary Rail Program Status as of December 2000*; UMTA/RTD, *Los Angeles Rail*, S-4-2.

116. Email from Marc Littman, Director of Public Relations, Los Angeles County Metropolitan Transportation Authority, July 21, 2003.

117. Phone call from Ed Scannell, MTA Public Relations, July 24, 2003.

118. MTDB, *East Urban Corridor, San Diego Region, Final Environmental Impact Statement*, San Diego Metropolitan Transit Development Board and San Francisco: Urban Mass Transportation Administration, U.S. Department of Transportation (1986), 2–37.

119. Bonsall, "A Bus For All Seasons," 11.

120. Korve Engineering et al., *Mid-City/Westside Transit Corridor Re-Evaluation/Major Investment Study*, submitted to MTA (Feb. 24, 2000), 5-2.

121. UMTA, *Memorandum for Approval of Capital Grant PA-03-0095, Amendment 14* (Mar. 1987), 2

122. Pickrell, *Urban Rail Transit Projects*, 33; Allen D. Biehler, "Exclusive Busway Versus Light Rail: A Comparison of New Fixed Guideways," *Light Rail Transit: New System Successes at Affordable Prices. Transport Research Board Special Report 221* (1988): 89–98.

123. RTD.

124. MTA, *Draft 2001 Long Range Transportation Plan*, 2–43; MTA.

125. *Los Angeles Times*, July 27, 2001; MTA, *Draft 2001 Long Range Transportation Plan*, 2–20.

126. *Los Angeles Times*, Oct. 20, 1985; Tri-Met, *Staff Recommendation*, 12; RT (Sacramento Regional Transit District), *Sacramento Light Rail Overview* (1987).

127. Richmond, *New Rail Transit Investments*, and Richmond, "A Whole-System Approach".

128. Mass Transit Administration, Maryland Department of Transportation, *Environmental Effects Report for the Central Light Rail Line* (1988), 2–11.

129. Alan M. Voorhees & Associates, Metro for Buffalo, *Transit Alternatives for the Buffalo–Amherst Corridor, Technical Report* (June 1976).

130. NTD.

131. Data supplied to author by NFTA.

132. NTD.

133. Bi-State Development Agency, *Quarterly Performance Indicators, Period Ending June, 1996* (St. Louis: 1996).

134. José A. Gómez-Ibáñez, "A Dark Side to Light Rail?" *Journal of the American Planning Association* (summer 1985): 342.

135. Mundle & Associates, Inc., *Review of Transportes Amarillo y Rosa*, prepared for the San Diego Metropolitan Transit Development Board (1986).

136. MTDB data.

137. DeLeuw Cather, *Comparative Analysis Study of Alternative Transit Systems South Hills Corridor*, prepared for the PAT, Pa. (Mar. 1976), II-3, II-5, II-26.

138. Because of errors in data collection/allocation, the MTA advised, in a November 1, 1996 memorandum to Supervisor Antonovich's office, that budgeted amounts represent the best indicators of actual costs for fiscal year 1995 (results shown for that fiscal year are according to a September 13, 1996, MTA response to a data request from Supervisor Antonovich's office, adjusted according to the information in the November 1, 1996, memorandum).

139. LACTC, *Final Environmental Impact Report*, III-86; Padron & Associates, *Blue Line FY 1991 Budget (Revised)*, letter from Manuel Padron to Sharon Neely, LACTC (Atlanta, Ga: June 26, 1990); RTD, *Proposed Annual Budget, Fiscal Year 1991* (June 7, 1990).

140. Calculated from national transit database data for the fiscal year 1995 included with MTA (Los Angeles County Metropolitan Transportation Authority), response to data request from Supervisor Antonovich's office (Sept. 13, 1996).

141. Korve Engineering, *Mid-City/Westside Transit Corridor Re-Evaluation*, 6-6.

142. Data based on NTD submissions. FY 2000 ridership data according to this methodology is more favorable to rail than data from the MTA's monthly methodology.

143. MTA.

144. Wachs, "Consumer Attitudes," 336.

145. RTD.

146. Alan F. Pegg, Tom Rubin, and Larry Schlegel, memorandum to SCRTD Board of Directors on "Fiscal Year 1990 Third Quarter Operating Budget Forecast and Year-to-Year Changes of Operating Revenues and Expenses" (Los Angeles: Southern California Rapid Transit District, Apr. 27, 1990).

147. Ibid., 8.

148. *Labor/Community Strategy Center et al. v. MTA et al.*, United States District Court for the Central district of California, Case No. CV 94-5936 TJH (MCx) (1994).

149. *U.S. District Court, Consent Decree in Labor/Community Strategy Center et al v. Los Angeles County Metropolitan Transportation Authority et al.* United States District Court for the Central District of California, Case No. CV 94-5936 TJH (MCx) (Oct. 29, 1996).

150. Churchman, *The Systems Approach*; Churchman, *Thought and Wisdom*.

Chapter 4

1. Guy Benveniste, *The Politics of Expertise*, 2d ed. (San Francisco: Boyd & Fraser, 1977), 43.

2. Martin Wachs, "Ethical Dilemmas in Forecasting for Public Policy," in *Ethics in Planning*, ed. Martin Wachs (New Brunswick, N.J.: Center for Urban Policy Research, 1985), 246.

3. John M. Mulvey, "Computer Modelling for State Use: Implications for Professional Responsibility" (paper given at meeting of the American Academy for the Advancement of Science, Detroit, Mich., May 1983).

4. Schön, *The Reflective Practitioner*, 44.

5. Richard J. Chorley and Peter Haggett, *Models in Geography* (London: Methuen, 1967), 23.

6. Ida R. Hoos, *Systems Analysis in Social Policy: A Critical Review*, Research Monograph no. 19 (London: Institute of Economic Affairs, 1969), 23.

7. See John F. Kain, "Deception in Dallas: Strategic Misrepresentation in Rail Transit Promotion and Evaluation," *Journal of the American Planning Association* (spring 1990).

8. Wachs, "Ethical Dilemmas in Forecasting for Public Policy," 257.

9. Wachs, "Ethical Dilemmas in Forecasting for Public Policy," 253.

10. Garry D. Brewer, *Politicians, Bureaucrats and the Consultant* (New York: Basic Books, 1973), 141.

11. Ibid., 143.

12. Ibid., 147.

13. William Ascher, *Forecasting: An Appraisal for Policy-Makers and Planners* (Baltimore: Johns Hopkins University Press, 1978), 202.

14. Jacques Ellul, *The Technological Society* (New York: Knopf, 1964), 134.

15. Joseph Weizenbaum, *Computer Power and Human Reason: From Judgment to Calculation* (San Francisco: W. H. Freeman, 1976), 239.

16. Ibid., 238.

17. Edward E. Leamer, "Let's Take the Con out of Econometrics," *American Economic Review* (Mar. 1983), 31–43.

18. Ibid., 41.

19. Ibid., 41.

20. Ibid., 42.

21. Franklin M. Fisher, "Multiple Regression in Legal Proceedings," *Columbia Law Review*, vol. 80, no. 3 (Apr. 1980), 702.

22. See Alasdair MacIntyre, "Utilitarianism and the Presuppositions of Cost-Benefit Analysis: An Essay on the Relevance of Moral Philosophy to the Theory of Bureaucracy," and Steven Kelman, "Cost-Benefit Analysis and Environmental, Safety, and Health Regulation: Ethical and Philosophical Considerations," in *Ethics in Planning*, ed. Martin Wachs (New Brunswick, N.J.: Center for Urban Policy Research, 1985), 216–45 for penetrating discussions of the assumptions of utilitarianism.

23. Robert M. Young, "Why are Figures so Significant? The Role and the Critique of Quantification," in *Demystifying Social Statistics*, ed. John Irvine, Ian Miles, and Jeff Evans (London: Pluto Press, 1979), 68.

24. Ibid., 69.

25. Peter R. Stopher and Arnim H. Meyburg, *Urban Transportation Modeling and Planning* (Lexington, Mass.: Lexington Books, 1975), 60.

26. UMTA, *A Detailed Description of UMTA's System for Rating Proposed Major Transit Investments* (Washington, D.C.: May 1984), 1.

27. Ibid., 4.

28. Stopher and Meyburg, *Urban Transportation Modeling and Planning*, 60.

29. Ibid., 60.

30. Ibid., 18.

31. Caltrans, *Long Beach to Los Angeles Light Rail Transit Feasibility Study* (Los Angeles: Caltrans District 7-Public Transportation Branch, Oct. 1981), 32.

32. Ibid., 33.

33. PBQD, "Investigate Potential Ridership for Baseline Los Angeles–Long Beach Light Rail Project," Working Paper no. 7, part of *The Los Angeles to Long Beach Light Rail Project and Evaluation of Other Rapid Transit Opportunities Study*. Prepared for Los Angeles County Transportation Commission (Los Angeles, 1982).

34. Ibid., 2–3.

35. SCAG, *Los Angeles–Long Beach Light Rail Transit Project, Patronage Estimation and Impacts* (Mar. 1984).

36. SCAG, *Regional Transportation Modeling System, Description and Assumptions of Travel Demand Models* (Los Angeles, Feb. 1984), 170.

37. Ibid., 172.

38. Stopher and Meyburg, *Urban Transportation Modeling and Planning*, 61.

39. SCAG, *Regional Transportation Modeling System, Description and Assumptions of Travel Demand Models* (Los Angeles, Feb. 1984), 174.

40. Peter Hall, *Great Planning Disasters*, American ed. (Berkeley: University of California Press, 1982), 6.

41. Andrew M. Hamer, *The Selling of Rail Rapid Transit* (Lexington, Mass.: Lexington Books, 1976), 248.

42. SCAG, *Regional Transportation Modeling System, Description and Assumptions of Travel Demand Models*, 174.

43. SCAG, *Los Angeles–Long Beach Light Rail Transit Project, Patronage Estimation and Impacts*, 2-15.

44. Ibid., 2-15.

45. Ibid., 2-15.

46. SCAG, *Regional Transportation Modeling System, Description and Assumptions of Travel Demand Models*, 174.

47. Stopher and Meyburg, *Urban Transportation Modeling and Planning*, 109.

48. SCAG, *Los Angeles–Long Beach Light Rail Transit Project, Patronage Estimation and Impacts*, 2-6.

49. Ibid., 2-7.

50. SCAG, *Regional Transportation Modeling System, Description and Assumptions of Travel Demand Models*, 180.

51. See discussion in Stopher and Meyburg, *Urban Transportation Modeling and Planning*.

52. See George Lakoff and Mark Johnson, *Metaphors We Live By* (Chicago: University of Chicago Press, 1980).

53. *Los Angeles Times*, Oct. 20, 1985

54. Stopher and Meyburg, *Urban Transportation Modeling and Planning*, 175.

55. SCAG, *Regional Transportation Modeling System, Description and Assumptions of Travel Demand Models*, 181.

56. Ibid., 181.

57. Stopher and Meyburg, *Urban Transportation Modeling and Planning*, 61.

58. Ibid., 61.

59. Ibid., 62.

60. SCAG, *Los Angeles–Long Beach Light Rail Transit Project, Patronage Estimation and Impacts*, 2-5.

61. Ibid., 2-5.

62. Ibid., 2-5–2-6.

63. Ibid., 2-14.

64. Ibid., 2-11.

65. Deloitte/Kellogg Joint Venture, *Memorandum Re: Start-up Review Report No. 2* from Rodney J. Dawson to Ed McSpedon, LACTC (Los Angeles, Sept. 15, 1989).

66. SCAG, *Los Angeles–Long Beach Light Rail Transit Project, Patronage Estimation and Impacts*, 6-2.

67. Ibid., 6-1.

68. Ibid., 6-2.

69. Ibid., 2-24.

70. Don H. Pickrell, *Urban Rail Transit Projects: Forecast Versus Actual Ridership and Costs. Final Report*, prepared for Office of Grants Management, UMTA. Report DOT-T-91-04 (Oct. 1990), 74.

71. LACTC, *The Long Beach–Los Angeles Rail Transit Project, Draft Environmental Impact Report* (May 1984).

72. Michel Godet, *The Crisis in Forecasting and the Emergence of the "Prospective" Approach: With Case Studies in Energy and Air Transport*, translated from the French by J. D. Pearse and Harry K. Lennon (New York: Pergamon Policy Studies 15, 1979), 15.

73. John F. Kain, "Deception in Dallas: Strategic Misrepresentation in Rail Transit Promotion and Evaluation," *Journal of the American Planning Association* (spring 1990): 184–96.

74. Ridership according to National Transit Database report. Ridership according to alternative MTA monthly data methodology was 57, 584 average weekday riders.

75. See Jonathan E. D. Richmond, *New Rail Transit Investments*, and Jonathan E. D. Richmond, "A Whole-System Approach," 131–79.

Chapter 5

1. Stanislav Andreski, *Social Science as Sorcery* (London: Andre Deutsch, 1972), 127.

2. Alan M. Voorhees & Associates, *Metro for Buffalo. Transit Alternatives for the Buffalo–Amherst Corridor, Technical Report* (June 1976), 2.

3. NFTA.

4. Tri-Met, *East Side Transit Operations* (Dec. 1977), 37; Tri-Met, *Staff Recommendation to the Tri-Met Board of Directors on the Banfield Transitway Project* (Aug. 1978), 23.

5. UMTA/SACOG, *Draft Alternatives Analysis/Environmental Impact Statement/Environmental Impact Report on Prospective Interstate Substitution Transportation Improvements in Northeast Sacramento*, California (1981), 4-23; See Jonathan E. D. Richmond, "A Whole-System Approach," 155–58.

6. MTA, response to data request from Supervisor Antonovich's office (Sept. 13, 1996).

7. MTA, *Facts At A Glance* (Feb. 1996).

8. Paul Taylor, interview, Aug. 14, 1985.

9. RTD, *Short Range Transit Plan: Guideway Plan, Fiscal Years 1991 through 1995* (1990), II-14.

10. Edward McSpedon, letter to Jonathan Richmond (July 3, 1990).

11. *Los Angeles Times*, Jan. 21, 1991

12. SCAG, *Los Angeles–Long Beach Light Rail Transit Project, Patronage Estimation and Impacts* (Mar. 1984); Paul Taylor, interview, Aug. 14, 1985.

13. SCAG, "Re-Training the Region: Rail Systems Prepare to Make a Comeback in L.A." *Crossroads, The Southern California Transportation Magazine* (1985): 12.

14. Ibid.

15. Churchman, *Thought and Wisdom*, 132.

16. Edward E. Leamer, "Let's Take the Con out of Econometrics," *American Economic Review* (Mar. 1983): 31–43.

17. Andrew M. Hamer, *The Selling of Rail Rapid Transit* (Lexington, Mass.: Lexington Books, 1976).

18. Don H. Pickrell, *Urban Rail Transit Projects: Forecast Versus Actual Ridership and Costs, Final Report*, Prepared for Office of Grants Management, UMTA, Report DOT-T-91-04 (Oct. 1990), 5.

19. A possibility Pickrell acknowledges in his later journal article: "A Desire Names Streetcar: Fantasy and Fact in Rail Transit Planning," *Journal of the American Planning Association*, vol. 58, no. 2 (spring, 1992): 159.

20. Altshuler, *The Urban Transportation System*, ix.

Chapter 6

1. Brodsly, *L.A. Freeway*, 84; Robert M. Fogelson, *The Fragmented Metropolis*, 132.

2. Brodsly, *L.A. Freeway*, 84–85.

3. Los Angeles Traffic Commission, *A Major Traffic Street Plan for Los Angeles*, prepared by Frederick Law Olmsted Jr., Harland Bartholomew, and Charles Henry Cheney (Los Angeles, 1924).

4. Brodsly, *L.A. Freeway*, 88.

5. Mark S. Foster, "The Model-T, The Hard Sell, and Los Angeles' Urban Growth: The Decentralization of Los Angeles during the 1920s," *Pacific Historical Review* 44 (Nov. 1975): 165.

6. Los Angeles County Regional Planning Commission, *A Comprehensive Report on the Regional Plan of Highways: Section 4, Long Beach–Redondo Area* (Los Angeles, 1931), 42.

7. Fogelson, *The Fragmented Metropolis*, 143–46; Sir Peter Hall, *Cities in Civilization* (New York: Pantheon Books, 1998), 816.

8. Hall, *Cities in Civilization*, 816; see also Scott L. Bottles, *Los Angeles and the Automobile: The Making of the Modern City* (Berkeley and Los Angeles: University of California Press, 1987), 206, 213.

9. Brodsly, *L.A. Freeway*, 112, 97.

10. H. Marshall Goodwin Jr., "From Dry Gulch to Freeway." *Southern California Quarterly* 47 (summer 1965): 93–94.

11. From letter of transmittal, Automobile Club of Southern California (1937).

12. Brodsly, *L.A. Freeway*, 98.

13. Los Angeles County Regional Planning Commission, *Master Plan of Highways* (Los Angeles, 1941), 33.

14. Brodsly, *L.A. Freeway*, 105; Los Angeles County Regional Planning Commission, *Freeways for the Region* (Los Angeles, 1943), 12; Ibid., 26–27.

15. Brodsly, *L.A. Freeway*, 115.

16. Hall, *Cities in Civilization*, 830.

17. Fogelson, *The Fragmented Metropolis*, 185.

18. Brodsly, *L.A. Freeway*, 116.

19. Altshuler, *The Urban Transportation System*, 26, 27, 28–31.

20. Goodwin, "From Dry Gulch to Freeway," 324–46.

21. Brodsly, *L.A. Freeway*, 136.

22. *Electric Railway Journal* 66 (July–Dec. 1925): 1087.

23. Laurence R. Veysey, *The Pacific Electric Railway Company, 1910–1953, A Study in the Operations of Economic, Social and Political Forces Upon American Local Transportation.* (Los Angeles, 1953), 113. Available through Los Angeles County Metropolitan Transportation Authority Library.

24. Ibid., 328.

25. Kelker, De Leuw & Co., *Report and Recommendations on a Comprehensive Rapid Transit Plan for the City and County of Los Angeles* (Chicago, 1925), 1.

26. Ibid., 3, 4.

27. Ibid., 5.

28. Ibid., 6, 7.

29. Ibid., 25.

30. City Club of Los Angeles, *Report on Rapid Transit*, supplement to *City Club Bulletin* 30 (Jan. 1926), 4.

31. Brodsly, *L.A. Freeway*, 154.

32. *Los Angeles Times*, Feb. 1 1926.

33. Brodsly, *L.A. Freeway*, 155.

34. Donald M. Baker, *A Rapid Transit System for Los Angeles, California* (Los Angeles, 1933).

35. Veysey, *The Pacific Electric Railway Company*, 171.

36. Los Angeles Transportation Engineering Board, *A Transit Program for the Los Angeles Metropolitan Area* (Los Angeles, 1939).

37. RTAG, *Rail Rapid Transit Now!, LIVE where you like, WORK where you please, An Immediate Program by The Rapid Transit Action Group*, Los Angeles Chamber of Commerce-Coordinator (Feb., 1948). Sy Adler, *Understanding the Dynamics of Innovation in Urban Transit, Center for Transit Research and Management Development*, Portland State University, Portland, Ore. U.S. DOT report UMTA-OR-11-003-86-1 (Washington D.C.: U.S. Department of Transportation, 1986), 10.

38. RTAG, *Rail Rapid Transit Now!* 1

39. Ibid., 5, 11.

40. Adler, *Understanding the Dynamics of Innovation in Urban Transit*, 11–12, and Adler, "The Transformation of the Pacific Electric Railway," 72.

41. See Adler, *Understanding the Dynamics of Innovation in Urban Transit*, 13–14, and Adler, "The Transformation of the Pacific Electric Railway," 72–74.

42. *Santa Monica Evening Outlook*, April 18, 1949.

43. California Assembly Interim Committee on Public Utilities and Corporations, *A Preliminary Report on Rapid Transit for the Los Angeles Area, California Legislature* (Sacramento, 1950); Veysey, *The Pacific Electric Railway Company*, 273.

44. Los Angeles Central City Committee and the Los Angeles City Planning Department, *Los Angeles Centropolis 1980: Economic Survey* (Los Angeles, 1960), 4–5.

45. Ibid., 5; Ibid., 35–36.

46. Altshuler, *The Urban Transportation System*, 48.

47. Michael N. Danielson, *Federal-Metropolitan Politics and the Commuter Crisis* (New York: Columbia University Press, 1965), 27.

48. Symes/Dilworth Alliance, Resolution, reprinted in *Hearings on S. 345, Subcommittee on Housing, Senate Committee on Banking and Currency* (Washington, D.C.: U.S. Government Printing Office, 1961), 78–79.

49. James A. Dunn Jr., *Driving Forces: The Automobile, Its enemies, and the Politics of Mobility* (Washington, D.C.: Brookings Institution Press, 1998), 89.

50. Alan Altshuler, "The Politics of Urban Mass Transportation," paper presented at the Annual Meeting of the American Political Science Association (New York City, Sept. 4–7, 1963), 28–29.

51. David W. Jones Jr., *Urban Transit Policy: An Economic and Political History* (Englewood Cliffs, N.J.: Prentice-Hall, 1985), 117.

52. Ibid., 118.

53. Danielson, *Federal-Metropolitan Politics and the Commuter Crisis*, 138–39.

54. Jones, *Urban Transit Policy*, 120–21.

55. IPA, *Urban Transportation and Public Policy*, Report prepared for U.S. Department of Commerce and the Housing and Home Finance Administration (New York, 1961). Also available as Lyle C. Fitch, *Urban Transportation and Public Policy* (San Francisco: Chandler, 1964).

56. Ibid., VII-2.

57. "Message on Transportation" (1962).

58. Altshuler, "The Politics of Urban Mass Transportation," 42.

59. Danielson, *Federal-Metropolitan Politics and the Commuter Crisis*, 177.

60. Peter Hall, *Great Planning Disasters*, 109.

61. Stephen Zwerling, *Mass Transit and the Politics of Technology: A Study of BART and the San Francisco Bay Area* (New York: Praeger Publishers, 1974), 21.

62. Frederick M. Wirt, *Power in the City: Decision Making in San Francisco* (Berkeley: University of California Press, 1974), 191.

63. Lee Shipnuck and Dan Feshbach, "Bay Area Council: Regional Powerhouse." *Pacific Research and World Empire Telegram* 4 (Nov./Dec. 1972): 6.

64. Richard Allen Sundeen Jr., "The San Francisco Bay Area Council: An Analysis of A Non-Governmental Metropolitan Organization" (master's thesis, University of California, Berkeley, 1963), 71, 90–100; John C. Bollens, *The Problem of Government in the San Francisco Bay Region* (Berkeley: Bureau of Public Administration, University of California, 1948), 115.

65. Zwerling, *Mass Transit and the Politics of Technology*, 22.

66. PBHM, *Regional Rapid Transit. Report to the San Francisco Bay Area Rapid Transit Commission* (Jan. 5, 1956), 1.

67. Zwerling, *Mass Transit and the Politics of Technology*, 24.

68. PBHM, *Regional Rapid Transit*, 3.

69. PBTB, *The Composite Report: Bay Area Rapid Transit* (San Francisco, 1962), 82–83.

70. Robert J. Bazell, "Rapid transit: A real alternative to the auto for the Bay Area?" *Science* 171 (Mar. 19, 1971): 1126.

71. J. Allen Whitt, *Urban Elites and Mass Transportation: The Dialectics of Power* (Princeton: Princeton University Press, 1982), 59–60.

72. Metropolitan Transportation Commission, *A History of the Key Decisions in the Development of Bay Area Rapid Transit* (San Francisco: Jossey-Bass, 1973), 42.

73. Hall, *Great Planning Disasters*, 124. See also Metropolitan Transportation Commission, *A History of the Key Decisions in the Development of Bay Area Rapid Transit*, 11–29.

74. Albert M. Liston, "Regional Rapid Transit Development in the San Francisco Bay Area" (master's thesis, Sacramento State College, 1970), 45.

75. Ebasco Services, *Bay Area Rapid Transit Economic Study for the San Francisco Bay Area Rapid Transit District* (San Francisco: June 1961), 25–26.

76. PBTB, *The Composite Report*, 80.

77. K. M. Fong, "Rapid Transit-Decision-Making: The Income Distribution Impact of San Francisco Bay Area Rapid Transit" (bachelor's thesis, Harvard University, 1976).

78. Hall, *Great Planning Disasters*, 127–28.

79. Andrew M. Hamer, *The Selling of Rail Rapid Transit*, 179, 180.

80. Adler, *Understanding the Dynamics of Innovation in Urban Transit*, 44.

81. SCAG, *Status of Regional Transportation Planning and Coordination in Southern California* (1968).

82. Hamer, *The Selling of Rail Rapid Transit*, 180.

83. DMJM, *Planning and Economic Considerations Affecting Transportation in the Los Angeles Region* (Los Angeles: Southern California Rapid Transit District, 1965), I-4, I-5.

84. RTD, *Planning and Economic Considerations Affecting Transportation in the Los Angeles Region*, DMJM (1965), V-9.

85. Whitt, *Urban Elites and Mass Transportation*, 84.

86. Citizens Advisory Council on Public Transportation, *Improving Public Transportation in Los Angeles* (Los Angeles: 1967), 94, 20–21, 24.

87. Whitt, *Urban Elites and Mass Transportation*, 86.

88. *Los Angeles Times*, Aug. 8, 1968

89. Whitt, *Urban Elites and Mass Transportation*, 90.

90. Hamer, *The Selling of Rail Rapid Transit*, 192.

91. Debra Kagan, David Newman, Stan Sander, and Bonnie Woltoncroft, *Mass Transit in Los Angeles: An Analysis* (Claremont, Calif.: Program in Public Policy Studies of the Claremont Colleges, June, 1972), 24.

92. RTD, *Rapid Transit for Los Angeles, Summary Report of Consultants' Recommendations*, prepared for the RTD and The Citizens of The Los Angeles Region (July 1973).

93. Ibid., 30.

94. Hamer, *The Selling of Rail Rapid Transit*, 198–99.

95. Ibid., 200.

96. Kaiser Engineering and DMJM, *Technical Study of Alternative Transit Corridors and Systems, Plan Refinement, Volume I: Engineering and Planning, Phase III.* (Los Angeles: Southern California Rapid Transit District, March, 1974), III-12, III-14.

97. Hamer, *The Selling of Rail Rapid Transit*, 202.

98. SCAG, *Southern California Regional Transit System Plan: An Element of the Southern California Association of Governments Critical Decisions Program.* Los Angeles (July 11, 1974), 25, 28.

99. Whitt, *Urban Elites and Mass Transportation*, 97, 99.

100. *Los Angeles Times*, November 4, 1974.

101. Jones, *Urban Transit Policy*, 126.

102. Ibid., 127.

103. George M. Smerk, *The Federal Role in Urban Mass Transportation.* (Bloomington and Indianapolis: Indiana University Press, 1991), 114, 115.

104. Altshuler, *The Urban Transportation System*, 36.

105. Ibid., 37.

106. James A. Dunn, Jr, *Miles to Go: European and American Transportation Policies* (Cambridge: MIT Press, 1981), 83.

107. Altshuler, *The Urban Transportation System*, 38.

108. Jones, *Urban Transit Policy*, 127.

109. Kagan et al. *Mass Transit in Los Angeles*, 18.

110. Senate Committee on Public Utilities, Transit, and Energy, Southern California Rapid Transit Subcommittee, *Hearings*, California Legislature, July 25, 1975.

111. *Los Angeles Times*, Aug. 14, 1975.

112. Adler, *Understanding the Dynamics of Innovation in Urban Transit*, 53

113. *Reader*, March 14, 1980.

114. Gina Shaffer, "Warding Off Opposition," *Valley Trader*, July 11–17, 1980.

115. Baxter Ward, *The Sunset Coast Line: Route of the New Red Cars*, prepared through office of Supervisor Baxter Ward by county departments and others (Los Angeles, 1976), 6, 11.

116. Peter L. Shaw, Keith G. Baker, Mel D. Powell, and Renee Simon, *Los Angeles County Transportation Commission Public Policy Impact Study (September, 1978–August, 1979)*, Final Report (Long Beach: Institute for Transportation Policy and Planning, Bureau of Governmental Research and Services, Center for Public Policy and Administration, California State University. U.S. Department of Transportation, UMTA, available through National Technical Information Service, Springfield, Va. Nov. 1979), 95.

117. Ibid., 46.

118. Ibid., 1.

119. Ibid., 46.

120. LACTC, *Staff Budget Memorandum* (Apr. 26, 1978).

121. Ibid., AB 1246, Sec. 130001, (c).

122. Ibid., Sec. 130250.

123. Ibid., Sec. 130257.

124. Peter L. Shaw and Renee Simon, *Los Angeles County Transportation Commission Public Policy Impact Study (September, 1979–August, 1980)*, Final Report (Long Beach: Institute for Transportation Policy and Planning, Bureau of Governmental Research and Services, Center for Public Policy and Administration, California State University. U.S. Department of Transportation, UMTA, available through National Technical Information Service, Springfield, Virginia, Apr. 1981), 22.

125. Cited in Shaw et al, *Los Angeles County Transportation Commission Public Policy Impact Study* (September, 1978–August, 1979), 54.

126. Caltrans, *Freeway Transit Element of the Regional Transit Development Plan for Los Angeles County*, District 7, Transit Branch (Aug. 1978).

127. LACTC, *Minutes* (Jan. 10, 1979), 2.

128. Baxter Ward, *Sunset, Ltd.* (Los Angeles: Board of Supervisors, County of Los Angeles, Office of Supervisor Baxter Ward, 1978).

129. LACTC, *Transcript* of LACTC Meeting (July 11, 1979), 4.

130. Shaw et al, *Los Angeles County Transportation Commission Public Policy Impact Study* (September, 1978–August, 1979), 69.

131. LACTC, *Minutes* (Aug. 8, 1979), 3.

132. Shaw et al. *Los Angeles County Transportation Commission Public Policy Impact Study* (September, 1978–August, 1979), 70, 110.

133. LACTC, *Staff Report on Transit System and Financing Alternatives for Los Angeles County* (Jan. 23, 1980), 53–54.

134. LACTC, *Minutes*, Commission Meeting (Mar. 26, 1980).

135. *Reader*, 14 Mar. 1980.

136. LACTC, *Minutes*, Commission Meeting (Aug. 6, 1980).

137. Adler, *Understanding the Dynamics of Innovation in Urban Transit*, 58.

138. LACTC, *Minutes*, Commission Meeting (Aug. 6, 1980).

139. Reported in LACTC, *Transcript of Commission Meeting* (Aug. 20, 1980), 4.

140. Ibid., 4–5.

141. Ibid., 9.

142. Ibid., 10.

143. Ibid., 17.

144. *Los Angeles Times*, Sept. 9, 1980.

145. LACTC, *Minutes*, Commission Meeting (Aug. 6, 1980), 17.

146. Ibid., 18.

147. Ibid., 21.

148. Interview, *Los Angeles Times*, Aug. 21, 1980.

149. LACTC, *Minutes*, Commission Meeting, (Aug. 20, 1980), 22.

150. Ibid., 23, 24, 26.

151. Ibid., 27.

152. Ibid., 28.

153. Ibid., 28–29, 31.

154. Ibid., 36, 40.

155. *Century City News*, Sept. 24, 1980.

156. *Valley News*, Sept. 29, 1980.

157. *Culver City News*, Nov. 13, 1980.

158. *Los Angeles Herald-Examiner*, Nov. 21, 1980.

159. *Los Angeles Herald-Examiner*, Dec. 24, 1980.

160. Quoted in Ray Hebert, "Prop A: From Bleak to Bright in LA," *Mass Transit* (Sept. 1982): 38.

161. California Assembly Committee on Transportation, "Light Rail Transit in Southern California—Return of the Red Cars," *Transcript of Proceedings*, Public Hearing, California Legislature, Long Beach City Hall, Long Beach, Aug. 14, 1981.

162. Ibid., 4, 5.

163. Ibid., 6.

164. Ibid., 7.

165. Ibid., 10.

166. Ibid., 49.

167. Ibid., 62.

168. Ibid., 62.

169. Ibid., 65–66.

170. Ibid., 157.

171. White House 1981, 5.3–5.4.

172. LACTC, *Staff Report on a Rapid Transit Development Strategy for Los Angeles County* (Sept. 9, 1981), 3.

173. LACTC, *Minutes* (Sept. 23, 1981).

174. Assembly Committee on Transportation, *Light Rail Program for Los Angeles County*, California Legislature (1981).

175. LACTC, *Minutes* (Oct. 13, 1981), 3.

176. LACTC, *Minutes* (Oct. 28, 1981).

177. LACTC, *Rapid Transit Development in Los Angeles County: Staff Recommendations* (Mar. 17, 1982).

178. Ibid.

179. LACTC, *Memorandum* from John Zimmerman, Chairman, Ad Hoc Rapid Transit Committee, to LACTC members and alternates on "March 17 Actions of Ad Hoc Rapid Transit Committee" (Mar. 19, 1982), 1.

180. LACTC, *Transcript* of Commission meeting (Mar. 24, 1982), 7–8.

181. Ibid., 8–9.

182. Ibid., 12–13.

183. Ibid., 15.

184. Memo from Burke Roche to Kenneth Hahn (June 12, 1982).

185. *Los Angeles Times*, Mar. 28, 1985.

186. Smerk, *The Federal Role in Urban Mass Transportation*, 216.

187. UMTA/East-West Gateway Coordinating Council, *St. Louis Central/Airport Corridor. St. Louis City and County, Missouri. East St. Louis and St. Clair County, Illinois. Alternatives Analysis & Draft Environmental Impact Statement for Major Transit Capital Investments* (May, 1984), S-12, 38–39.

188. David M. Kennedy, *UMTA and the New Rail Start Policy*, Kennedy School of Government Case Program, Harvard University (1984), 7.

189. Smerk, *The Federal Role in Urban Mass Transportation*, 254–55.

190. UMTA/RTD, *Los Angeles Rail Rapid Transit Project Metro Rail, Final Environmental Impact Statement, December* (1983).

191. Brian D. Taylor and Eugene J. Kim, *The Politics of Rail Planning: A Case Study of the Wilshire Red Line in Los Angeles*, working paper, UCLA Institute of Transportation Studies (Feb. 15, 1999), 11.

192. Ibid., 24; *Los Angeles Times*, Aug. 8, 1986.

193. Taylor and Kim, *The Politics of Rail Planning*, 21.

194. LACTC, *30-Year Integrated Transportation Plan. Report to U.S. Department of Transportation* (Washington D.C.: Federal Transit Administration, 1992), 4–37.

195. MTA, *Transportation for the 21st Century, A Plan for Los Angeles County* (Mar. 1995), 55–59.

196. U.S. District Court, *Consent Decree in Labor/Community Strategy Center el al vs. Los Angeles County Metropolitan Transportation Authority et al.* United States District Court for the Central district of California, Case No. CV 94-5936 TJH (MCx) (Oct. 29, 1996), 3.

197. Ibid.

198. Gordon Linton, letter from U.S. Federal Transit Administrator to Larry Chairman, Chair, MTA,(Jan. 6, 1997), 1–2.

199. MTA, *Metro Rail Red Line Recovery Plan*, Submitted to the U.S. Department of Transportation, Federal Transit Administration (Jan. 15, 1997). Larry Zarian, letter from Chair, Los Angeles County Metropolitan Transportation Authority to Gordon Linton, U.S. Federal Transit Administrator (Jan. 15, 1997).

200. Gordon Linton, letter from U.S. Federal Transit Administrator to Larry Chairman, Chair, Los Angeles County Metropolitan Transportation Authority (Apr. 9, 1997), 1.

201. Ibid., 2–4.

202. Leslie Rogers, letter from Regional Administrator, U.S. Department of Transportation, Federal Transit Administration, Region IX, San Francisco, to Linda Bohlinger, Interim Chief Executive Officer, MTA (1997), 1.

203. MTA, *Restructuring Plan. Analysis and Documentation of the MTA's Financial and Managerial Ability to Complete North Hollywood Rail Construction and Meet the Terms of the Bus Consent Decree* (May 15, 1998), 5–6.

204. Gordon Linton, Statement of Federal Transit Administrator Gordon J. Linton Regarding Acceptance of Recovery Plan submitted by Los Angeles County Metropolitan Transportation Authority (July 2, 1998).

205. MTA/Booz Allen, *Regional Transit Alternatives Analysis, Study Results & Appendix*, Attachment No. 1 to MTA Board Report (Nov. 9, 1998).

206. MTA, *Draft 2001 Long Range Transportation Plan for Los Angeles County* (Feb. 2001), 1–13; Ibid., 1–14.

Chapter 7

1. Graham T. Allison, *Essence of Decision: Explaining the Cuban Missile Crisis* (Boston: Little, Brown & Co., 1971), 10.

2. Edward de Bono, *The Use of Lateral Thinking* (Harmondsworth, Mddx: Penguin Books, 1967), 13.

3. Philip Morrison, "Knowing Where You Are: A First Essay Towards Crossing Critical Barriers," in *Critical Barriers Phenomenon in Elementary Science*, by Maja Apelman, David Hawkins, and Philip Morrison (Grand Forks: Center for Teaching and Learning, University of North Dakota, 1985), 81.

4. Ibid., 82.

5. Ibid., 82–83.

6. Christopher Alexander, "A City is Not a Tree," *Architectural Forum* 122 (1965): Pt. 2, 60.

7. John D. Steinbruner, *The Cybernetic Theory of Decision* (Princeton: Princeton University Press, 1974), 92.

8. Friedrich Nietzsche, *The Will to Power* (New York: Random House, 1969), 14–15.

9. Steinbruner, *The Cybernetic Theory of Decision*, 95–96.

10. de Bono, *The Use of Lateral Thinking*, 71.

11. James R. Pomerantz and Michael Kubovy, "Perceptual Organization: An Overview," in *Perceptual Organization*, ed. Michael Kubovy and James R. Pomerantz (Hillsdale, N.J.: Lawrence Erlbaum Associates, 1981), 438.

12. James R. Pomerantz, "Perceptual Organization in Information Processing," in *Perceptual Organization*, ed. Michael Kubovy and James R. Pomerantz (Hillsdale, N.J.: Lawrence Erlbaum Associates, 1981), 160.

13. Julian E. Hochberg, *Perception*, 2d ed. (Englewood Cliffs, N.J.: Prentice-Hall, 1978), 140.

14. Pomerantz, "Perceptual Organization in Information Processing," 162.

15. Roger N. Shepard, "Psychophysical Complementarity," in *Perceptual Organization*, 285–86.

16. Pomerantz and Kubovy, "Perceptual Organization: An Overview," 437.

17. Martin Wachs and Joseph L. Schofer, "Abstract Values and Concrete Highways," *Traffic Quarterly* 23, no. 1 (Jan. 1969): 138.

18. Paul Ricoeur, *The Philosophy of Paul Ricoeur, An Anthology of His Work*, ed. by Charles E. Reagan and David Stewart (Boston: Beacon Press, 1978), 37.

19. Ludwig Von Bertalanffy, "On the Definition of the Symbol," in *Psychology and the Symbol, An Interdisciplinary Symposium*, ed. by Joseph R. Royce (New York: Random House, 1965), 27.

20. Murray Edelman, *The Symbolic Uses of Politics* (Urbana and Chicago: University of Illinois Press, 1964), 31.

21. Susanne K. Langer, *Philosophy in a New Key: A Study in the Symbolism of Reason, Rite and Art*, 3d ed. (Cambridge: Harvard University Press, 1957).

22. Ibid., 59.

23. Ibid., 60–63.

24. Edward de Bono, *The Mechanism of Mind* (Harmondsworth, Mddx.: Penguin Books, 1969), 112.

25. Frederick J. Hacker, "Psychology and Psychopathology in Symbolism," in *Psychology and the Symbol, An Interdisciplinary Symposium*, ed. Joseph R. Royce (New York: Random House, 1965), 82.

26. Von Bertalanffy, "On the Definition of the Symbol," 42.

27. Philip Wheelwright, *Metaphor and Reality* (Bloomington: Indiana University Press, 1962), 94.

28. Hacker, "Psychology and Psychopathology in Symbolism," 81.

29. Judith Innes de Neufville, "Symbol and Myth in Public Choice, The Case of Land Policy in the United States," *Working Paper* 359 (Berkeley: Institute for Urban & Regional Development, University of California, 1981), 7.

30. Roland Bartel, *Metaphor and Symbols* (Urbana, Il.: National Council of Teachers of English, 1983), 71–72.

31. Hacker, "Psychology and Psychopathology in Symbolism," 80.

32. Von Bertalanffy, "On the Definition of the Symbol," 34.

33. Bartel, *Metaphor and Symbols*, 72.

34. George Lakoff, *Women, Fire, and Dangerous Things. What Categories Reveal about the Mind* (Chicago: University of Chicago Press, 1987), 77.

35. George Lakoff and Mark Johnson, *Metaphors We Live By* (Chicago: University of Chicago Press, 1980), 36.

36. Lakoff, *Women, Fire, and Dangerous Things*, 78.

37. Ibid., 77.

38. Martin Foss, *Symbol and Metaphor in Human Experience* (Princeton: Princeton University Press, 1949), 13.

39. Ibid., 16.

40. Lakoff and Johnson, *Metaphors We Live By*, 36.

41. Ibid., 37.

42. Lakoff, *Women, Fire, and Dangerous Things*, 78–79; Richard Rhodes, "Semantics in Rela-

tional Grammar," *Proceedings of the Thirteenth Annual Meeting of the Chicago Linguistic Society* (Chicago: Chicago Linguistic Society, 1977).

43. Hacker, "Psychology and Psychopathology in Symbolism," 80.

44. Cited in Wheelwright, *Metaphor and Reality*, 95.

45. Murray Edelman, *The Symbolic Uses of Politics* (Urbana and Chicago: University of Illinois Press, 1964), 32.

46. See figure 7-2, from *Newsweek*, July 27 1987, 11.

47. Michael Polanyi and Harry Prosch, *Meaning* (Chicago: University of Chicago Press, 1975), 72–73.

48. Kenneth E. Boulding, *The Image. Knowledge in Life and Society* (Ann Arbor: University of Michigan Press, 1956), 6.

49. Mary Schaldenbrand, "Metaphoric Imagination: Kinship Through Conflict," in *Studies in the Philosophy of Paul Ricoeur*, ed. Charles E. Reagan (Athens: Ohio University Press, 1979), 60.

50. Boulding, *The Image*, 111.

51. Langer, *Philosophy in a New Key*, 145–46.

52. Maurice Merleau-Ponty, *The Primacy of Perception* (Evanston, Ill.: Northwestern University Press, 1964), 60.

53. Jean-Paul Sartre, *The Psychology of Imagination* (London: Methuen, 1972), 8.

54. Ibid., 133.

55. Susanne K. Langer, "On Cassirer's Theory of Language and Myth," in *The Philosophy of Ernst Cassirer*, ed. Paul Arthur Schilpp (Evanston, Ill.: The Library of Living Philosophers, 1949), 386.

56. de Neufville, "Symbol and Myth in Public Choice," 1.

57. Boulding, *The Image*, 84.

58. George A. Miller, "Images and Models, Similes and Metaphors," in *Metaphor and Thought*, ed. Andrew Ortony (Cambridge: Cambridge University Press, 1979), 204.

59. Lakoff and Johnson, *Metaphors We Live By*, 59.

60. Miller, "Images and Models, Similes and Metaphors," 205.

61. Boulding, *The Image*, 53.

62. Sartre, *The Psychology of Imagination*, 177.

63. Langer, "On Cassirer's Theory of Language and Myth," 386; Sartre, *The Psychology of Imagination*, 200.

64. John D. Steinbruner, *The Cybernetic Theory of Decision*, 115.

65. Boulding, *The Image*, 18.

66. Sartre, *The Psychology of Imagination*, 160.

67. de Bono, *The Mechanism of Mind*, 224.

68. Schaldenbrand, "Metaphoric Imagination," 66.

69. Langer, *Philosophy in a New Key*, 139.

70. Ibid., 139.

71. Donald A. Schön, *Displacement of Concepts* (London: Tavistock Publications, 1963), 58–64.

72. Polanyi and Prosch, *Meaning*, 78.

73. Max Black, *Models and Metaphors, Studies in Language and Philosophy* (Ithaca: Cornell University Press, 1962), 40.

74. Lakoff and Johnson, *Metaphors We Live By*, 44–45.

75. Schön, *Displacement of Concepts*, 60; Ted Peters, "Metaphor and the Horizon of the Unsaid," *Philosophy and Phenomenological Research* 38 (Mar. 1978): 356.

76. Lakoff and Johnson, *Metaphors We Live By*, 14–21.

77. Ibid., 20.

78. Paul Ricoeur, *The Rule of Metaphor, Multi-disciplinary Studies in the Creation of Meaning in Language* (Toronto: University of Toronto Press, 1977), 252.

79. Ricoeur, *The Philosophy of Paul Ricoeur*, 132; George E. Yoos, "A Phenomenological Look at Metaphor," *Philosophy and Phenomenological Research* 37, no. 1 (Sept. 1977): 84.

80. Schön, *Displacement of Concepts*, 104.

81. Lakoff and Johnson, *Metaphors We Live By*, 4.

82. Ibid., 5.

83. Michael J. Reddy, "The Conduit Metaphor—A Case of Frame Conflict in our Language about Language," in *Metaphor and Thought*, ed. Andrew Ortony (Cambridge: Cambridge University Press, 1979).

84. Ibid., 308.

85. Schön, *Displacement of Concepts*, 42.

86. Schön, "Generative Metaphor: A Perspective on Problem-Setting in Social Policy," in *Metaphor and Thought*, ed. Andrew Ortony (Cambridge: Cambridge University Press, 1979), 255.

87. Ibid., 264–65.

88. Colin Murray Turbayne, *The Myth of Metaphor*, 2d ed. (Columbia: University of South Carolina Press, 1970), 15; 6. Schön, "Generative Metaphor," 265; Ibid., 266

89. Schön, "Generative Metaphor," 262–63.

90. Ibid., 263, 265.

91. Ibid., 265–66.

92. Ibid., 267; Lawrence M. Hinman, "Nietzsche, Metaphor, and Truth," *Philosophy and Phenomenological Research* 43, no. 2 (Dec. 1982): 198–99.

93. Schön, "Generative Metaphor," 266–67.

94. Turbayne, *The Myth of Metaphor*, 27.

95. Schön, "Generative Metaphor," 266.

96. Ibid., 267.

97. Ibid., 267.

98. Earl R. MacCormac, *Metaphor and Myth in Science and Religion* (Durham, N.C.: Duke University Press, 1976), 103.

99. Ernst Cassirer, *An Essay on Man* (New Haven: Yale University Press, 1944), 81; Ibid., 77; Ernst Cassirer, "Mythical Thought," *The Philosophy of Symbolic Forms*. vol. 2 (New Haven: Yale University Press, 1955), 36. Translation here by Langer, "On Cassirer's Theory of Language and Myth," 397.

100. MacCormac, *Metaphor and Myth in Science and Religion*, XVII.

101. de Bono, *The Mechanism of Mind*, 191.

102. Ibid., 192; Wheelwright, *Metaphor and Reality*, 131; Turbayne, *The Myth of Metaphor*, 59.

103. de Neufville "Symbol and Myth in Public Choice," 2–3.

104. Alan W. Watts, *Myth and Ritual in Christianity* (New York: Macmillan, 1954), 7.

105. Lakoff and Johnson, *Metaphors We Live By*, 97.

106. Black, *Models and Metaphors*, 40.

107. Altshuler, *The Urban Transportation System*

Chapter 8

1. Brodsly, *L.A. Freeway*, 2.

2. Ibid., 4.

3. Joan Didion, *Play It As It Lays* (New York: Bantam Books, 1970).

4. Brodsly, *L.A. Freeway*, 5.

5. Ibid., 43.

6. Brodsly, *L.A. Freeway*, 51–53.

7. Lakoff and Johnson, *Metaphors We Live By*, 59.

8. Kelker, De Leuw & Co., *Report and Recommendations on a Comprehensive Rapid Transit Plan for the City and County of Los Angeles* (Chicago, 1925), 4.

9. RTAG, *Rail Rapid Transit Now!, LIVE where you like, WORK where you please, An Immediate Program by The Rapid Transit Action Group*, Los Angeles Chamber of Commerce-Coordinator (Feb. 1948), 4.

10. Peter Hall, *Great Planning Disasters*, 122–23.

11. California Assembly Committee on Transportation, "Light Rail Transit in Southern California."

12. California Legislature, "Light Rail Transit in Southern California," 20.

13. Paul Moyer, newscaster, KABC-TV, Los Angeles, July 20, 1980.

14. California Legislature, "Light Rail Transit in Southern California," 16.

15. RTAG, *Rail Rapid Transit Now!*, 1.

16. R. L. Bacchus to Supervisor Kenneth Hahn, Mar. 11, 1982.

17. Gina Shaffer, "Warding Off Opposition."

18. LACTC, *Transcript* of Commission Meeting held Mar. 24 1982, 6.

19. Robert A. Johnston and Daniel Sperling, *Politics and Technical Uncertainty in Transportation Investment Analysis* (Davis Cal: Division of Environmental Studies and Department of Civil Engineering, University of California, Oct. 15, 1986), 22.

Chapter 9

1. Douglas A. Hart, *Strategic Planning in London: The Rise and Fall of the Primary Road Network* (Oxford: Pergamon Press, 1976).

2. Patrick Abercrombie, *Town and Country Planning* (London: Oxford University Press, 1933), 27.

3. C. B. Purdom, *How Shall We Rebuild London?* (London: Dent, 1945).

4. Patrick Abercrombie, *Planning in Town and County. An Inaugural Lecture in the Department of Town Planning, School of Architecture*, University College (London: Hodder & Stoughton, 1937), 43; Hart, *Strategic Planning in London,* 69.

5. Los Angeles Traffic Commission, *A Major Traffic Street Plan for Los Angeles*, prepared by Frederick Law Olmsted Jr., Harland Bartholomew, and Charles Henry Cheney (Los Angeles, 1924).

6. Mark Johnson, *The Body in the Mind: The Bodily Basis of Meaning, Imagination and Reason.* (Chicago: University of Chicago Press, 1987), 130.

7. *Oxford English Dictionary*, 2d ed. vol. 3 (Oxford: Clarendon Press, 1989), 726.

8. Johnson, *The Body in the Mind*, 74.

9. Ibid., 75–76.

10. George Lakoff and Zoltan Kovecses, "The Cognitive Model of Anger Inherent in American English," *Berkeley Cognitive Science* Report No. 10 (May 1983).

11. Johnson, *The Body in the Mind*, 88–89.

12. Ibid., 89.

13. Donald A. Schön, *Displacement of Concepts.*

14. Ibid., 119–20; 121.

15. Ibid., 122.

16. *Los Angeles Times*, November 5, 1981.

17. *Los Angeles Times*, July 1, 1984.

18. RTAG, Foreword to *Rail Rapid Transit Now!*

19. Debra Kagan et al., *Mass Transit in Los Angeles: An Analysis* (Claremont, Calif.: Program in Public Policy Studies of the Claremont Colleges, 1972).

20. California Assembly Committee on Transportation, "Light Rail Transit in Southern California," 96–97.

21. For the origin of this term, see Hart, *Strategic Planning in London,* 59.

22. Interview, *Los Angeles Times*, Nov. 1, 1985.

23. Press release, Mar. 24, 1982.

24. Editorial, *Los Angeles Times* Oct. 26, 1980.

25. News broadcast, KABC-TV, Los Angeles, Mar. 24, 1982.

26. KNBC-TV, Los Angeles, Oct. 31, 1985.

27. LACTC meeting, Mar. 24, 1982.

28. Cal Hamilton, Director of Planning, City of Los Angeles, July 13, 1984, quoted in LACTC, *The Long Beach–Los Angeles Rail Transit Project, Final Environmental Impact Report*, Mar 1985, III-157.

Chapter 10

1. Noel T. Braymer, *Uncle Jam International*, July 1985.
2. House Subcommittee on Transportation and Commerce of the Committee on Interstate and Foreign Commerce, *Hearings* from April 1979.
3. Baxter Ward, *Reader* (Mar. 14, 1980).
4. California Assembly Committee on Transportation, "Light Rail Transit in Southern California—Return of the Red Cars," 74–75, 79.
5. Dennis Gabor, *Innovations: Scientific, Technological and Social* (New York and Oxford: Oxford University Press, 1970), 8.
6. Arnold Pacey, *The Culture of Technology* (Cambridge: MIT Press, 1983), 84–85.
7. KCOP-TV, Los Angeles, Oct. 21, 1985.
8. Sigmund Freud, "Three Contributions to the Theory of Sex," in *The Basic Writings of Sigmund Freud*, ed. and trans. by A. A. Brill (New York: Random House, 1938), 600.
9. Gina Shaffer, "Warding Off Opposition."
10. RTAG, *Rail Rapid Transit Now!*, 4.
11. Baxter Ward, *The Sunset Coast Line: Route of the New Red Cars*, 7.
12. Ibid., 55, 57, 59, 9.
13. Baxter Ward, *Sunset, Ltd.*, 162–67.
14. Marcia Brandwynne, KNXT-TV, Los Angeles, April 28, 1980.
15. Herbert Kupferberg, "Bus Talk," *Parade*, Dec. 14, 1980.
16. Ward, *The Sunset Coast Line*, 7, 24.
17. Shaffer, "Warding Off Opposition."
18. California, "Light Rail Transit in Southern California," 67.
19. Ward, *The Sunset Coast Line*, 1.
20. *Los Angeles Times*, May 1, 1986.
21. Interview, *Los Angeles Times* Nov. 11, 1984.
22. California, "Light Rail Transit in Southern California," 57.
23. Ibid.
24. KABC-TV, Los Angeles, July 30, 1985.
25. KABC-TV, Los Angeles, July 30, 1980.
26. KTTV-TV, Los Angeles, Oct. 31, 1985.
27. KHJ-TV, Los Angeles, Oct. 31, 1985.
28. METRO, *Final Environmental Impact Statement, Downtown Seattle Transit Project* (June 1985), foreword, 3.
29. Robert A. Johnston, and Daniel Sperling, *Politics and Technical Uncertainty in Transportation Investment Analysis*. (Davis: Division of Environmental Studies and Department of Civil Engineering, University of California, Oct. 15, 1986), 27.
30. *Los Angeles Times*, Oct. 20, 1985.
31. *Los Angeles Times*, April 13, 1982.
32. Raymond Firth, *Symbols: Public and Private* (Ithaca: Cornell University Press, 1973), 427; Joseph M. Conforti, "The Cargo Cult and the Protestant Ethic as Conflicting Ideologies: Implications for Education," *The Urban Review* 21, no. 1 (1989): 4.
33. Firth, *Symbols*, 200.
34. Ibid., 427.
35. Conforti, "The Cargo Cult," 4, 6.

Chapter 11

1. John McCone, *Violence in the City—An End or a Beginning? A Report by the Governor's Commission on the Los Angeles Riots* (Dec. 2, 1965), 75.
2. Pat Adler, "Watts: A Legacy of Lines."
3. McCone, *Violence in the City*, 76–77.
4. Ibid., 2, 1.

5. Ibid., 4.

6. Ibid., 67–68.

7. Robert M. Fogelson, "White on Black: A Critique of the McCone Commission Report on the Los Angeles Riots," *The Los Angeles Riots* (New York: Arno Press & The New York Times, 1969), 115; Ibid., 143.

8. Robert Blauner, "Whitewash over Watts: The Failure of the McCone Commission Report," *The Los Angeles Riots* (New York: Arno Press & The New York Times, 1969), 169.

9. Rustin, "The Watts "Manifesto" and the McCone Report," *The Los Angeles Riots* (New York: Arno Press & The New York Times, 1969), 150.

10. Fogelson, "White on Black," 131–32.

11. Los Angeles County and City Human Relations Commissions, *McCone Revisited: A Focus on Solutions to Continuing Problems in South Central Los Angeles. Report on a Public Hearing Jointly Sponsored by the Los Angeles County and City Human Relations Commissions* (Jan. 1985), 11.

12. Ibid., 2; Ibid., 8; Ibid., 9.

13. Ibid., 2.

14. Anastasia Loukaitou-Sideris and Tridib Banerjee, "The Blue Line Blues: Why the Vision of Transit Village May Not Materialize Despite Impressive Growth in Transit Ridership," *Journal of Urban Design* 5, no. 2 (2000): 104, 106.

15. Ibid., 104.

16. State of California Business and Transportation Agency, *Transportation-Employment Project: South Central and East Los Angeles, Final Report* (Aug. 1971), v–vi.

17. Ibid., v.

18. RTD, *Rapid Transit for Los Angeles, Summary Report of Consultants' Recommendations. Prepared for the Southern California Rapid Transit District and The Citizens of The Los Angeles Region* (July 1973), 35.

19. *Los Angeles Times*, October 20, 1985.

20. *Labor/Community Strategy Center et al. vs. MTA et al.*, United States District Court for the Central district of California, Case No. CV 94-5936 TJH (MCx) (1994).

21. Keith Killough, "Labor/Community Strategy Center et al. vs. Los Angeles County Metropolitan Transportation Authority et al," Deposition, United States District Court for the Central district of California, Case No. CV 94-5936 TJH (MCx) (1994), 6–7.

22. Tom Rubin and James E Moore II, "Why Rail Will Fail: An Analysis of the Los Angeles County Metropolitan Transportation Authority's Long Range Plan," *Policy Study* no. 209 (Los Angeles: Reasons Foundation, July 1996).

23. State of California Department of Finance, *Los Angeles County Demographic Projection for 1992*, Demographic Research Unit (Nov. 1993).

24. *Los Angeles Times*, 25 Apr. 1985; *Los Angeles Times*, 28 March 1985.

25. Sedway Cooke Associates, *Technical Report to the Long Beach–Los Angeles Rail Transit Project Environmental Impact Statement/Report: Land Use and Development Impacts* (1984), III-19–III-20.

26. Keyser Marston Associates, Katz, Hollis, Coren & Associates, and Jefferson Associates, *Long Beach–Los Angeles Rail Transit Project, Internal Memo 3.4: Evaluation of Development Potential* (1983), 31; Ibid., 31–32; Ibid., 33.

27. Anastasia Loukaitou-Sideris and Tridib Banerjee, "There's no There There: Or Why Neighborhoods Don't Readily Develop Near Light-Rail Transit Stations," *Access* no. 9 (University of California Transportation Center, fall 1996): 5, and Loukaitou-Sideris and Banerjee, "The Blue Line Blues," 114.

28. Loukaitou-Sideris and Banerjee, "There's no There There," 5, and Loukaitou-Sideris and Banerjee, "The Blue Line Blues," 111, 116–22.

29. Loukaitou-Sideris and Banerjee, "There's no There There," 5–6.

30. Glen E. Holt, "The Changing Perception of Urban Pathology: An Essay on the Development

of Mass Transit in the United States," in *Cities in American History*, edited by Kenneth J. Jackson and Stanley K. Schultz (New York: Knopf, 1972), 324–43.

31. Gilmore Simms, "The Philosophy of the Omnibus," *Godey's Lady's Book* 23, (September 1841): 105; and Holt, "The Changing Perception of Urban Pathology," 326.

32. Holt, "The Changing Perception of Urban Pathology," 329.

33. *Reader* (Mar. 14 1980).

34. *Los Angeles Times*, Apr. 25, 1985.

35. Ibid.

36. Loukaitou-Sideris and Banerjee, "There's no There There," 2.

Chapter 12

1. Churchman, *The Systems Approach*, 157.

2. Schön, *Beyond the Stable State*, 123–24; 128.

3. J. Allen Whitt, *Urban Elites and Mass Transportation: The Dialectics of Power.*

Chapter 13

1. Churchman, *The Systems Approach*, 36.

2. Murray Edelman, *The Symbolic Uses of Politics*, 32.

Epilogue

1. Churchman, *The Systems Approach*, 157.

References

Note: When a source for data is given in the text as a particular agency, this means that the data was provided by staff at that agency in the form of raw data, spreadsheets, or other nonpublished formats. Such data is often used internally, but may not have been verified for external distribution or intended to represent the agency's official position. No further reference is made to such data in the following list.

Abercrombie, Patrick. *Town and Country Planning.* London: Oxford University Press, 1933.
———. *Planning in Town and County.* An Inaugural Lecture in the Department of Town Planning, School of Architecture, University College. London: Hodder & Stoughton, 1937.
Ackoff, Russell L. *The Art of Problem Solving.* New York: Wiley, 1978.
Adler, Pat. "Watts: A Legacy of Lines." *Westways* 58 (Aug. 1966).
Adler, Sy. *Understanding the Dynamics of Innovation in Urban Transit. Center for Transit* Research and Management Development, Portland State University, Portland, Ore. US DOT report UMTA-OR-11-003-86-1. Washington, D.C.: U.S. Department of Transportation, 1986.
———. "The Transformation of the Pacific Electric Railway: Bradford Snell, Roger Rabbit, and the Politics of Transit in Los Angeles." *Urban Affairs Quarterly* 27, no. 1 (Sept. 1991): 51–86.
Alexander, Christopher. "A City is Not a Tree." *Architectural Forum* 122 (1965): Part 1: No. 1, 58–61 & Part 2: No. 2, 58–62.
Algers, S., S. Hansen, and G. Tegner. "Role of Waiting Time, Comfort, and Convenience in Modal Choice for Work Trip." *Transportation Research Record No. 534.* Washington, D.C.: Transportation Research Board, 1975.
Allison, Graham T. *Essence of Decision: Explaining the Cuban Missile Crisis.* Boston: Little, Brown & Co., 1971.
Altshuler, Alan. "The Politics of Urban Mass Transportation." Paper presented at the annual meeting of the American Political Science Association, New York City, Sept. 4–7, 1963.
———. *The Urban Transportation System: Politics and Policy Innovation.* Cambridge: MIT Press, 1979.
Andreski, Stanislav. *Social Science as Sorcery.* London: Andre Deutsch, 1972.
Arrington, G. B. "Beyond the Field of Dreams: Light Rail and Growth Management in Portland." *Seventh National Conference on Light Rail Transit.* Vol. 1, pp. 42–51. Washington, D.C.: Transportation Research Board, 1995.
Ascher, William. *Forecasting: An Appraisal for Policy-Makers and Planners.* Baltimore: Johns Hopkins University Press, 1978.
Automobile Club of Southern California. *Traffic Survey.* Los Angeles, 1937.

Bail, Eli. *From Railway to Freeway: Pacific Electric and the Motor Coach.* Glendale, Calif.: Interurban Press, 1984.

Baker, Donald M. *A Rapid Transit System for Los Angeles, California.* Los Angeles, 1933.

Baltimore Metropolitan Council. *Central Light Rail Survey. Volume 1: Methodology and Results.* Cornelius Nuworsoo, project manager. Task Report 94-2, 1994.

Bartel, Roland. *Metaphor and Symbols.* Urbana, Ill.: National Council of Teachers of English, 1983.

Bazell, Robert J. "Rapid Transit: A Real Alternative to the Auto for the Bay Area?" *Science* 171 (Mar. 19, 1971): 1125–1128.

Benveniste, Guy. *The Politics of Expertise.* 2d ed. San Francisco: Boyd & Fraser, 1977.

Biehler, Allen D. "Exclusive Busway Versus Light Rail: A Comparison of New Fixed Guideways." *Light Rail Transit: New System Successes at Affordable Prices. Transport Research Board Special Report 221* (1988): 89–98.

Bi-State Development Agency. *Quarterly Performance Indicators, Period Ending June, 1996.* St. Louis, 1996.

Black, Max. *Models and Metaphors, Studies in Language and Philosophy.* Ithaca: Cornell University Press, 1962.

Blauner, Robert. "Whitewash over Watts: The Failure of the McCone Commission Report." In *The Los Angeles Riots,* edited by Robert M. Fogelson. New York: Arno Press & The New York Times, 1969.

Board of Public Utilities. *Annual Report,* 1926.

Bollens, John C. *The Problem of Government in the San Francisco Bay Region.* Berkeley: Bureau of Public Administration, University of California, 1948.

Bonsall, John. "A Bus For All Seasons." Paper presented at seminar on *The Canadian Experience: Making Transit Work in the Golden Gate Corridor,* co-sponsored by the Golden Gate Bridge, Highway and Transportation District and the Canadian Consulate General, San Rafael, Calif., Oct. 3, 1985.

Bottles, Scott L. *Los Angeles and the Automobile. The Making of the Modern City.* Berkeley and Los Angeles: University of California Press, 1987.

Boulding, Kenneth E. *The Image. Knowledge in Life and Society.* Ann Arbor: University of Michigan Press, 1956.

———. "The Ethics of Rational Decision." *Management Science* 12, no. 6 (1966): B-161–B-169.

Bradbury, Ray. *Dandelion Wine.* Garden City, N.Y.: Doubleday, 1957.

Brewer, Garry D. *Politicians, Bureaucrats and the Consultant.* New York: Basic Books, 1973.

Brodsly, David. *L.A. Freeway: An Appreciative Essay.* Berkeley: University of California Press, 1981.

Bronner, Ethan. "The Wearing of His Uniform Is Also an Issue." *Boston Globe,* (July 9, 1987): 14.

Brubaker, Rogers. *The Limits of Rationality: An Essay on the Social and Moral Thought of Max Weber.* London: George Allen & Unwin, 1984.

Buckstein, Steve. *Top Ten Light Rail Myths,* Jan. 30 1996. Available at: http://www.cascadepolicy.org/transit/topten.htm.

California Legislature. *A Preliminary Report on Rapid Transit for the Los Angeles Area.* California Assembly Interim Committee on Public Utilities and Corporations. Sacramento, 1950.

———. Senate Committee on Public Utilities, Transit, and Energy. Southern California Rapid Transit Subcommittee. *Hearings,* July 25, 1975.

———. "Light Rail Transit in Southern California-Return of the Red Cars." *Transcript of Proceedings.* Public Hearing before the California Assembly Committee on Transportation, Long Beach City Hall, Long Beach, Aug. 14, 1981.

———. Assembly Committee on Transportation. Light Rail Program for Los Angeles County. 1981.

California Railroad Commission. *Decision No. 33088,* May 14, 1940, 631–32.

———. *Decision No. 33984,* Mar. 11, 1941. 379.

———. *Decision No. 44161,* May 9, 1950. 661–62.

Caltrans (California Department of Transportation). *Freeway Transit Element of the Regional Transit Development Plan for Los Angeles County,* District 7, Transit Branch, Aug. 1978.

————. *Long Beach to Los Angeles Light Rail Transit Feasibility Study.* Los Angeles: Caltrans District 7—Public Transportation Branch, Oct. 1981.

Cassirer, Ernst. *An Essay on Man.* New Haven: Yale University Press, 1944.

————. *The Philosophy of Symbolic Forms.* Vol. 2, *Mythical Thought.* New Haven: Yale University Press, 1955.

Cervero, Robert. "Light Rail Transit and Urban Development." *Journal of the American Planning Association* (spring 1984): 133–47.

Charles, John A. "The Mythical World of Transit Oriented Development." *Policy Perspective No. 1019.* Portland, Ore.: Cascade Policy Institute, 2001.

Chorley, Richard J., and Peter Haggett. *Models in Geography.* London: Methuen, 1967.

Churchman, C. West. *The Design of Inquiring Systems: Basic Concepts of Systems and Organization.* New York: Basic Books, 1971.

————. *The Systems Approach and Its Enemies.* New York: Basic Books, 1979.

————. *Thought and Wisdom.* Seaside, Calif.: Intersystems Publications, 1982.

Citizens Advisory Council on Public Transportation. *Improving Public Transportation in Los Angeles.* Los Angeles, 1967.

City of Beaverton. *Mayor's Report, June/July 2001.* Available at www.ci.beaverton.or.us/mayor/welcome_june_july01.html.

City Club of Los Angeles. *Report on Rapid Transit.* Supplement to *City Club Bulletin.* Jan. 30 1926.

City of Toronto Planning Board. *Air Rights and Land Use Adjacent to the Bloor–Danforth Subway.* June 1963.

————. *Report in a Proposal to Rezone Lands Along the East/West Subway Route to Permit High Rise Apartments.* Oct.12, 1971.

Conforti, Joseph M. "The Cargo Cult and the Protestant Ethic as Conflicting Ideologies: Implications for Education." *The Urban Review* 21, no. 1 (1989).

Crain, John and Associates. *First-year Report: San Bernardino Freeway Express Busway Evaluation.* Menlo Park, Calif.: John Crain and Associates, 1974.

Crain & Associates, Inc. *The Martin Luther King East Busway in Pittsburgh, PA. Final Report, October 1987.* UMTA/TSC Evaluation Series. U.S. Department of Transportation, Urban Mass Transportation Administration. UMTA Technical Assistance Program, 1987.

Crump, Spencer. *Ride the Big Red Cars: How Trolleys Helped Build Southern California.* Corona del Mar, Calif.: Trans-Anglo Books, 1970.

————. *Henry Huntington and the Pacific Electric.* Corona del Mar, Calif.: Trans-Anglo Books, 1978.

Cunneen, Michael J. *MAX, Lies and Videotapes: How "Oregonians for Roads and Rails" is Lying to Oregonians to Sell the Measure 32 Gravy Train.* Portland, Ore. Monograph. n.d.

Danielson, Michael N. *Federal-Metropolitan Politics and the Commuter Crisis.* New York: Columbia University Press, 1965.

de Bono, Edward. *The Use of Lateral Thinking.* Harmondsworth, Mddx.: Penguin Books, 1967.

————. *The Mechanism of Mind.* Harmondsworth, Mddx.: Penguin Books, 1969.

DeLeuw Cather. *Comparative Analysis Study of Alternative Transit Systems South Hills Corridor.* Prepared for the Port Authority of Allegheny County, Pennsylvania, Mar. 1976.

Deloitte/Kellogg Joint Venture. *Memorandum Re: Start-up Review Report No. 2* from Rodney J. Dawson to Ed McSpedon, LACTC, Los Angeles, Sept. 15 1989.

de Neufville, Judith Innes. "Symbol and Myth in Public Choice, The Case of Land Policy in the United States." *Working Paper 359,* Berkeley: Institute or Urban & Regional Development, University of California, 1981.

Didion, Joan. *Play It As It Lays.* New York: Bantam Books, 1970.

DMJM (Daniel, Mann, Johnson and Mendenhall). *Planning and Economic Considerations Affecting Transportation in the Los Angeles Region.* Los Angeles, Southern California Rapid Transit District, 1965.

Dror, Yehezkel. *Public Policymaking Reexamined*. Scranton, Pa.: Chandler Publishing Company, 1968.

Dunn, James A. Jr. *Miles to Go: European and American Transportation Policies*. Cambridge: MIT Press, 1981.

———. *Driving Forces: The Automobile, Its Enemies, and the Politics of Mobility*. Washington, D.C.: Brookings Institution Press, 1998.

Dyckman, John W. "Riding the Sunset Coast Line." In *Transportation Alternatives for Southern California*. Edited by Peter Gordon and Ross D. Eckert. Conference Proceedings of a Symposium, The Institute for Public Policy Research, Center for Public Affairs, University of Southern California, Apr. 12, 1976. Also available at http://www-rcf.usc.edu/~pgordon/transit.html.

Ebasco Services. *Bay Area Rapid Transit Economic Study for the San Francisco Bay Area Rapid Transit District*. San Francisco, June 1961.

Edelman, Murray. *The Symbolic Uses of Politics*. Urbana and Chicago: The University of Illinois Press, 1964.

Ellul, Jacques. *The Technological Society*. New York: Knopf, 1964.

Firth, Raymond. *Symbols: Public and Private*. Ithaca: Cornell University Press, 1973.

Fisher, Franklin M. "Multiple Regression in Legal Proceedings." *Columbia Law Review* 80, no. 3 (Apr. 1980): 702–736.

Florman, Samuel C. *The Existential Pleasures of Engineering*. New York: St. Martin's Press, 1976.

Fogelson, Robert M. *The Fragmented Metropolis: Los Angeles 1850–1930*. Cambridge: Harvard University Press, 1967.

———. "White on Black: A Critique of the McCone Commission Report on the Los Angeles Riots." In *The Los Angeles Riots*. Edited by Robert M. Fogelson. New York: Arno Press & The New York Times, 1969.

Fong, K. M. "*Rapid Transit–Decision-Making: The Income Distribution Impact of San Francisco Bay Area Rapid Transit*." Bachelor's thesis, Harvard University, 1976.

Foss, Martin. *Symbol and Metaphor in Human Experience*. Princeton: Princeton University Press, 1949.

Foster, Mark S. "The Decentralization of Los Angeles During the 1920s". Doctoral dissertation, University of Southern California, 1971.

———. "The Model-T, The Hard Sell, and Los Angeles' Urban Growth: The Decentralization of Los Angeles during the 1920s." *Pacific Historical Review* 44 (Nov. 1975): 459–484.

Freud, Sigmund. "Three Contributions to the Theory of Sex." In *The Basic Writings of Sigmund Freud*, edited and translated by A. A. Brill. New York: Random House, 1938. Also available as Freud, Sigmund. *Three Contributions to the Theory of Sex*. New York: Dutton, 1962.

Friedmann, John, and Barclay Hudson. "Knowledge and Action: A Guide to Planning Theory." *AIP Journal* (Jan. 1974): 2–16.

Gabor, Dennis. *Innovations: Scientific, Technological and Social*. New York and Oxford: Oxford University Press, 1970.

Garrison, William L. "Urban Transportation and Land Use." In *Public Transportation: Planning, Operations and Management*, edited by George E. Gray, and Lester E. Hoel. Englewood Cliffs, N.J.: Prentice-Hall, 1979.

General Motors. "The Truth About 'American Ground Transport'—A Reply by General Motors." Part 4A, Appendix to Part 4, Hearings before the Subcommittee on Antitrust and Monopoly of the Committee on the Judiciary, United States Senate, 1974, A-107–A-127.

Glaab, Charles N. and A. Theodore Brown. *The History of Urban America*. London: Macmillan, 1967.

Godet, Michel. *The Crisis in Forecasting and the Emergence of the "Prospective" Approach with Case Studies in Energy and Air Transport*. Translated from the French by J. D. Pearse and Harry K. Lennon. New York: Pergamon Policy Studies 15, 1979.

Gómez-Ibáñez, José A. "A Dark Side to Light Rail?" *Journal of the American Planning Association* (summer 1985): 337–51.

Goodwin, H. Marshall, Jr. "From Dry Gulch to Freeway." *Southern California Quarterly* 47 (summer 1965): 73–102.

Gordon, Peter, and Ross D. Eckert, eds. *Transportation Alternatives for Southern California*. Conference Proceedings of a Symposium, The Institute for Public Policy Research, Center for Public Affairs, University of Southern California, Apr. 12, 1976. Also available at http://www-rcf.usc.edu/~pgordon/transit.html.

Gordon, Peter, and Harry W. Richardson. "Compactness or Sprawl: America's Future vs. The Present." Paper presented at the conference of the Association of Collegiate Schools of Planning, Atlanta, Nov. 2000. Available at: http://www.usc.edu/schools/sppd/lusk/research/papers/pdf/wp_2000_1008.pdf.

Gruen Associates, Inc. "Indirect Environmental Impacts." *TM 244-77*, BART Impact Program. Berkeley: Metropolitan Transportation Commission, 1977.

Hacker, Frederick J. "Psychology and Psychopathology in Symbolism." In *Psychology and the Symbol, An Interdisciplinary Symposium*, edited by Joseph R. Royce. New York: Random House, 1965.

Hall, Peter. *Great Planning Disasters*. American Edition. Berkeley: University of California Press, 1982.

Hall, Sir Peter. *Cities in Civilization*. New York: Pantheon Books, 1998.

Hallihan, Joseph P. "Relation of Rapid Transit to Community Development." *Electric Railway Journal* 73 (Sept. 14, 1929).

Hamer, Andrew M. *The Selling of Rail Rapid Transit*. Lexington, Mass.: Lexington Books, 1976.

Hart, Douglas A. *Strategic Planning in London: The Rise and Fall of the Primary Road Network*. Oxford: Pergamon Press, 1976.

Hartman Company. "The Round." Available at: http://www.hartmanco.com/about/awards/round.htm.

Hebert, Ray. "Prop A: From Bleak to Bright in LA." *Mass Transit* (Sept. 1982): 38–42.

Heenan, G. Warren. "The Economic Effect of Rapid Transit on Real Estate Development in Toronto." *The Appraisal Journal* (Apr. 1968): 213–224.

Henderson, C., and J. Billheimer. *Manhattan Passenger Distribution Project: Effectiveness of Midtown Manhattan System Alternatives*. Menlo Park, Calif.: Stanford Research Institute, 1972.

Henk, Russell H., Daniel E. Morris, and Dennis L. Christiansen. *An Evaluation of High-Occupancy Vehicle Lanes in Texas, 1994*. College Station: Texas Transportation Institute Rept. FHWA/TX-96/1353-3, 1995.

"Henry E. Huntington Dead." *Electric Railway Journal* (May 28, 1927).

Hilton, George W. "Rail Transport and the Pattern of Modern Cities: The California Case." *Traffic Quarterly* 3 (July 1967): 379–93.

———. "What Did We Give Up With The Big Red Cars?" In *Transportation Alternatives for Southern California*, edited by Peter Gordon and Ross D. Eckert. Conference Proceedings of a Symposium, The Institute for Public Policy Research, Center for Public Affairs, University of Southern California, Apr. 12, 1976. Also available at: http://www-rcf.usc.edu/~pgordon/transit.html.

Hilton, George W. and John F. Due. *The Electric Interurban Railway in America*. Stanford: Stanford University Press, 1960.

Hinman, Lawrence M. "Nietzsche, Metaphor, and Truth." *Philosophy and Phenomenological Research* 43, no. 2 (Dec. 1982): 179–99.

Hochberg, Julian E. *Perception*. 2nd ed.. Englewood Cliffs, N.J.: Prentice-Hall, 1978.

Holt, Glen E. "The Changing Perception of Urban Pathology: An Essay on the Development of Mass Transit in the United States." In *Cities in American History*. edited by Kenneth J. Jackson and Stanley K. Schultz. New York: Knopf, 1972.

Hoos, Ida R. *Systems Analysis in Social Policy. A Critical Review*. Research Monograph no. 19, London: Institute of Economic Affairs, 1969.

"How the New Gas Buses Have Improved Earnings." *Transit Journal* (Sept. 15, 1936): 337–341.

Howell Research Group. *Regional Transportation District Light Rail Passenger Survey,* May 1995.

Hunt, Rockwell D., and William S. Amert. *Oxcart to Airplane.* Los Angeles: Powell Publishing Co., 1929.

IPA (Institute for Public Administration). *Urban Transportation and Public Policy.* Report prepared for U.S. Department of Commerce and the Housing and Home Finance Administration, New York, 1961. Also available as Fitch, Lyle C. *Urban Transportation and Public Policy.* San Francisco: Chandler, 1964.

Jester, Norman J. "Preparations for 'Show Time': The Los Angeles Story." In *Light Rail Transit, New System Successes at Affordable Prices.* Special Report 221. Washington, D.C.: Transportation Research Board, 1989.

Johnson, Mark. *The Body in the Mind: The Bodily Basis of Meaning, Imagination and Reason.* Chicago: University of Chicago Press, 1987.

Johnston, Robert A., and Daniel Sperling. *Politics and Technical Uncertainty in Transportation Investment Analysis.* Davis: Division of Environmental Studies and Department of Civil Engineering, University of California, Oct. 15, 1986.

Jones, David W., Jr. *Urban Transit Policy: An Economic and Political History.* Englewood Cliffs, N.J.: Prentice-Hall, 1985.

Kagan, Debra; David Newman; Stan Sander; and Bonnie Woltoncroft. *Mass Transit in Los Angeles: An Analysis.* Claremont, Calif.: Program in Public Policy Studies of the Claremont Colleges, June 1972.

Kain, John F. "Deception in Dallas: Strategic Misrepresentation in Rail Transit Promotion and Evaluation." *Journal of the American Planning Association* (spring 1990): 184–196.

Kain, John F., and Zhi Liu. *Secrets of Success: How Houston and San Diego Transit Providers Achieved Large Increases in Transit Ridership.* Prepared for Federal Transit Administration, May 1995.

Kaiser Engineering, and DMJM (Daniel, Mann, Johnson and Mendenhall). *Technical Study of Alternative Transit Corridors and Systems, Plan Refinement, Volume I: Engineering and Planning, Phase III.* Los Angeles: Southern California Rapid Transit District, Mar. 1974.

Karr, Randolph. "Rail Passenger Service History of Pacific Electric Railway Company." Available at: Los Angeles County Metropolitan Transportation Authority Library, 1973.

Kelker, De Leuw & Co. *Report and Recommendations on a Comprehensive Rapid Transit Plan for the City and County of Los Angeles,* Chicago, 1925.

Kelman, Steven. "Cost-Benefit Analysis and Environmental, Safety, and Health Regulation: Ethical and Philosophical Considerations." In *Ethics in Planning.* Edited by Martin Wachs. New Brunswick, N.J.: Center for Urban Policy Research, 1985.

Kennedy, David M. *UMTA and the New Rail Start Policy.* Kennedy School of Government Case Program, Harvard University, 1984.

Keyser Marston Associates, Katz, Hollis, Coren & Associates, and Jefferson Associates. *Long Beach–Los Angeles Rail Transit Project, Internal Memo 3.4: Evaluation of Development Potential,* 1983.

Killough, Keith. Deposition, "*Labor/Community Strategy Center et al. vs. Los Angeles County Metropolitan Transportation Authority et al.*" United States District Court for the Central District of California, Case No. CV 94-5936 TJH (MCx), 1994.

Kirkman-Harriman. *The Kirkman-Harriman Pictorial and Historical Map of Los Angeles County, 1860 A.D.* Los Angeles, 1937.

Knight, Robert L., and Lisa L. Trygg. "Evidence of Land Use Impacts of Rapid Transit Systems." *Transportation* 6 (1977): 231–47.

Korve Engineering et al. *Mid-City/Westside Transit Corridor Re-Evaluation/Major Investment Study.* Submitted to Los Angeles County Metropolitan Transportation Authority, Feb. 24, 2000.

Kovach, Carol. "On Conducting an 'Impact' Study of a Rapid Transit Facility—The Case of Toronto." Presented to Joint Transportation Engineering Meeting, Montreal, July 15–19, 1974, *Meeting preprint MTL-23.*

Kramer, Fred A. "Policy Analysis as Ideology." *Policy Analysis* (Sept./Oct. 1975): 509–17.

Kuhn, Thomas S. *The Copernican Revolution.* Cambridge: Harvard University Press, 1957.

Kupferberg, Herbert. "Bus Talk." *Parade*, Dec. 14. 1980.

Labor/Community Strategy Center et al. vs. MTA et al. (Labor/Community Strategy Center et al. vs. Los Angeles County Metropolitan Transportation Authority et al). United States District Court for the Central District of California, Case No. CV 94-5936 TJH (MCx), 1994.

LACTC (Los Angeles County Transportation Commission). Staff Budget Memorandum. Apr. 26, 1978.

———. LACTC Minutes. Jan. 10, 1979.

———. Transcript of LACTC meeting. July 11, 1979.

———. LACTC Minutes. Aug. 8, 1979.

———. *Staff Report on Transit System and Financing Alternatives for Los Angeles County.* Jan. 23, 1980.

———. Minutes, Commission Meeting, Mar. 26, 1980.

———. Minutes, Commission Meeting, Aug. 6, 1980.

———. Transcript of Commission Meeting. Aug. 20, 1980.

———. *Staff Report on a Rapid Transit Development Strategy for Los Angeles County.* Sept. 9, 1981.

———. Minutes, Sept. 23, 1981.

———. Minutes, Oct. 13, 1981.

———. Minutes, Oct. 28, 1981.

———. *Rapid Transit Development in Los Angeles County: Staff Recommendations.* Los Angeles, Mar. 17, 1982.

———. Memorandum from John Zimmerman, Chairman, Ad Hoc Rapid Transit Committee, to LACTC members and alternates on "March 17 Actions of Ad Hoc Rapid Transit Committee." Mar. 19, 1982.

———. Transcript of Commission Meeting. Mar. 24, 1982.

———. *The Long Beach–Los Angeles Rail Transit Project, Draft Environmental Impact Report,* May 1984.

———. *The Long Beach–Los Angeles Rail Transit Project, Final Environmental Impact Report,* Mar. 1985.

———. *Metro Moves.* Feb. 1991.

———. *30-Year Integrated Transportation Plan.* Report to U.S. Department of Transportation, Washington D.C.: Federal Transit Administration, 1992.

Lakoff, George. *Women, Fire, and Dangerous Things. What Categories Reveal about the Mind.* Chicago: University of Chicago Press, 1987.

Lakoff, George, and Mark Johnson. *Metaphors We Live By.* Chicago: The University of Chicago Press, 1980.

Lakoff, George, and Zoltan Kovecses. "The Cognitive Model of Anger Inherent in American English." *Berkeley Cognitive Science Report* no. 10 (May 1983).

Landis, John, and Robert Cervero. "Middle Age Sprawl: BART and Urban Development." *Access* (University of California Transportation Center) no. 14 (summer 1999): 2–15.

Langer, Susanne K. "On Cassirer's Theory of Language and Myth." In *The Philosophy of Ernst Cassirer.* edited by Paul Arthur Schilpp. Evanston, Ill.: The Library of Living Philosophers, 1949.

———. *Philosophy in a New Key: A Study in the Symbolism of Reason, Rite and Art.* 3rd ed. Cambridge: Harvard University Press, 1957.

Larwin, Thomas F. "1988 Transportation Achievement Award (Operations). San Diego's Light Rail System: A Success Story." *ITE Journal* (Jan. 1989): 19–20.

Laski, Harold J. "Limitations of the Expert." *Chemtech* (Apr. 1974): 198–202. Originally published in 1930 in *Harpers.*

Lasswell, Harold. *Psychopathology and Politics.* New York: Viking Press, 1960.

Lave, Charles A. "The Mass Transit Panacea and Other Fallacies About Energy." *The Atlantic* (Oct. 1979): 39–43.

Leamer, Edward E. "Let's Take the Con out of Econometrics." *American Economic Review* (Mar. 1983): 31–43.

Lee, Douglass B., Jr. "Impacts of BART on Prices of Single-Family Residences and Commercial Property." *BART Impact Studies.* Berkeley: Institute of Urban and Regional Development, University of California, 1973.

Lee, Douglass B., Jr., and D. F. Wiech. "Market Street Study." *BART Impact Studies.* Berkeley: Institute of Urban and Regional Development, University of California, 1972.

Libicki, Martin C. "Land Use Impacts of Major Transit Improvements." Urban Analysis Program, Office of the Secretary of Transportation, Washington, D.C., Mar. 1975.

Linton, Gordon. Letter from U.S. Federal Transit Administrator to Larry Chairman, Chair, Los Angeles County Metropolitan Transportation Authority, Jan. 6, 1997.

———. Letter from U.S. Federal Transit Administrator to Larry Chairman, Chair, Los Angeles County Metropolitan Transportation Authority, Apr. 9, 1997.

———. *Statement of Federal Transit Administrator Gordon J. Linton Regarding Acceptance of Recovery Plan submitted by Los Angeles County Metropolitan Transportation Authority,* July 2, 1998.

Liston, Albert M. "Regional Rapid Transit Development in the San Francisco Bay Area." Master's thesis, Sacramento State College, 1970.

Los Angeles Central City Committee and the Los Angeles City Planning Department. *Los Angeles Centropolis 1980: Economic Survey.* Los Angeles, 1960.

Los Angeles County, and City Human Relations Commissions. *McCone Revisited: A Focus on Solutions to Continuing Problems in South Central Los Angeles. Report on a Public Hearing Jointly Sponsored by the Los Angeles County and City Human Relations Commissions,* Jan. 1985.

Los Angeles County Regional Planning Commission. *A Comprehensive Report on the Regional Plan of Highways: Section 4, Long Beach–Redondo Area.* Los Angeles, 1931.

———. *Master Plan of Highways.* Los Angeles, 1941.

———. *Freeways for the Region.* Los Angeles, 1943.

Los Angeles Metropolitan Parkway Engineering Committee. *Interregional, Regional, and Metropolitan Parkways.* Los Angeles, 1946.

Los Angeles Traffic Commission. *A Major Traffic Street Plan for Los Angeles.* Prepared by Frederick Law Olmsted Jr., Harland Bartholomew, and Charles Henry Cheney: Los Angeles, 1924.

Los Angeles Transportation Engineering Board. *A Transit Program for the Los Angeles Metropolitan Area,* Los Angeles, 1939.

Loukaitou-Sideris, Anastasia, and Tridib Banerjee. "There's no There There: Or Why Neighborhoods Don't Readily Develop Near Light-Rail Transit Stations." *Access* (University of California Transportation Center) no. 9 (fall 1996): 2–6.

———. "The Blue Line Blues: Why the Vision of Transit Village May Not Materialize Despite Impressive Growth in Transit Ridership." *Journal of Urban Design* 5, no. 2 (2000): 101–25.

MacCormac, Earl R. *Metaphor and Myth in Science and Religion.* Durham, N.C.: Duke University Press, 1976.

MacIntyre, Alasdair. "Utilitarianism and the Presuppositions of Cost-Benefit Analysis: An Essay on the Relevance of Moral Philosophy to the Theory of Bureaucracy." In *Ethics in Planning.* Edited by Martin Wachs. New Brunswick, N.J.: Center for Urban Policy Research, 1985.

Marcou, O'Leary & Associates. *Transit Impact on Land Development.* Washington, D.C., 1971.

Maritz Marketing Research. 1989.

Mass Transit Administration, Maryland Department of Transportation. *Environmental Effects Report for the Central Light Rail Line,* 1988.

McCone, John. *Violence in the City—An End or a Beginning? A Report by the Governor's Commission on the Los Angeles Riots,* Dec. 2, 1965.

McFadden, Daniel. "The Measurement of Urban Travel Demand." Working Paper no. 227. Berkeley: Institute of Urban and Regional Development, University of California, Berkeley, 1974.

McSpedon, Edward. "Building Light Rail Transit in Existing Rail Corridors—Panacea or Nightmare? The Los Angeles Experience." In *Light Rail Transit, New System Successes at Affordable Prices.* Special Report 221. Washington, D.C.: Transportation Research Board, 1989.

McSpedon, Edward. Letter to Jonathan Richmond, July 3, 1990.

McWilliams, Carey. *Southern California: An Island on the Land.* Santa Barbara and Salt Lake City: Peregrine Smith, 1973.

Merleau-Ponty, Maurice. *The Primacy of Perception.* Evanston, Ill.: Northwestern University Press, 1964.

METRO (Municipality of Metropolitan Seattle). *Final Environmental Impact Statement, Downtown Seattle Transit Project.* June 1985.

Metropolitan Transportation Commission. *A History of the Key Decisions in the Development of Bay Area Rapid Transit.* San Francisco: Jossey-Bass, 1973.

Miller, George A. "Images and Models, Similes and Metaphors." In *Metaphor and Thought.* Edited by Andrew Ortony. Cambridge: Cambridge University Press, 1979.

Miller, Gerald K., and Keith M. Goodman. *The Shirley Highway Express-Bus-on-Freeway Demonstration Project: First Year Results.* Washington, D.C.: Technical Analysis Division, National Bureau of Standards, 1972.

Morrison, Philip. "Knowing Where You Are: A First Essay Towards Crossing Critical Barriers." In *Critical Barriers Phenomenon in Elementary Science.* by Maja Apelman, David Hawkins, and Philip Morrison. Grand Forks: Center for Teaching and Learning, University of North Dakota, 1985.

MTA (Los Angeles County Metropolitan Transportation Authority). *Transportation for the 21st Century, A Plan for Los Angeles County.* Mar. 1995.

——. *Facts At A Glance,* Feb. 1996.

——. "The Metro Green Line Turns One." News release, Aug 12, 1996.

——. Response to data request from Supervisor Antonovich's office, Sept 13, 1996.

——. Memorandum to Supervisor Antonovich's office, Nov. 1, 1996.

——. *Metro Rail Red Line Recovery Plan,* Submitted to the U.S. Department of Transportation, Federal Transit Administration, Jan. 15, 1997.

——. *Restructuring Plan. Analysis and Documentation of the MTA's Financial and Managerial Ability to Complete North Hollywood Rail Construction and Meet the Terms of the Bus Consent Decree.* May 15, 1998.

——. *Executive Summary Rail Program Status as of December 2000,* Dec. 2000.

——. *Draft 2001 Long Range Transportation Plan for Los Angeles County,* Feb. 2001.

MTA/Booz Allen (Los Angeles County Metropolitan Transportation Authority/Booz-Allen & Hamilton, Inc.). *Regional Transit Alternatives Analysis, Study Results & Appendix,* Attachment No. 1 to MTA Board Report, Nov. 9, 1998.

MTDB (Metropolitan Transit Development Board). *East Urban Corridor. Alternatives Analysis/Environmental Impact Statement. Capital Cost Estimates. Technical Report.* Submitted to Urban Mass Transportation Administration, U.S. Department of Transportation, Sept. 24, 1984.

MTDB (San Diego Metropolitan Transit Development Board). *East Urban Corridor, San Diego Region, Final Environmental Impact Statement.* San Diego: San Diego Metropolitan Transit Development Board, and San Francisco: Urban Mass Transportation Administration, U.S. Department of Transportation, 1986.

MTDB (Metropolitan Transit Development Board). *The San Diego Trolley. South Line and East Line Summary,* Sept. 1995.

Mulvey, John M. "Computer Modelling for State Use: Implications for Professional Responsibility." Paper presented at meeting of the American Academy for the Advancement of Science, Detroit, Mich., May 1983.

Mundle & Associates, Inc. *Review of Transportes Amarillo y Rosa.* Prepared for the San Diego Metropolitan Transit Development Board, 1986.

Nash, Allan N., and Stanley J. Hille. "Public Attitudes Toward Transport Modes: A Summary of Two Pilot Studies." *Highway Research Record 233.* Washington D.C.: Highway Research Board, 1968.

National Analysis. *A Survey of Commuter Attitudes Toward Rapid Transit Systems.* Vol. 1. Washington, D.C.: National Capital Transportation Agency, 1963.

Nelson, Howard J. "The Spread of an Artificial Landscape over Southern California." *In Man, Time, and Space in Southern California,* special supplement to *Annals of the Association of American Geographers,* edited by William L. Thomas Jr., Sept. 1959.

NFTA (Niagara Frontier Transportation Authority, Buffalo). *Light Rail Rapid Transit System, Memorandum of Light Rail Capital Cost Estimates, Ridership Forecasts, Fleet Size and Running Times (Projected vs. Actual),* 1987.

Nietzsche, Friedrich. *The Will to Power.* New York: Random House, 1969.

OC Transpo. *Operating Statistics.* OC Transpo Research & Development, Ottawa, 1996.

Pacey, Arnold. *The Culture of Technology.* Cambridge: MIT Press, 1983.

Padron & Associates. *Blue Line FY 1991 Budget (Revised).* Letter from Manuel Padron to Sharon Neely, LACTC. Atlanta, Ga., June 26, 1990.

PAT (Port Authority of Allegheny County, Pittsburgh, Pennsylvania). *Light Rail and Busway Statistical Overview (Based on Fiscal Year 1995 Operations),* May 1996.

PBHM (Parsons, Brinkerhoff, Hall, and Macdonald). *Regional Rapid Transit. Report to the San Francisco Bay Area Rapid Transit Commission,* Jan. 5, 1956.

PBQD (Parsons, Brinckerhoff, Quade & Douglas). "Investigate Potential Ridership for Baseline Los Angeles–Long Beach Light Rail Project." Working Paper no. 7, part of The Los Angeles to Long Beach Light Rail Project and Evaluation of Other Rapid Transit Opportunities Study. Prepared for Los Angeles County Transportation Commission, Los Angeles, 1982.

————. *Summary Report, The Los Angeles-Long Beach Light Rail Project and Evaluation of Other Rapid Transit Opportunities.* Santa Ana, Calif., 1982.

PBTB (Parsons Brinckerhoff, Tudor and Bechtel). *The Composite Report: Bay Area Rapid Transit.* San Francisco, 1962.

Pegg, Alan F., Tom Rubin, and Larry Schlegel. Memorandum to SCRTD Board of Directors on "Fiscal Year 1990 Third Quarter Operating Budget Forecast and Year-to-Year Changes of Operating Revenues and Expenses." Los Angeles: Southern California Rapid Transit District, Apr. 27, 1990.

Peters, Ted. "Metaphor and the Horizon of the Unsaid." *Philosophy and Phenomenological Research* 38 (Mar. 1978): 355–69.

Peterson, Neil. Memorandum to LACTC Planning and Mobility Improvement Committee: "Interim Downtown METRO Blue Line Shuttle Update." Prepared by Alan E. Patashnick, Sept. 14, 1990. Los Angeles: Los Angeles County Transportation Commission.

Pickrell, Don H. *Urban Rail Transit Projects: Forecast Versus Actual Ridership and Costs. Final Report.* Prepared for Office of Grants Management, Urban Mass Transportation Administration, U.S. Department of Transportation. Report DOT-T-91-04, Oct. 1990.

————. "A Desire Names Streetcar: Fantasy and Fact in Rail Transit Planning," *Journal of the American Planning Association* 58, no. 2 (spring 1992): 158–76.

Polanyi, Michael, and Harry Prosch. *Meaning.* Chicago: The University of Chicago Press, 1975.

Pomerantz, James R. "Perceptual Organization in Information Processing." In *Perceptual Organization.* Edited by Michael Kubovy and James R. Pomerantz. Hillsdale, N.J.: Lawrence Erlbaum Associates, 1981.

Pomerantz, James R. and Michael Kubovy. "Perceptual Organization: An Overview." In *Perceptual Organization.* Edited by Michael Kubovy and James R. Pomerantz. Hillsdale, N.J.: Lawrence Erlbaum Associates, 1981.

Post, Bob. Interoffice Memorandum to R. Higbee, D. Porter, and G. Arrington on the subject of "U.S. DOT Banfield Visit." Tri-Met, Sept. 16, 1988.

Primack, Phil. "Death of the Hiawatha." *The Washington Monthly* (Dec. 1979): 16–23.

Purdom, C. B. *How Shall We Rebuild London?* London: Dent, 1945.

Quarmby, David A. "Choice of Travel Mode for the Journey to Work." *Journal of Transport Economics and Policy* 1 (1967): 273–314.

Rapid Transit Action Group. *Rail Rapid Transit Now!, LIVE where you like, WORK where you please, An Immediate Program by The Rapid Transit Action Group, Los Angeles Chamber of Commerce—Coordinator,* Feb. 1948.

Reddy, Michael J. "The Conduit Metaphor—A Case of Frame Conflict in our Language about Language." In *Metaphor and Thought.* Edited by Andrew Ortony. Cambridge: Cambridge University Press, 1979.

Rhodes, Richard. "Semantics in Relational Grammar," *Proceedings of the Thirteenth Annual Meeting of the Chicago Linguistic Society.* Chicago: Chicago Linguistic Society, 1977.

Richmond, Jonathan E. D. "Perpetuum Mobile-AMTRAK: The Original Sin". Master's thesis, MIT, Sept. 1981.

———. "The Costly Lure of the Bullet Train—Legislature Acts Quickly and Blindly on an Untested Scheme." *Los Angeles Times.* Sept. 8, 1982.

———. "Mental Barriers to Learning and Creativity in Transportation Planning." *Carolina Planning* 13, no. 1 (fall 1987): 42–53.

———. "Theories of Symbolism, Metaphor and Myth, and the Development of Western Rail Passenger Systems or Penis Envy in Los Angeles." Paper presented at the Conference of the Association of Collegiate Schools of Planning, Portland, Ore., Oct. 7, 1989.

———. "Introducing Philosophical Theories to Urban Transportation Planning or Why All Planners Should Practice Bursting Bubbles." *Systems Research* 7, no. 1 (1990): 47–56.

———. "Introducing Philosophical Theories to Urban Transportation Planning." In *Planning Ethics: A Reader in Planning Theory, Practice, and Education,* edited by Sue Hendler. New Brunswick, N.J.: Center for Urban Policy Research, 1995.

———. "Simplicity and Complexity in Design for Transportation Systems and Urban Forms," *Journal of Planning Education and Research* 17, no. 3 (spring 1998): 220–230.

———. *New Rail Transit Investments—A Review,* Taubman Center for State and Local Government, John F. Kennedy School of Government, Harvard University, June 1998.

———. "The Mythical Conception of Rail Transit in Los Angeles." *Journal of Architectural and Planning Research* 15, no. 4. (winter 1998): 294–320.

———. "A Whole-System Approach to Evaluating Urban Transit Investments," *Transport Reviews* 21, no. 2 (Apr.–June 2001): 141–179.

Ricoeur, Paul. *The Rule of Metaphor, Multi-disciplinary studies in the creation of meaning in language.* Toronto: University of Toronto Press, 1977.

———. *The Philosophy of Paul Ricoeur, An Anthology of His Work.* Edited by Charles E. Reagan and David Stewart. Boston: Beacon Press, 1978.

Robinson, W. W. *Los Angeles: From the Days of the Pueblo.* San Francisco: Chronicle Books, for the California Historical Society, 1981.

Rogers, Leslie. Letter from Regional Administrator, U.S. Department of Transportation, Federal Transit Administration, Region IX, San Francisco, to Linda Bohlinger, Interim Chief Executive Officer, Los Angeles County Metropolitan Transportation Authority, 1997.

RT (Sacramento Regional Transit District). *Sacramento Light Rail Overview,* 1987.

RTD (Southern California Rapid Transit District). *Planning and Economic Considerations Affecting Transportation in the Los Angeles Region.* Daniel, Mann, Johnson, and Mendenhall, 1965.

———. *Rapid Transit for Los Angeles, Summary Report of Consultants' Recommendations.* Prepared for the Southern California Rapid Transit District and The Citizens of The Los Angeles Region, July 1973.

———. *Line Performance Trends Report.* (Data as of Dec. 19, 1985).

———. *Short Range Transit Plan: Guideway Plan, Fiscal Years 1991 through 1995.* (1990).

———. *Proposed Annual Budget, Fiscal Year 1991.* June 7, 1990.

Rubin, Tom. Memorandum to Alan Pegg, Art Leahy, Al Perdon, John Richeson, and Gary Spivack on "Bus vs. Rail Costs." Los Angeles: Southern California Rapid Transit District, Aug. 21, 1990.

———. "The Watts 'Manifesto' and the McCone Report" In *The Los Angeles Riots.* edited by Robert M. Fogelson. New York: Arno Press & The New York Times, 1969.

Rubin, Tom, and James E. Moore II. "Why Rail Will Fail: An Analysis of the Los Angeles County Metropolitan Transportation Authority's Long Range Plan." *Policy Study No. 209.* Los Angeles: Reasons Foundation, July 1996.

SANDAG (San Diego Association of Governments). *1990 San Diego Regional Transit Survey, Volume 2.* Oct. 1991.

Sanyal, Bish. "The Make-Believe World of the Calcutta Metro-Rail." *The Statesman* (July 12, 1987): 4–5.

Sartre, Jean-Paul. *The Psychology of Imagination.* London: Methuen, 1972.

SCAG (Southern California Association of Governments). *Status of Regional Transportation Planning and Coordination in Southern California,* 1968.

———. *Southern California Regional Transit System Plan: An Element of the Southern California Association of Governments Critical Decisions Program.* Los Angeles, July 11, 1974.

———. Transcript of SCAG Executive Committee Regional Transportation Plan Workshop. Sept. 1, 1983.

———. *Regional Transportation Modeling System, Description and Assumptions of Travel Demand Models.* Los Angeles, Feb. 1984.

———. *Los Angeles–Long Beach Light Rail Transit Project, Patronage Estimation and Impacts.* Mar. 1984.

———. "Re-Training the Region: Rail Systems Prepare to Make a Comeback in L.A." *Crossroads, The Southern California Transportation Magazine* (1985): 8–13.

Schaldenbrand, Mary. "Metaphoric Imagination: Kinship Through Conflict." In *Studies in the Philosophy of Paul Ricoeur.* Edited by Charles E. Reagan. Athens: Ohio University Press, 1979.

Schön, Donald A. *Displacement of Concepts.* London: Tavistock Publications, 1963.

———. *Beyond the Stable State.* New York: W. W. Norton, 1971.

———. "Generative Metaphor: A Perspective on Problem-Setting in Social Policy." In *Metaphor and Thought,* edited by Andrew Ortony. Cambridge: Cambridge University Press, 1979.

———. *The Reflective Practitioner: How Professionals Think in Action.* New York: Basic Books, 1983.

Sedway Cooke Associates. *Technical Report to the Long Beach-Los Angeles Rail Transit Project Environmental Impact Statement/Report: Land Use and Development Impacts,* 1984.

Shaffer, Gina. "Warding Off Opposition." *Valley Trader,* July 11–17, 1980.

Shaw, Peter L., Keith G. Baker, Mel D. Powell, and Renee Simon. *Los Angeles County Transportation Commission Public Policy Impact Study (September, 1978–August, 1979). Final Report.* Long Beach: Institute for Transportation Policy and Planning, Bureau of Governmental Research and Services, Center for Public Policy and Administration, California State University. U.S. Department of Transportation, Urban Mass Transportation Administration, Nov. 1979. Available through National Technical Information Service, Springfield, Va.

Shaw, Peter L., and Renee Simon. *Los Angeles County Transportation Commission Public Policy Impact Study (September, 1979–August, 1980), Final Report.* Long Beach: Institute for Transportation Policy and Planning, Bureau of Governmental Research and Services, Center for Public Policy and Administration, California State University. U.S. Department of Transportation, Urban Mass Transportation Administration, Apr. 1981. Available through National Technical Information Service, Springfield, Va.

Shepard, Roger N. "Psychophysical Complementarity." In *Perceptual Organization.* Edited by Michael Kubovy and James R. Pomerantz. Hillsdale, N.J.: Lawrence Erlbaum Associates, 1981.

Shipnuck, Lee, and Dan Feshbach. "Bay Area Council: Regional Powerhouse." *Pacific Research and World Empire Telegram* 4 (Nov./Dec. 1972): 3–11.

Simms, Gilmore, "The Philosophy of the Omnibus." *Godey's Lady's Book* 23 (Sept. 1841).

Smerk, George M. *The Federal Role in Urban Mass Transportation.* Bloomington and Indianapolis: Indiana University Press, 1991.

Snell, Bradford C. "American Ground Transport, A Proposal for Restructuring the Automobile, Truck, Bus, and Rail Industries." Part 4A, Appendix to Part 4, Hearings before the Subcommittee on Antitrust and Monopoly of the Committee on the Judiciary, United States Senate, 1974, A-1–A-103.

State of California Business and Transportation Agency. *Transportation-Employment Project: South Central and East Los Angeles, Final Report*, Aug. 1971.

State of California Department of Finance. *Los Angeles County Demographic Projection for 1992*. Demographic Research Unit, Nov. 9, 1993.

Stauffer, J. R. "Opportunities for Profits in De Luxe Bus Operation." *Electric Railway Journal* (Feb. 1930): 91–6.

Steinbruner, John D. *The Cybernetic Theory of Decision*. Princeton: Princeton University Press, 1974.

Stocks, Carl W. "Improving the Bus to Increase Its Usefulness." *Electric Railway Journal* (June 14, 1930): 387–90.

Stopher, Peter R., and Arnim H. Meyburg. *Urban Transportation Modeling and Planning*. Lexington, Mass.: Lexington Books, 1975.

"Street Cars Coming Back." *Electric Traction* (Oct. 1925).

Sundeen, Richard Allen, Jr. "The San Francisco Bay Area Council: An Analysis of A Non-Governmental Metropolitan Organization." Master's Thesis, University of California, Berkeley, 1963.

Symes/Dilworth Alliance. Resolution, reprinted in *Hearings on S. 345*. Subcommittee on Housing, Senate Committee on Banking and Currency. Washington, D.C.: U.S. Government Printing Office, 1961.

Taylor, Brian D., and Eugene J. Kim. *The Politics of Rail Planning: A Case Study of the Wilshire Red Line in Los Angeles*. Working Paper, UCLA Institute of Transportation Studies, Feb. 15, 1999.

Texas Transportation Institute. *Houston High-Occupancy Vehicle Lane Operations Summary*. Dec. 1996.

Transportation Research Board. *Highway Capacity Manual*, Special Report 209, 1985.

Treutlein, Theodore E. "Los Angeles, California: The Question of the City's Original Spanish Name." *Southern California Quarterly* 55 (spring 1973): 1–8.

Tri-Met (Tri-County Metropolitan Transportation District of Oregon). *East Side Transit Operations*. Dec. 1977.

———. *Staff Recommendation to the Tri-Met Board of Directors on the Banfield Transitway Project*. Aug. 1978.

———. *Portland's Metropolitan Area Express: Forecast Versus Actual Ridership and Costs*. Public Services Division. Unnumbered cover and three text sheets, June 1990.

———. *Tri-Met Five Year Plan FY 93-97, Phase I. Foundation Document No. 1. Comprehensive Service Analysis*, Sept. 1991.

———. Letter with comments on draft of Richmond. June 12, 1998.

Turbayne, Colin Murray. *The Myth of Metaphor*. 2d ed. Columbia: University of South Carolina Press, 1970.

UMTA (Urban Mass Transportation Administration, U.S. Department of Transportation). *A Detailed Description of UMTA's System for Rating Proposed Major Transit Investments*. Washington, D.C., May 1984.

———. *Memorandum for Approval of Capital Grant PA-03-0095, Amendment 14*, Mar. 1987.

UMTA/East-West Gateway Coordinating Council (U.S. Department of Transportation, Urban Mass Transportation Administration and East-West Gateway Coordinating Council, St. Louis). *St. Louis Central/Airport Corridor. St. Louis City and County, Missouri. East St. Louis and St. Clair County, Illinois. Alternatives Analysis & Draft Environmental Impact Statement for Major Transit Capital Investments*, May 1984.

UMTA/RTD. *Los Angeles Rail Rapid Transit Project Metro Rail, Final Environmental Impact Statement*, Dec. 1983.

———. *Los Angeles Rail Rapid Transit Project Metro Rail, Final Supplemental Environmental Impact Statement/Subsequent Environmental Impact Report*, July 1989.

UMTA/SACOG (U. S. Department of Transportation, Urban Mass Transportation Administration, and Sacramento Area Council of Governments). *Draft Alternatives Analysis/Environmental Impact Statement/Environmental Impact Report on Prospective Interstate Substitution Transportation Improvements in Northeast Sacramento, California*, 1981.

U.S. Congress. Hearings before the House Subcommittee on Transportation and Commerce of the Committee on Interstate and Foreign Commerce, Apr. 1979.

U.S. District Court. Consent Decree in *Labor/Community Strategy Center el al vs. Los Angeles County Metropolitan Transportation Authority et al.* United States District Court for the Central District of California, Case No. CV 94-5936 TJH (MCx), Oct. 29, 1996.

Vance, James E., Jr. "California and the Search for the Ideal." *Annals of the Association of American Geographers* 62 (1972): 185–210.

Veysey, Laurence R. "The Pacific Electric Railway Company, 1910–1953, A Study in the Operations of Economic, Social and Political Forces Upon American Local Transportation". Available at Los Angeles County Metropolitan Transportation Authority Library, (1953).

Von Bertalanffy, Ludwig. "On the Definition of the Symbol." In *Psychology and the Symbol, An Interdisciplinary Symposium*, edited by Joseph R. Royce. New York: Random House, 1956.

Voorhees, Alan M. & Associates. *Metro for Buffalo. Transit Alternatives for the Buffalo–Amherst Corridor. Technical Report*, June 1976.

Wachs, Martin. "Consumer Attitudes Toward Transit Service: An Interpretive Review." *Journal of the American Institute of Planners* (Jan. 1976): 96–104.

———. "The Case for Bus Rapid Transit in Los Angeles." In *Transportation Alternatives for Southern California*, edited by Peter Gordon and Ross D. Eckert. Conference Proceedings of a Symposium, The Institute for Public Policy Research, Center for Public Affairs, University of Southern California, Apr. 12, 1976. Also available at http://www-rcf.usc.edu/~pgordon/transit.html.

———. "Autos, Transit, and the Sprawl of Los Angeles: The 1920s." *Journal of the American Planning Association* 50, no. 4 (summer 1984): 297–310.

———. "Ethical Dilemmas in Forecasting for Public Policy." In *Ethics in Planning*, edited by Martin Wachs. New Brunswick, N.J.: Center for Urban Policy Research, 1985.

———. "Policy Implications of Recent Behavioral Research in Transportation Demand Management." *Journal of Planning Literature* 5, no. 4 (May 1991): 333–41.

Wachs, Martin, and Joseph L. Schofer. "Abstract Values and Concrete Highways." *Traffic Quarterly* 23, no. 1 (Jan. 1969): 133–55.

Ward, Baxter. *The Sunset Coast Line: Route of the New Red Cars.* Prepared through office of Supervisor Baxter Ward by county departments and others. Los Angeles, 1976.

———. Sunset, Ltd. Los Angeles: Board of Supervisors, County of Los Angeles, Office of Supervisor Baxter Ward, 1978.

Warman, Cy. *Tales of an Engineer, with Rhymes of the Rail.* New York: C. Scribner's Sons, 1895.

Warner, A. T. "Development of the Bus for Mass Transportation." *Electric Railway Journal* (Dec. 1930): 745–46.

Watts, Alan W. *Myth and Ritual in Christianity.* New York: Macmillan, 1954.

Webber, Melvin M. *The BART Experience—What Have We Learned?* Monograph no. 26. Berkeley: Institute of Urban and Regional Development, University of California, Berkeley, Oct. 1976.

Weizenbaum, Joseph. *Computer Power and Human Reason: From Judgment to Calculation.* San Francisco: W. H. Freeman, 1976.

Wells, William R. *Rapid Transit Impact on Suburban Planning and Development, Perspective and Case Study.* Department of Industrial Engineering, Stanford University, 1973.

Wheelwright, Philip. *Metaphor and Reality.* Bloomington: Indiana University Press, 1962.

White, Andrew D. *A History of the Warfare of Science with Theology in Christendom.* New York: Appleton, 1896.

Whitt, J. Allen. *Urban Elites and Mass Transportation: The Dialectics of Power.* Princeton: Princeton University Press, 1982.

Wirt, Frederick M. *Power in the City: Decision Making in San Francisco.* Berkeley: University of California Press, 1974.

Wohl, Martin. "The Case for Rapid Transit: Before and After the Fact." In *Transportation Alternatives for Southern California,* edited by Peter Gordon and Ross D. Eckert. Conference Proceedings of a Symposium, The Institute for Public Policy Research, Center for Public Affairs, University of Southern California, Apr. 12, 1976. Also available at http://www-rcf.usc.edu/~pgordon/transit.html.

White House. *America's New Beginning: A Program for Economic Recovery,* Washington, D.C., 1981.

Wohlwill, David E. *Development Along a Busway. A Case Study of Development Along the Martin Luther King, Jr. East Busway in Pittsburgh.* Port Authority of Allegheny County, Pittsburgh, Pa., June 1996.

Yoos, George E. "A Phenomenological Look at Metaphor." *Philosophy and Phenomenological Research* 37, no. 1 (Sept. 1977): 78–88.

Young, Robert M. "Why are Figures so Significant? The Role and the Critique of Quantification." In *Demystifying Social Statistics.* edited by John Irvine, Ian Miles, and Jeff Evans. London: Pluto Press, 1979.

Zarian, Larry. Letter from Chair, Los Angeles County Metropolitan Transportation Authority to Gordon Linton, U.S. Federal Transit Administrator, Jan. 15, 1997.

Zwerling, Stephen. *Mass Transit and the Politics of Technology: A Study of BART and the San Francisco Bay Area.* New York: Praeger Publishers, 1974.

Biographical Note

"One of the nation's leading experts on urban transportation," according to the *St. Paul Pioneer Press*, Jonathan Richmond addresses controversial questions about the types of transportation services that should be provided, and how they should be supplied, with a focus on revealing unstated assumptions and developing understandings of complex phenomena which promise to contribute to better decision-making.

An undergraduate at the London School of Economics, he continued as a Fulbright Scholar to the Center for Transportation Studies at MIT, from which he received masters and doctoral degrees. He is currently conducting research on developing country transportation problems and developing and teaching new coursework for developing country transportation professionals as a visiting fellow at Asian Institute of Technology, Thailand. Previously, he has held positions at the University of North Carolina, Chapel Hill, UCLA, and as New South Wales Department of Transport Visiting Professor of Transport Planning at the Institute of Transport Studies, University of Sydney. His first book, *The Private Provision of Public Transport*, was researched while a fellow at the A. Alfred Taubman Center for State and Local Government, John F. Kennedy School of Government, Harvard University.

Jonathan Richmond has also served as policy advisor to the Chair of the Los Angeles County Transportation Commission, and conducted work on strategic transportation planning for the World Bank. He has broad interests in planning theory and ethics, with a particular concern for the ethical design of systems of evaluation in planning, and for developing systems of government and management that learn and act in ethical and well-informed ways.

With interests in reaching popular audiences, Richmond writes in an accessible style and has contributed opinion articles to the *Los Angeles Times* and other major newspapers, as well as delivered public testimony to legislative bodies. In his spare time, Richmond's obsessions are travel and music. He has climbed sand dunes under the Saharan sun, been granted a prisoner-led tour of Bolivia's central jail, crossed Madagascar by taxi brousse and truck, and hitch-hiked across Tibet, arriving in the shadow of Everest as Buddhist monks sounded their evensong. Awarded MIT's Wiesner Award for contributions to the arts, he was the *Christian Science Monitor*'s Boston-based classical music critic as well as Arts Editor of MIT's *The Tech* while a graduate student, and continues his relationship with *The Tech* as an advisor.

Index

Bay Area Rapid Transit (BART) system: in business development, 67–68; business promotion of, 167–68; capacity of, 267; comfort of, 53; declining ridership of, 164–68; design of, 45; Disneyland effect of, 330; in downtown development, 67–69; feeder services in, 294; funding of, 169; imagery of, 382; Lafayette Station of, *165*; in reducing freeway congestion, 61, 62; speed of, 257; technological virtuosity of, 314; transbay tube of, 167; transfers on, 49

BCB Group, 69

Beaverton, urban development in, 69

beliefs, mythology and, 243–45

Benbow, Richard, 354, 359

Bernardi, Ernani, 273, 334, 386

Biehler, Allen D., 76

Bilbray, Brian, 259, 314, 316, 339

Billheimer, J., 47

Billions for Buses protests, 371–72

Black, Max, 235–36

Blauner, Robert, 349

Blue Line, *37*; all-bus alternative to, 77; attractive features of, 55; capital costs of, 73–75, 89–90; in community renewal, 345–46; conversion of, 50–51; daily passenger travel miles of, 87–88; early performance of, 32; end-to-end travel time on, 51; farebox recover ratio of, 78; first, *256*; at July 1990 opening, *34*; in Los Angeles–Long Beach, 373–89; Los Angeles as terminal site of, 302–3; minority ridership of, 355–57, 359–60; no-transfer trip times on, 49–50; opening of, *400*, 401–10; operating costs of, 83–84; operating performance of, 81–82; perceived benefits of to Watts, 352–60; project funding for, 3; ridership on, 41: among minority groups, 355–57, 359–60; costs of, 32–33, 86–87; forecasts for, 40, 140; increasing, 129; romance of tunnel opening on, 323, 324; security costs of, 82; service description of, 33–38; station in, *387*; subsidy per passenger mile cost of, 85; Train Number 1 of, *400*; transfers on, 49; in Watts, 352

Blue Line Final Environmental Impact Statement, 81–82

Blyth-Zellerbach Committee, 167

body-as-machine metaphor, 276–77, 279–80; for Los Angeles transit system, 282

Bohlinger, Linda, 209–10

Bonsall, John, 71

Boulding, Kenneth, 1, 230

Bradbury, Ray, 311–12

Bradley, Mayor Tom, 172, 175, 180–81, 185, 206, 208, 297, 301, 377–78, 402, *404*; at opening of Blue Line, 406; on Proposition A, 191

Bradshaw, Thornton, 172

Bragg, Felicia, 350–51, 366–67, 368–69

Brandwynne, Marcia, 249, 321, 328; gas addiction coverage of, 295–96

Brathwaite-Burke, Yvonne, 336

Braymer, Noel T., 310

Brazil, high-capacity bus system of, 335–38, 393

Brewer, Garry D., 94

Brinckerhoff, Parsons, 73–74, 100, 355

Broadway-Spring Couplet light rail service, 121–22

Brodsly, David, 11, 152, 153–54, 249–50, 252

Bronner, Ethan, 227

Buffalo rail system: versus bus service performance, 78; costs of, 71; forecasting as political tool for, 139; pedestrian mall in, 42–43; ridership of, 42

Burke, Julian, 183, 189, 192, 210–11

bus fares, 4, 38, 51–52; reduction of, 85, 202–3; subsidies for, 185, 373; in Watts, 349

bus route 56, 49–50

bus services (*see also* All-bus alternative model; Rapid Bus system): to areas where most needed, 50–51; assumptions about, 119–20; attracting ridership for, 55–61; augmentation of, 89–90; in balanced transportation system, 290; as best alternative, 179, 181; Blue Line and, 33; capacity of, 266–67; capital costs of, 71, 76–77; characteristics of, 44–61; circulation metaphor for, 385–86; comfort of, 53–55, 257–59; complementary network of, 33, 119–20; congested traffic and, 64–65, 263–64, 298; conversion of railways to, 20–21; crime on, 260–61; crosstown, 51; curb lanes for, 64–65; demand for funding of, 371–72; direct-to-downtown, 43–44, 51; dirty image of, 382; door-to-door service of, 48; efficiency of, 267–71; in energy conservation, 66; evaluation of, 43–44; expanded, 57–58; experience of, 381; express, 48, 50, 56; farebox recover ratios of, 78–79; feeder lines in, 25, 79–80, 255, 292–94; flexibility of, 25–26, 282; grid service of, 89; high-capacity, 60, 335–38; high-speed, 336–37; imagery of, 274; improvements of, 24–25, 59–60, 209; integrating metaphors of, 275–308; lack of federal subsidies for, 160–61; lack of sex appeal of, 322; in Long Beach corridor, *35*; luxury coaches in, 258; mid-corridor ridership of, 355; for minority communities, 355–57; most desired improvements in, 54t; no-transfer trip times on, 49–50; operating and maintenance costs of, 78–85; operating performances of, 79–80; opportunity costs of, 85–90; Pacific Electric shift to, 26–27; passenger travel miles of, 88–89; patronage estimation for, 100; at peak hour, *264*; pollution and, 265–66; queuing effects of stops of, 64; in reducing freeway congestion, 61–65; ridership studies of, 40; security of, 259–60; in South Central Los Angeles, 354; speed of compared to train, 256–57; stops in, 115; subsidy per passenger mile cost of, 84–85; transfers in, 117; in urban development, 70; "userfriendliness" of, 55; versus rail service benefits, 4–5, 252–73; versus Red Cars, 24–26; in Western cities, 4–5

Bush, George, 402, 408

business community: Proposition A and, 189–91; supporting BART development, 167–68

busways/lanes, 56–57, 60, 63, 64–65, 85, 301, 314 (*see also* HOV lanes); costs of, 77, 88; development of, 179; for high-capacity buses, 336; impact of on congestion, 63; operating performances of, 81; priority, 56–58; in urban development, 70

Calcutta metro-rail system, 338, 346, 371

California (*see also* Southern California; *specific city*): freeway and expressway system of, 153–54; road-building in, 150–53

California Assembly Bill 1246, 178

California Assembly Interim Committee on Public

Dallas, ridership forecasting in, 128
Dana, Deane, 74, 192, 195, 196, 199–202, 288, 302, 408
Daniel, Mann, Johnson and Mendenhall (DMJM) report, 168–69
Davis, Aubry, 340
de Bono, Edward, 215, 217, 243–44
de Neufville, Judith, 226–27, 244, 380
De Witt, Mayor, 406–7
death penalty model, 146; econometric measurement of efficacy of, 96–97
decision-making: forecasts in legitimizing, 140–42; myths in, 5; rationality in, 7–8
deduction, common sense, 215–16
DeLeuw Cather engineering firm, 166
demand assessments, 100–101
demand forecasting: alternatives analysis and, 98–101; for Long Beach line, 98–130; techniques and problems in, 91–98; validation of, 102
demographic data, compilation of, 103–4
Denver: bus versus light rail ridership in, 44; farebox recover ratios in, 78
Depression, Great, 23–24, 30
depth cues, 217–18
development: public transportation impact on, 67–70; trips generated by, 106
Didion, Joan, 250
diesel buses, 19–20, 66. See also electric buses
diesel engine technology, 28, 30
diesel fuel pollution, 265–66
disease metaphor, 277; escape valve metaphor and, 297–300; logic of, 306, 307; in transportation system breakdown, 281–84
Disneyland: experience of, 329–41, 410; monorail at, 219, 381–82
dispersed lifestyle, 248, 379. See also low-density residential development; population dispersal
disutility, 112–13
Dixon, Julian, 207
Donley, Roy: on buses versus trains, 261, 265, 266; on mass transit in Los Angeles, 340; on religion of rails, 325–26, 384, 386–87; on "sex appeal" of trains, 315, 321, 332; on speed of trains, 255–56
door-to-door services, 48
Dotterer, Steve, 258, 339
downtown: direct service to, 51; office development in, 67–70
Downtown Seattle Transit Project, 335
Drew, Joe, 118, 209
Drier, David, 205
driver costs, 82
driver efficiency, 269–70t
driver image, 267–71
drug addiction metaphor. See addiction metaphor; gas addiction
Due, John, 21
Dunn, James, 161
Dyckman, John, 47
Dyer, John, 203, 205–6, 355

ease of use, 55
East Busway, 58, 70, 76, 77
East Coast rail transit systems, 4
econometric models, 96–97
economic analysis, 397

Edelman, Ed, 187, 189, 192, 201–2, 401–6; at opening of Blue Line, 407–8
Edelman, Murray, 220–21, 227, 374, 392
Edgerton, Wallace, 138, 252, 266, 282, 300, 351–52
efficiency, trains versus buses, 267–71
electric buses, 296, 335, 382. See also diesel buses
El Monte Busway, 301; appeal of, 314; impact of on congestion, 63; ridership of, 56–57, 262
Elder, Dave, 257, 265, 271, 313
Elder, Ed, 193–94, 195
Ellul, Jacques, 95
emotions: balance of, 285–86; images reflecting, 232–33; symbols representing, 384
employment, 394; connection to, 370–71; distribution of, 105, 107–8; growth forecasts for, 104–6; public transportation and, 352–57; trip generation and, 107–8
energy consumption, reducing, 65–66, 373
energy crisis, 179–80, 191; metaphor in understanding of, 242
environment, uncertainty-controlled, 1–2
environmental factors, 61–66
Environmental Impact Assessment, 101, 203
environmentalism: in anti-highway movement, 173–74; bus ridership and carpools encouraged by, 56; in freeway development opposition, 17; grassroots, 173
equilibrium, restoring, 285–86, 297–98, 307, 385
errors of convenience, 128
errors of optimism, 128
escape valve metaphor, 277, 297–300, 385–86, 408–9; logic of, 306–7
Estrada, Carmen, 138, 300–301, 304
ethics, of forecast modeling, 131–35, 145–48
Euler, 131
European rail system: high technology of, 333–34
European rail systems, 330–31, 383; personal experience of, 330–31
evolution metaphor, 300–301, 386
experience: in cognitive process, 215–16; power of, 334–35; of rails, 330–31, 383; role of, 328–29; supporting rail myth, 380–81; translated from one setting to another, 381–82
Exposition Corridor, cost-effectiveness of, 76, 90

face, symbolizing person, 225–26
facilities maintenance costs, 82
farebox recovery ratios: for Blue Line, 81–82; of MTA bus system versus rail system, 87; of rail versus bus services, 78; of Red Line, 84
fares: for buses versus rails, 51–52, 119; flat, 43; reduction of, 85–86; ridership and, 83, 86; structuring of for light rail systems, 42–43
Federal Aid Highway Act (1973), 17; supporters of, 174
Federal Highway Trust Fund, 27
Federal Mass Transportation Act of 1987, 206–7
Federal policies, 160–64
Federal Urban Renewal Program, constitutionality of, 240
feeder systems, 292–94
Ferraro, John, 138
Fiedler, Bobbi, 208
Filer, Maxcy, 138, 294, 305, 342–43, 358, 364–65, 366, 372, 387–88, 392–93, 397–98

265; population dispersal and, 271; public
control of, 30–31; romance and sex appeal of,
316, 322; route of, 16–17; as solution to gas
addiction, 295–96; speed of, 378; tunnel for, 45;
"untimely" death of, 194–95, 248, 327–28,
379–80; in urban development, 70; in Watts
community, 345, 346–47, 364, 369
Red Line, 35–36, 37; capital costs of, 76, 89–90;
design in, 200; downtown stops of, 48; financial
performance of, 84; at peak hours, 38; proposal
for, 205, 206–7; recovery plan for, 209–210;
ridership of, 42; ridership reporting for, 41; route
of, 207; service connecting points of, 59; time and
fare savings on, 51–52; transfer fees for, 43;
transfers on, 49
Reddy, Michael, 238–39
Reed, Christine, 219, 326, 333, 340–41, 387, 403; on
bus security, 260; on forecast modeling, 134, 136,
137, 144; on light rail system, 144, 198; on "sex
appeal" of light rail, 304, 309
reflective rationality, 8, 9, 395, 396–97
Regional Planning Commission, Los Angeles
County: *Freeways for the Region* of, 152; in street
planning, 150
regional statistical areas, trip patterns between, 45,
46–47
Regional Transportation Modeling Plan, 103
Regional Transportation Plan (RTP) system,
ridership forecasts for, 101–2
regression analysis, 118
religion of trains, 325–28
Remy, Ray, 184–85, 187, 197
la Reyna de los Angeles settlers, 13
Rhodes, Richard, 226
Richardson, Harry, 69
Richmond, Jonathan E. D., 39, 40, 70
Richmond, Rick, 141, 142, 198–99, 265, 303, 353–54,
358
Ricoeur, Paul, 220, 237
ridership: evaluations of, 40–44; fares and, 83, 86;
peak versus off-peak, 119; reasons for decline in,
166
ridership forecast modeling, 98–101, 391–92; failure
of, 126–30; negative, 142–45; overprediction in,
122–25; population increases versus improved
services in, 121; reliability of, 135–36
right-of-way, cost of new, 77
Riordan, Richard, 335
road systems (*see also* freeway systems; highways):
balancing with rail system, 287–91; capacity of,
64–65; circulation metaphor for, 277–78; funding
of, 149–50; local government in construction of,
149–50
Robenhymer, Bob, 257, 298–99, 316, 339
Roberts, Dan, 263, 290
Roberts, George, 159
Robertson, Bill, 189, 191, 205
Roche, Burke, 144, 184, 189, 191, 251, 263, 289, 297,
331
Roche, Pat, 364
Rogers, Leslie, 210
romance, of rails, 375–76
RTAG proposal (1948), 376
RTD. *See* Rapid Transit District (RTD), Southern
California

Rubley, Russ, 182–83, 195, 197, 202
Rudin, Anne, 339
rural immigrants, Los Angeles, 13–14
Russ, Ed, 179, 184, 186, 332
Russell, Pat, 135, 184, 302, 357
Rustin, 349

Sacramento rail system, 292t; costs of, 78; forecasting
as political tool in, 139; investment trends in, 42
St. Louis: fare structures in, 79; improved image of,
204; rail ridership studies in, 40, 42
Sales tax revenue, 182–83; funding Blue Line, 3
San Bernardino Freeway, busway on, 56–57, 63
San Bernardino line, 24
San Diego: balanced transportation system in, 287;
Blue Line ridership in, 40; bus versus light rail
ridership in, 44; farebox recovery ratios in, 78;
light rail system in, 248, 332–33, 339, 378; cost of,
193; investment trends in, 42; Orange Line in, 40;
transit operating performances in, 79; trolley in,
72, 340
San Fernando Valley Bus Rapid Transit Project,
212–13; capital costs of, 77
San Fernando Valley: East-West rail line of, 208;
express bus services from, 44
San Fernando Valley-Long Beach corridor, 175
San Francisco (*see also* BART): Community Renewal
Program model of, 94
San Francisco Bay Area Rapid Transit Commission,
166
San Francisco Bay Regional Council, 165
San Jose rail system, 42
Sanborn, Blake, 138, 141–42, 265
Santa Monica, Rapid Bus in, 88
Sanyal, Bish, 338, 341, 346, 371
Sartre, Jean-Paul, 230–31, 232–33
Sato, Eunice, 136, 138, 255, 282, 330, 334, 386–87
SCAG-82 Growth Forecast Policy, 104–5, 106;
socioeconomic data of, 121–22
Schabarum, Peter, 189, 191, 192, 287, 330
Schiff, Adam, 211, 213
Schneider, Ronald, 184
Schofer, Joseph, 218
Schön, Donald A., 2–3, 91, 234–35, 236, 286–87, 376;
generative metaphor concept of, 239–42
scientific analogues, 93–94
Seattle (*see also* METRO): bus tunnel project in, 393;
mass transit system of, 335; METRO Transit
Committee of, 337
security: costs of, 82; imagery of, 383; of trains versus
buses, 259–60
Sedway Cooke Associates report, 359
self-determination, 388
Senate Banking, Currency and Urban Affairs
Committee, 161
Senate Commerce Committee Subcommittee on
Transportation, 161
service characteristics, rail versus bus, 44–61
sex appeal, technology as, 316–21
Shaffer, Gina, 325
Shaw, Peter L., 178
Shaw-Johnson, Frieta, 364
Shepard, Roger, 218
Sherman, Moses, 15
Shirley Highway buses, 56

Shoup, Paul, 24
Siegel, Steve, 288
signs, symbols and, 221–22
Silver, Tom, 211
simplicity, 216–18
Sims, Jim, 137, 142, 293, 303
Singer, Edgar, 8
60 Minutes, metrorail story on, 209
Smerk, George M., 173
Snell, Bradford, 247; conspiracy theory of, 17–21. *See also* General Motors
Snell Report, 31, 377
social change, 392–93
social decision-making, 93
social integration: light rail service and, 370; symbolism of trains in, 360–69
social organism, 276
social policy: generative metaphor in, 239–42; transportation solutions in, 392–93, 394
social processes, scientific analogues for, 93–94
socioeconomic data: gathering of, 103–4; in ridership analysis, 121–22; in trip distribution, 110–11
Southern California: balanced transportation system in, 286–91; circulation metaphor in, 281–84; highway development in, 153–54; Pacific Electric railway lines in, *18*
Southern California Association of Governments (SCAG), 6–7; creation of, 168; Development Guide Program, 104–5; Long Beach corridor demand modeling, 101–26; model of: assumptions in, 132–35; responsibility of modeler in, 131–35; opposing proposed Los Angeles rapid transit system, 172; pollution impact study of, 66; Regional Transportation Plan of, 99, 177; report of: inadequacy of, 125–26; reactions to negative findings of, 142–45; on traffic congestion, 63; ridership forecasts of versus Caltrans studies, 129–30
Southern California Transportation Action Committee, 191
speed: focus on, 253–57; imagery of, 376–77, 408–9; symbols of, 381–82, 383
Spence, Ted, 310–11, 315–16, 339
Standard Oil of California, conspiracy conviction of, 20
Stanger, Richard, 135, 143, 303, 310–11, 358; on balanced transportation system, 287, 293, 294; on driver efficiency, 269t
Stanley, Ralph, 207
State Transportation Improvement Program (STIP), 197
statistical analysis, 97
status symbol, rail system as, 339–41
Stauffer, J.R., 54
Steinbruner, John, 216–17
Stoddard, Joseph, 328
Stopher, Peter, demand forecasting of, 98–99, 103, 104–5, 107–8, 113–14, 132–33, 141, 147
storytelling, 245
strangulation metaphor, 282–83, 284t
streetcar system: abandonment of, 26–27; bus competition with, 24–26; conspiracy to eliminate, 17–21; in Los Angeles, 14–17; revival of, 5; route of, 4
Struiksma, Ed, 290, 321, 339

subsidy costs, 78–85
subsidy per passenger mile, 84–85; of Blue Line, 87–88
suburban communities, 172
subway systems (*see also* Wilshire subway): "christening" of terminal in, 247; congestion and, 334; downtown stops of, 48; early development of, 154; services of, 35–36
Sunset Coast Line, 176–77 (*see also* Ward, Baxter); proposal for, 182, 377–78; romance of trains and proposal for, 322; scaled-down version of, 180–81; sex appeal of, 318–21
Sunset Ltd. proposal, 318–21
surface imagery, 382
surface language, 242, 246
Surface Transportation Assistance Act (1983), 206
"Sweep-in" approach, 8–9
symbolic action, 341–42
symbolic logic, 380
symbolism: of community renewal, 345–72; in Los Angeles rail service myth, 245–47; of rail system, 380–89; of technology, 309–44; in understanding transportation needs, 398
symbols, 220; abstraction and, 225–26; as blueprints, 223–24; boundaries of, 223; conveying values, 224; culturally established patterns of, 222–23; example of, 227–28; as gateways, 222, 382–83; in human nature, 226–27; images as, 231–32; internal logic of, 246–47; interpreting, 229; metaphor and, 229, 234–42; in myth, 243; signs and, 221–22; in specific contexts, 225–26; thought and, 220–21; in understanding systems, 390–91
synechdoche, 252–53
system impact, ignoring, 146
systems analysis, 98–101, 390
Szabo, Barna, 141–42, 144, 183, 185, 260, 271–72, 304–5, 339

Tajuata, 346–47
Taylor, Paul, 111, 116, 142–43, 268, 333, 358–59
technical analysis: role of, 131; to support assumptions, 147; value and reliability of, 135–37
technological virtuosity, 313–15; phallic symbolism of, 316; sex appeal of, 315–21
technology: displacing Red Cars, 21–31; emotional attachment to, 384; evolution and displacement of, 11–31; fixed evolutionary path of, 301; focus on, 391; imagery of, 385; religion of, 325–28; sex appeal of, 253, 393–94; symbolic meanings of, 309–310
Teele, Arthur, 196
thought: non-reflective, 233–34; symbols and, 220–21
Tijuiana Trolley, 333
tire rationing, end of, 27
Tiresias, 1, 275
Toronto, downtown development in, 67–68
traffic: bumper-to-bumper, 251–52, 297; canalization of, 278–79; choking metaphor for, 297–98; flow of: circulation metaphor for, 276–84, 278–79, 306; metaphors for, 384–85; public transportation impact on, 61–65
Traffic Commission, Los Angeles, 149–50
trains. *See* light rail systems; rail systems; subway systems; trolleys